RELIGION
The Social Context

Fourth Edition

MEREDITH B. McGUIRE

Trinity University

Wadsworth Publishing Company

I(T)P® An International Thomson Publishing Company

Belmont, CA • Albany, NY • Bonn • Boston • Cincinnati • Detroit • Johannesburg • London • Madrid
Melbourne • Mexico City • New York • Paris • Singapore • Tokyo • Toronto • Washington

Sociology Editor: Eve Howard
Assistant Editor: Deirdre McGill
Marketing Manager: Michael Dew
Project Editor: Karen Garrison
Print Buyer: Karen Hunt
Cover: Craig Hanson
Compositor: Thompson Type
Printer: Quebecor/Fairfield

For more information, contact Wadsworth Publishing Company, 10 Davis Drive, Belmont, CA
94002, or electronically at http://www.thomson.com/wadsworth.html

International Thomson Publishing Europe
Berkshire House 168-173
High Holborn
London, WC1V 7AA, England

Thomas Nelson Australia
102 Dodds Street
South Melbourne 3205
Victoria, Australia

Nelson Canada
1120 Birchmount Road
Scarborough, Ontario
Canada M1K 5G4

International Thomson Publishing GmbH
Königswinterer Strasse 418
53227 Bonn, Germany

International Thomson Editores
Campos Eliseos 385, Piso 7
Col. Polanco
11560 México D.F. México

International Thomson Publishing Asia
221 Henderson Road
#05-10 Henderson Building
Singapore 0315

International Thomson Publishing Japan
Hirakawacho Kyowa Building, 3F
2-2-1 Hirakawacho
Chiyoda-ku, Tokyo 102, Japan

International Thomson Publishing
Southern Africa
Building 18, Constantia Park
240 Old Pretoria Road
Halfway House, 1685 South Africa

Library of Congress Cataloging-in-Publication Data

McGuire, Meredith B.
        Religion, the social context / Meredith B.
McGuire. — 4th ed.
                p.   cm.
        Includes bibliographical references (p. 331)
        and indexes.
        ISBN 0-534-50572-4 (alk. paper)
        1. Religion and sociology.   I. Title.
BL60.M36 1997
306.6—dc20   96-36182

This book is dedicated to

Rachel Rebecca McGuire
(July 2, 1970–October 10, 1995)

and to all other victims of a violent world.

Contents

Preface

I n this fourth edition, I continue to emphasize that religion exists in a social context, is shaped by that social context, and, in turn, often influences it. This book is ideally suited to be a core of essays around which instructors can build a course and assign other readings to complement their own emphasis; it does not attempt to cover every issue in the sociology of religion, nor is it meant to be the only text used for a course. My primary audience consists of undergraduates taking a first course in sociology of religion; this material should prove useful to seminary and graduate students and to interested general readers as well.

Each chapter focuses on a particular issue in the sociology of religion, presenting how sociologists approach these issues and critically summarizing research relevant to them. In most cases, I also suggest new approaches and advocate new lines of research that seem to me to be especially promising. This book is therefore prospective as well as retrospective: In addition to being a textbook, it is a systematic essay designed to move the field forward.

Chapter 1 examines different approaches to defining religion: different approaches to the subject result in entirely different ways of framing research and interpreting the religious situation in the modern world. The new edition gives expanded attention to religious ritual and religious experience. *Chapter 2* discusses how religion provides both meaning and belonging as links between the individual and social group. Several chapters include an extended example in addition to the main explicative portion. The purpose of such examples is to illustrate a sociological interpretation of relevant religious phenomena. The Extended Application in Chapter 2, on millenarianism and dualism in contemporary social movements, has been expanded and updated to help interpret such current movements as Islamic and Christian fundamentalism, the Christian Patriot movement, the Shining Path, Branch Davidians, and many other recent religious movements. Other chapters include extended examples on women's religion (Chapter 4), emerging religious movements in the United States (Chapter 5), the conflict in Northern Ireland (Chapter 6), African-American religion (Chapter 7), and religion and healing (Chapter 8).

Some of these essays reflect interpretations of phenomena on which I have done extensive empirical research. All the extended examples represent a more personal stance than the discussions in the main text and, as such, invite the reader to encounter the sociological issues intellectually and perhaps disagree. These interpretations are intended to illustrate useful sociological concepts in a stimulating and provocative way and to show how sociological theories are relevant to interesting aspects of social life. Several of these interpretive essays are far from neutral; my sympathies clearly lie with the exploited and oppressed (specifically women, minorities, natives of colonized lands, and victims of political or economic suppression), and my personal concern is to explore the role of religion in their oppression and liberation.

Chapter 3 looks at the individual's religion: how it is shaped, supported, developed, and changed over the lifetime. This edition includes new material on the role of religion in self-identity in the context of modern pluralistic societies; new cross-cultural material on conversion to and disengagement from religious groups has also been added. *Chapter 4* has been significantly changed to take into account changes in the social location of religion and corresponding patterns of religiosity. As in previous editions, Chapter 4 distinguishes official and nonofficial religion; the new edition also looks at recent studies suggesting that the relevant levels of analysis for official religion have shifted from denominational and church affiliation to other organizational levels, especially the local congregation. Similarly, the discussion of nonofficial religion has been amplified by numerous recent studies of popular and syncretic religious expressions. These developments require a rethinking of the nature of religiosity in both official and nonofficial modes. Chapter 4 suggests some appropriate research methods and better strategies for studying official and nonofficial religion and religiosity.

Chapter 5 continues to use a basic theoretical model of religious organization and dynamics, but the fourth edition includes several new diagrams and examples to clarify for students the distinctions implicit in the typology. Instructors will note that some of the terminology has been changed toward this end, and there is greater emphasis on the usefulness of thinking about religious groups and their dynamics using these models.

Writing this fourth edition, I have been impressed with the growing value of the sociology of religion as a "window" for understanding many far-reaching sociopolitical developments in the modern world. Religion has figured prominently in recent events in the Middle East, Latin America, and Eastern Europe. In the United States, religion has exhibited considerable political assertiveness and various religious groups have articulated important social-moral issues facing the nation, while others have raised the specter of violent and antidemocratic extremism. *Chapter 6* continues to examine how religion is both a source of social cohesion and social conflict. New material includes themes of ethnicity, group identities, nationalism, and public religion; sadly, the conflict in Northern Ireland continues (after nearly two decades) to be a relevant and timely case for the chapter's Extended Application.

The Application in *Chapter 7,* on religion and social change, returns to the theme that was developed in the first edition of this book: African-American religion. The Latin American examples that were the focus in editions two and three are now incorporated, wherever possible, in the main body of the chapter. Much relevant new material has been published about African-American religion in recent years, so this new Extended Application covers much fascinating descriptive analysis, as well as aptly illustrating the theories explained in the chapter.

This edition includes extensive, current international and cross-cultural material, ending with a focus on religion in a global perspective, *Chapter 8.* New material emphasizes the importance of interpreting religion in the context of the "world as a single place"—the globalization of economic, political, communicative, and cultural features of modern social life. Not only do I believe such cross-cultural understanding is an important part of a broad education, but also I dare to hope that greater understanding and an appreciation of the global setting of our individual lives will promote peace. At the same time, I have deliberately chosen a disproportionate number of brief examples from religion in the United States and other Western, industrialized societies, because most students more readily comprehend these examples without needing extensive historical and cultural background explanation.

Ending each chapter is an annotated list of recommended readings—articles, books, and selections from recent books of readings. This edition is the first to include a condensed *Glossary* of important terms used in the text. Two appendices suggest further useful materials. *Appendix A* outlines major resources for a literature search in sociology of religion. Students preparing term papers or theses will find these resources essential. Several useful new bibliographies and encyclopedias are now available. The bibliography of this text itself reflects an extensive and reasonably current literature review. *Appendix B* includes a new feature, an *Observation Guide,* to help students who are undertaking their first field observation of a religious gathering. Though observing religious groups can be a very good way to learn the sociological approach to religion, I have found that students need explicit guidance about what and how to observe in order to get the most out of their experiences. Instructors should feel free to modify this Guide to fit the specific situations they have assigned for visits. Appendix B also includes recommended film resources for teaching sociology of religion. I have found good films to be almost essential in teaching sociology of religion because they provide concise glimpses into religious situations beyond the often limited experiences of students.

With gratitude I acknowledge those who have given thoughtful suggestions for this fourth edition, as well as all whose assistance made possible three earlier editions. Thanks to Eileen Barker, Jay Demerath, Phillip Hammond,

Benton Johnson, John Martin, Jim Richardson, Ole Riis, and Sheryl Tynes. Also, very welcome were the comments of reviewers Roger G. Branch, Georgia Southern University; Richard Cramer, University of North Carolina; David P. Efroymson, La Salle University; Geoffrey Kapenzi, Babson College; Larry Poston, Nyack College; and Darren E. Sherkat, Vanderbilt University. Special thanks to Jim Spickard, who read the entire revised manuscript and offered insightful suggestions throughout. Thanks also to Demetria Lafkas for her careful assistance in preparing the manuscript. The continuing lively response from my students at Trinity University, as well as at Montclair State University and many other universities where I have lectured, has been particularly useful. Finally, I want to give special thanks to my family, Jim, Daniel, Rachel, Kieran, Janaki, and Dmitri, for their encouragement, understanding, and help in keeping the whole project in perspective.

1

The Sociological Perspective on Religion

R eligion is one of the most powerful, deeply felt, and influential forces in human society. It has shaped people's relationships with each other, influencing family, community, economic, and political life. Religious beliefs and values motivate human action, and religious groups organize their collective religious expressions. Religion is a significant aspect of social life, and the social dimension is an important part of religion.

Sociologists are interested in studying religion primarily for two reasons. First, religion is very important to many people. Religious practices are important parts of many individuals' lives. Religious values influence their actions, and religious meanings help them interpret their experiences. Sociologists seek to understand the meaning of religion to believers themselves. Second, religion is an important object for sociological study because of its influence on society and society's impact on religion. Analysis of this dynamic relationship requires examining the interdependence of religion and other aspects of society. Often this means questioning taken-for-granted ways of understanding social action. From the earliest foundations of the discipline, sociologists have sought to understand the larger society through examining religion and its influence.

THE RELIGIOUS CONTOURS OF A COMMUNITY

Before probing religious phenomena in depth, we might note the broad contours of religion as they appear in one community.[1] This highlighting of religion suggests some of the questions a sociologist might ask in surveying, analyzing, and interpreting the place of religion in contemporary society. For example, what are some of the religious contours of San Antonio, Texas, where I live and work?

Conveniently for this example, the city is named for a Catholic saint, Anthony, who was patron of the eighteenth-century Franciscan Mission San Antonio de Valero (better known today as the Alamo). Catholicism has always been a central religion in San Antonio, but there is also a long history of religious and ethnic diversity within local Catholic religious expression. Large clusters of immigrant peoples brought their Catholic faith from other places, like Spain's Canary Islands, Germany, Ireland, France, and Czechoslovakia, and these groups expressed their faith differently. Not only ethnic differences but also class and status differences shaped local Catholicism. In the nineteenth century, the economically comfortable children attending the convent school encountered a different version of their religion than did the poor laborers whose religion was mediated by folk saints, mystery plays, pilgrimages, and a rich popular lore.

---

1. The metaphor of religious contours was suggested by Phillip Hammond.

Many decades later, despite homogenizing experiences of education, economic mobility, intermarriage, and changes within their church internationally, many of these Catholics still participate in different styles of worship and religiosity, in different languages, gestures, and idioms, often celebrating very different aspects of the same world religion. A sociologist would want to know more: What does it mean to believers themselves to be Catholics in San Antonio today? How does their religion help them make sense of their lives? How does it influence their everyday behavior, such as decisions in the workplace or use of leisure time? How, too, does their religion influence their political and moral behavior? Do they experience a sense of belonging to a community of fellow believers? Which fellow believers are experienced as "us," and how does that community, in turn, shape religious expression?

The rough contours of religion in San Antonio also show that Catholicism no longer holds the allegiance of all Latinos (mostly Mexican-Americans), who constitute the ethnic majority of the city's populace. The city abounds with numerous small but active Spanish-speaking Pentecostal and evangelical Protestant churches. Separate but similar in style of worship are many Pentecostal and evangelical groups serving the city's minority populations of Anglos (whites) and African-Americans. Many of these growing congregations are nondenominational, neither affiliated with nor under the supervision of any larger denominational body. Some of these churches practice a much more enthusiastic form of worship than do mainstream religious groups.

Effervescent religiosity (including such expressions as speaking in tongues, faith healing, and exorcism) has long fascinated sociologists and anthropologists, who would ask questions like: What is the social nature of these extraordinary religious experiences? For example, how does the group mediate healing to the sick person? How does the group generate and channel such fervor? What are some of the effects of this kind of religiosity on members' everyday lives? How do these groups recruit and convert new members? How are some of the social class, educational, and other background characteristics of members linked with these groups' appeal? What are the effects on group structure, beliefs, and practices of being so independent of external denominational affiliation?

San Antonio's Southern Baptist churches exemplify the fact that even religious groups identifying with mainstream denominations may find that affiliation problematic. In recent years, this mainstream denomination has been wracked with disagreement and uncertainty about the proper role of the national denominational body: Does the faction that controls the national Convention have the right to demand member congregations to conform to its particular version of Baptist belief and practice? What seems like an organization struggle, however, involves some of this religion's core values and beliefs. Southern Baptists are the most numerous Protestant denomination in the city, and local leaders are also prominent in the national struggle. Each time the national conference is held, news of Southern Baptists is a front-page story—indeed, often headlines—in the city's major newspapers.

Sociologists would be fascinated to learn more about this denomination, locally and nationally: What happens (at any level of organization) when one segment of the membership tries to establish an orthodoxy (i.e., boundaries around what one should believe or practice in order to be "one of us")? How do religious groups variously respond to societal changes and modernity? Does change within a religious group occur from the top down or from the bottom up? How does the stance of this denomination affect its larger influence in society and its members' lives?

While orthodoxy is historically rather alien to Baptists, it is both familiar and valued among many hierarchical religions. Besides Roman Catholics and Episcopalians, many ancient hierarchical religious groups are nestled into the religious contours of this city. Making news headlines, for example, was the visit of the Patriarch to the local Coptic church, Saint Anthony the Great. Coptic Christianity dates to the first-century evangelization of Egypt, and today the vast majority of adherents of this faith live in the Mideast. The Coptic enclave in this U.S. city is very much a church in diaspora (i.e., a people settled far from their ancestral homelands), as are the congregations of Lebanese Maronites, Greek Orthodox, and other ethnoreligious minorities.

Surveying these small but significant outlines on the religious horizon of the city, the sociologist would ask, among other questions: How is religion linked with ethnic and language identity? How are those identities preserved in the face of pressures for integration into the larger culture, for intermarriage, and for bilingual or English-only education and mass media? How do religious groups so steeped in tradition respond to modern challenges to traditional norms (such as efforts to redefine women's roles or to democratize decision making in the church itself)? Similar questions would apply to non–Christian traditions active in San Antonio, such as the Jews, Hindus, and Muslims.

Developments in San Antonio's American Moslem Mission mirror national changes. A visitor to this highly committed congregation would immediately notice the balance of U.S.-born black members, together with Mideastern Muslims. The Black Muslim movement was once virulently intolerant of other races and religions; it was also relatively unconnected with Islam as a world religion. Malcolm X was one of the first Black Muslim leaders to embrace a larger vision of Islam, and one faction of the movement followed his lead to greater tolerance and intense study of Islam, becoming part of the Sunni branch of the international faith.

Observing these U.S.-born blacks, with their command of Koranic scriptures and their commitment to the strictures of Muslim religious practice, the sociologist might ask: How does this religious commitment constitute a form of protest against the values and attitudes of the dominant U.S. culture? How does religion figure into racism and other intolerance in U.S. society? How do sectarian enclaves, like the Black Muslims, recruit and maintain commitment? How have the beliefs and practices of this group influenced the socioeconomic situation of its members?

Besides representatives of recognizable world religions, the religious contours of San Antonio include many emerging religious movements. Numerous meditation groups gather to practice eclectic combinations of Eastern meditation techniques, body disciplines, and healing rituals. One New Age group, for example, borrows elements from Yoga meditation, Reiki (a Japanese "new" religion) healing, aromatherapy, pagan and Christian symbolism, and modern psychology. Many of its members simultaneously remain active in their Christian denominational churches. Another group is limited to women members, who seek to create and express a nonpatriarchal religion, focusing on goddess imagery derived from ancient Celtic lore but interwoven with their own new symbols and rituals.

Particularly fascinating to the sociologist-observer is that these emerging religions are frequently found among the relatively well-educated, economically comfortable, young adult or middle-aged members of the community—precisely those who were, according to earlier predictions, most likely to become nonreligious altogether. What do these emerging forms of religion suggest about the place of religion in modern society? What is the basis of their appeal to these segments of the community? How is it possible to put together a "new" religion? How potent are changes in ritual and symbolism for producing long-lasting transformations of people's identities and way of life?

Although these emerging religious movements may not advertise themselves as alternatives to traditional religions, the religious quality of their beliefs and practices is clearly one of their functions. Less obviously religious is the vast range of quasi-religious or parareligious groups in the community. To analyze fully the religious contours of San Antonio, one might include a study of some of these groups that function very much like religion, although whether they are actually religion is a matter of definitional debate (described shortly). One of the oldest parareligious groups in the city is the Ku Klux Klan (KKK), a militant and occasionally violent secret society organized against nonwhites and non-Christians. In the early decades of this century, the Klan paraded openly in full hooded regalia in the streets of San Antonio, asserting their local "muscle" and celebrating their group identity in photographs posed before the Alamo. The rituals, symbols, and group identity of parareligions like the Klan are also worth sociological analysis.

And what should the sociologist of religion make of the quasi-religious rituals and symbols by which the nation is celebrated and understood? The religious quality of the awe inspired by the Alamo, for example, is interesting. Visitors to this "shrine of American liberty" are moved by the image of martyrs, "sacred" places, and objects arranged solemnly within the crumbling walls of the old mission to bring to collective memory certain values of the nation. The Alamo inspires collective memory among U.S. tourists and many local Anglos, despite the conveniently forgotten historical likelihood that the heroes of the Alamo were fighting for the status of Texas within Mexico (rather than for union with the United States) and were motivated by economic self-interest as much as (or more than) political freedom.

This kind of religiousness about the nation and its symbols (discussed further in Chapter 6) is of considerable interest to sociologists, who ask: How do complex, modern societies generate, maintain, and express cohesion? How do our myths and rituals involve an ongoing selective remembering? How do these beliefs and practices shape what kind of a nation we want to be and serve as a basis for national self-criticism? How are civil religious ideas and fervor sometimes used to legitimate the interests of one group over another? How are they used in political manipulation?

Is religion losing its influence? I once challenged a class to use instances from this community to persuade me that the answer was not an unqualified yes. They presented extensive data about the vitality of church-oriented religiosity in the city, such as the size of church membership rolls and number of people attending services regularly. The most impressive evidence, however, was of the influence of religious expressions not overtly identified with religious organizations. One student described a middle-aged Anglo couple who chose to live, work, and raise their children in a Latino barrio, near one of the worst public-housing tracts in the city. Their daily efforts to help organize the poor for better education, health, and community services were expressions of their religious convictions, even though they did not identify their unpaid work with any specific religious organization. Another instance was a young, college-educated Mexican-American, whose religious commitment was his main reason for devotion to the cause of voter registration (in a state that makes exercising the vote extremely difficult for poor citizens). Yet another example was a group of women who have formed a spiritual and psychological support group for battered women. Instances such as these are surely significant, if less visible, parts of the religious contours of the community.

This rough sketch of just a few parts of the religious situation in one city illustrates that a deeper analysis of the role of religion is essential for an understanding of the whole society. Sociological analysis can give us an interpretive grasp of the everyday meanings, potent traditions, dynamism, and change of religion today.

THE NATURE OF THE SOCIOLOGICAL PERSPECTIVE

Because religion is so intensely personal and reverently held, some people have difficulty grasping a **sociological perspective** on it. The sociological perspective is a way of looking at religion that focuses on the human (especially social) aspects of religious belief and practice. Religion is both individual and social. Even the most intensively subjective mystical experience is given meaning through socially available symbols and has value partly because of culturally established interpretations of such experiences. A personal religious experience such as conversion is voluntary and subjective, yet is situated in social circumstances and given meaning by social conventions. When the individual communicates that religious experience, the symbols used to interpret it are socially determined.

Two characteristics of the sociological perspective separate it from nonscientific approaches to religion: It is *empirical* and *objective*. As much as possible, sociologists base their interpretations on empirical evidence. They seek to verify their images and explanations of social reality by experimental or experiential evidence. And they look for generalizations about a larger societal, historical, and cross-cultural picture, continually asking: Of what larger phenomenon is this particular situation an example? Objectivity means that sociological interpretations of religion do not attempt to evaluate, accept, or reject the content of religious belief. Sociological researchers set aside, for purposes of the study, their personal opinions about religion and try to be as objective as possible in observing and interpreting the religious phenomenon under study. From a sociological perspective, one religion is not superior to another. Indeed, the sociological perspective does not even presume the merits of religious over nonreligious approaches.

These aspects of the sociological perspective on religion may discomfort students who find their cherished beliefs and practices dispassionately treated as objects of study. It may be disturbing to have one's own religion treated as comparable to other religions and not as superior or uniquely true. The perspectives of the religious believer and the sociologist are necessarily quite different, but neither perspective represents the whole reality of religion.

Let's use the analogy of another object of attention. In looking at a flower, for example, several people could view the same one, yet each employ a different perspective. A botanist could employ a scientific perspective, analyzing in great detail the flower's physical properties, comparing it with other plants, and classifying the flower accordingly. An artist could interpret the flower onto the canvas, abstracting those visual qualities that convey its image. A mystic could use the flower as a focal point for meditation, perhaps experiencing a sense of oneness with the flower and the rest of the natural environment. And a child could examine the flower to consider its suitability for a daisy chain. The point is that none of these perspectives has a monopoly on the reality of the flower; what is discovered from the perspective of the botanist does not disconfirm what the artist's perspective has revealed.

Similarly, the religious perspective on human life often produces a very different picture of that reality than does a sociological perspective. What is relevant to the religious believer may be irrelevant or inadmissible evidence to the sociologist. What is central to the sociologist may be irrelevant or uninteresting to the religious believer. The reality perceived from the sociological perspective cannot disprove that of the religious believer; it is not possible for a sociologist to prove, for example, whether or not a religious prophecy or vision is from God. Furthermore, the perspective of the religious believer does not disprove that of the sociologist. If, for example, a religious "revelation" contradicts sociological evidence, the fact that the believer considers the "revelation" to be from God does not disprove the sociological observation.

These two different perspectives are, however, often difficult to reconcile. The sociological perspective, by definition, lacks a key religious quality—faith. The believer accepts certain beliefs and meanings on faith. Faith implies taking certain meanings or practices for granted, implicitly trusting, not questioning. By contrast, the sociologist does not take the believer's meanings for granted but takes them as an object of study.

The sociological perspective sometimes implies that people belong to religious groups for reasons other than the truth value of the belief system. For example, sociologists have observed that upper-class persons are likely to belong to different Christian denominations than lower-class persons. In 1990, the median reported household income for Baptists was $20,600 and for Nazarenes, $21,600, compared to $33,000 for Episcopalians and $34,800 for Unitarians. Similarly, the percent of college graduates among Baptists and Nazarenes was 10.4 and 12.5 percent, compared to 52.6 and 52.7 percent for Episcopalians and Unitarians (Kosmin and Lachman, 1993:260). Unless the very circumstance of affiliation with one of these latter denominations bestows greater wealth and education, we must conclude that social class, as well as belief, influences people's choice of denomination.

The more causality that believers attribute to supernatural sources, the less their interpretation can be reconciled with a sociological perspective. The very fact of treating certain interactions like conversion as human behavior, as an object of sociological study, is often incompatible with the basic beliefs of some religious groups. Sociology must necessarily "bracket" (i.e., methodologically set aside) the crucial religious question: To what extent is this action *also* from God? This does not mean that sociology treats religious behavior and experience as "merely" human. Important dimensions of religion may not be accessible to sociological interpretation. Nevertheless, whatever else they may be, religious behavior and experience are also human and are therefore proper subjects for sociological research and understanding. With these limitations of the sociological perspective in mind, let us explore some definitions of religion within this perspective.

DEFINING RELIGION

The purpose of a sociological definition is to bring order to a vast array of social phenomena. The definition of any concept establishes (somewhat arbitrarily) boundaries around those phenomena to be considered as instances of that concept. Thus, to focus our attention on phenomena to be considered "religion," we must establish a working definition of religion.

Try writing your own sociological definition of religion. It should be broad enough to include all kinds of religion but narrow enough to exclude what is similar to, yet not the same as, religion. One way to construct your definition is to think of specific instances of "religion" or "nonreligion" and see whether your definition includes or excludes them. For example, does your definition include the religions of Asia or Africa? Does your definition use terms (e.g., *church* or *god*) that, unless clarified, may apply only to some religions? Is your definition narrow enough? Does it distinguish between religious commitment and other commitments such as allegiance to a social club, an ethnic group, or family? There are several phenomena that only some sociologists define as religion (e.g., magic, superstition, witchcraft, astrology, spiritualism). Does your definition account for these phenomena? Why? How do the phenomena of nationalism, Marxism, atheism, or humanism fit into your definition? Does your definition include or exclude them? Why? Do you consider psychotherapy, sports, or rock music to be "religion"? Some sociological definitions of religion are so broad as to include all of these phenomena; others exclude some or all of them. We will examine some of these definitions and some arguments for excluding or including certain characteristics within a definition of religion.

Definitions of religion are an issue of serious debate, not just academic wrangling, in sociology. How one defines religion shapes one's explanation of its role in society. Different definitions of religion result in different interpretations of issues such as social change (see Chapter 7), modernity (see Chapter 8), and nonchurch religion (see Chapter 4). Indeed, part of the problem in determining a definition of religion that is satisfactory for analytical purposes is that the issue of what is "properly religious" is a continuing controversy in modern societies. In recent years, for example, the courts have examined several cases in which their verdicts hinged on deciding whether a given phenomenon was or was not a valid instance of religion. If the definitive boundaries of religion are not very firm for everyday social and legal purposes—as they are not—then producing a neat sociological definition may be counterproductive and misleading (Fenn, 1978:29).

Definition as Strategy

It is useful to approach sociological definitions as strategies rather than as "truths." A definitional strategy narrows the field under consideration and suggests ways of thinking about it. Definitions can be evaluated according to how useful they are for a given task. What interpretive tasks does any given definition help to accomplish? Pragmatically, then, it is sometimes helpful to use different, even opposing definitional strategies to approach a phenomenon.

Two major strategies used by sociologists of religion are **substantive** and **functional definitions**. Substantive definitions try to establish what religion *is;* functional definitions describe what religion *does.* These approaches can be illustrated by analogy with definitions of the concept *chair.* A substantive definition might state that a chair is an object of furniture that usually has four legs and a back; the definition might add further physical details to distinguish a chair from a sofa, bench, or toilet. A functional definition of *chair* might state that it is a seat, usually for one person. This functional definition is somewhat broader, allowing objects that various cultures use as seats but that may have no legs at all. Defining *chair* is easier than defining *religion,* however, because one can point to various physical objects called *chair* and then derive a set of distinguishing characteristics from observing them. Religion does not have clear-cut physical properties, nor are its characteristics readily ascertained and agreed on. Indeed, religious groups themselves disagree about religion's essential nature, partially accounting for much diversity of belief systems.

Substantive Definitions

A substantive definition defines what religion *is*. It attempts to establish categories of religious content that qualify as religion and other categories specified as nonreligion. Melford Spiro offers a straightforward substantive definition of *religion:* "An institution consisting of culturally patterned interaction with culturally postulated superhuman beings" (1966:96). By "institution," this definition refers to socially shared patterns of behavior and belief. All institutions include beliefs, patterns of actions, and value systems; the critical feature of religion is that the beliefs, patterns of action, and values refer to "superhuman beings" (Spiro, 1966:98).

Spiro's definition of religion is a good example of a sociological definitional strategy because all of the categories in the definition—"institution," "culturally patterned," "culturally postulated"—are sociologically relevant. His explication of the concept of "superhuman beings" is also sociologically significant because it emphasizes the sense of *power.* "Superhuman beings" are defined as those having greater power than humans, beings who can help or hurt humans but can be influenced by human action (Spiro, 1966:98). Power is one of the most important concepts in the sociology of religion, and a definition that emphasizes power can be useful. Other substantive definitions of religion use similar concepts, including "nonhuman agencies," "supernatural realm," "super-empirical reality," "transcendent reality," and "sacred cosmos."

Besides these theoretical approaches to religion, many empirical studies of institutional religion implicitly use substantive definitions for a different reason: They are easier to adapt to survey research. A somewhat oversimplified example is seen in the questionnaire that asks the respondent's religion, then lists the following alternative choices: "Protestant," "Catholic," "Jew," "other," "none." The underlying assumption of such a question is a substantive definition of religion as a specific Western religious institution.

The major advantage of substantive definitions is that they are more specific than functional ones. They are more explicit about the content of religion. Substantive definitions tend to be narrower and neater than functional definitions; using them, one can specify whether a phenomenon is or is not religion. Substantive definitions also tend to correspond more closely than functional definitions to commonsense notions of religion because they are generally based on Western—especially Christian—ideas about reality. For example, the distinction between *natural* and *supernatural* is a product of Western thinking, such as the elaborate medieval cosmographies (i.e., pictures of the universe) that classified natural and supernatural beings on numerous levels. Although hierarchies of archangels and seraphim may not be relevant in contemporary society, the division of natural from supernatural still seems familiar.

Substantive definitions are appropriate for studying religion in relatively stable societies, which present few problems with issues of social change and cross-cultural applicability. Substantive definitions are problematic precisely because they are historically and culturally bound, based on what was considered religion in one place and time. Because of their basis in Western historical experience, substantive definitions are often too narrow to account for non-Western religious phenomena. Substantive definitions are sometimes deceptively neat. Without specifying the functions of supernatural beings, for example, it is impossible to distinguish the gods from ghosts, Santa Claus, and the tooth fairy.

Some sociologists identify religion with church-oriented religiosity; relatively few non-Western societies, however, have formal organizations like churches. Does this mean that such societies therefore have no religion? If supernatural, nonempirical, or nonrational entities do not figure importantly in a society, does that society lack religion? Confucianism (the state religion of China from about 200 B.C. until the early twentieth century A.D.) is essentially a set of principles of order, especially regarding social relationships surrounding authority and kinship. It does not include the worship of any gods, although nature and ancestors receive much ritual reverence. If supernaturalism is a key criterion of a substantive definition, then Confucianism and some strains of Buddhism (in which there is no deity and no dualistic conception of the nature of being) would not be considered religions. A current also exists within contemporary Christian thinking toward leveling this-worldly and other-worldly distinctions (e.g., the theologies of Pannenberg, Metz, Solle, and Moltmann). Would Christianity without supernaturalism cease to be a religion?

Substantive definitions have difficulty accounting for religious change. If religion is identified only in terms of religious expression in one historical period, any change from that form of expression looks like nonreligion. Many theories of religious change, for example, start from a notion of a time when people were "really religious." In Christian history, the thirteenth century is usually identified as the period when religion was a powerful force in the entire society and thoroughly interwoven with other aspects of life—work, education, politics, family, and so on (note, however, that recent historical scholarship reveals "there never was an 'age of faith'" [Obelkevich, 1979:6]). If one equates that image of religion with "real religion," any change from that pattern is viewed as a trend toward nonreligion ("secularization"). A parallel problem occurs in studies of the family. If *family* is defined in terms of its historical manifestations (e.g., concrete blood or marriage ties), many contemporary living arrangements do not qualify as family. Do two unmarried people and their offspring constitute a family? What about a single woman and her adopted children? What about six unmarried persons sharing sexual partners and collectively caring for their offspring? Because cultures constantly and sometimes rapidly change, it is difficult to create a substantive definition that applies through time. On the other hand, it is difficult to document historical changes in religion's place in society if we lack a sufficiently specific definition of religion. If we compare, for example, the situation of religion in the French Revolution with the religious situation of the American Revolution, we must be able to identify exactly which aspects of social life we mean by *religion*.

As a definitional strategy, substantive definitions have advantages: They are more specific and amenable to empirical studies of religion. On the other hand, they tend to be more historically and culturally bound than functional definitions. Substantive definitions, in short, produce a very different interpretation of social change than functional definitions (for essays promoting substantive definitional approaches, see Berger, 1967:175–178; Robertson, 1970:34–51; Spiro, 1966).

Functional Definitions

A functional definition of religion emphasizes what religion *does* for the individual and social group. Accordingly, religion is defined by the social functions it fulfills. The content of religious belief and practice is less important for this definitional strategy than the consequences of religion.

Clifford Geertz's definition of religion is a useful example of a functional definition: "A *religion* is: (1) a system of symbols which acts to (2) establish powerful, pervasive, and long-lasting moods and motivations in men by (3) formulating conceptions of a general order of existence and (4) clothing these conceptions with such an aura of factuality that (5) the moods and motivations seem uniquely realistic" (1966:4).

This definition emphasizes several sociologically important concepts. The most important element is the provision of meaning because the establishing of shared meanings (i.e., symbols) is an essentially social event. The definition also accounts for social structural and social psychological functions through the concepts of *moods, motivations,* and *factuality* (which also relate to the notion of *institution,* as Spiro uses it). The distinguishing features of this definition are the "conceptions of a general order of existence" and their realism. According to Geertz, people interpret events and experiences as meaningful by linking them with a larger sense of *order.* This larger sense of order is perceived from a religious perspective as entirely real—even more real than mundane events and experiences. Empirically this distinction means that the content of religious beliefs and practices does not matter so long as it serves to symbolize some transcendent order to believers (Geertz, 1966:12ff.).

Some of the functions identified by various sociological definitions of religion include the provision of ultimate meaning, the attempt to interpret the unknown and control the uncontrollable, personification of human ideals, integration of the culture and legitimation of the social system, projection of human meanings and social patterns onto a superior entity, and the effort to deal with ultimate problems of human existence. Some of these functions are described further in later chapters.

One distinction used in many functional definitions is the social attribution of the **sacred.** Whereas the natural/supernatural distinction of some substantive definitions refers to the intrinsic quality of the *object* of worship, the sacred/profane distinction refers to the *attitude* of worshippers. The realm of the sacred refers to that which a group of believers sets apart as holy and protects from the "profane" by special rites and rules (Durkheim, 1965:62). The sacred is regarded as especially powerful and serious. For example, a communion wafer may be regarded as nothing but a piece of bread by nonbelievers, but Christian worshippers regard it as special and treat it differently from ordinary bread. Natural objects such as candles, beads, books, water, oil, and wood can be regarded as sacred. Thus, nonsupernatural cultural systems (e.g., nationalism) could be viewed as sacred systems because of the attitude of their followers. Nationalist groups often treat the state and its symbols—flags, national holidays, shrines—as sacred.

Functional definitions of religion include all that substantive definitions identify as religion, but they are usually much broader. Both substantive and functional approaches would define the phenomena of Calvinism, Roman Catholicism, Methodism, Mormonism, and Reform Judaism as "religion." The inclusiveness of some functionalist approaches, however, extends to some phenomena that substantivists identify as nonreligion: ideologies, ethos, value systems, worldviews, interpersonal relations, leisure activities, voluntary associations, and so on. Geertz (1966:13) points out that his definition of religion would include, for example, golf—not if a person merely played it with a passion, but rather if golf were seen as symbolic of some transcendent order.[2] Functional definitions often include as "religion" phenomena such as nationalism, Maoism, Marxism, psychologism, spiritualism, and even atheism. The religious qualities of less comprehensive human activities, such as sports, art, music, and sex, are incorporated into some functionalist definitions as well.

The primary advantage of a functionalist definitional strategy is its breadth. Functional definitions tend to be better than substantive definitions for encompassing cross-cultural, transhistorical, and changing aspects of religion. Functional definitions encourage the observer to be sensitive to the religious quality of many social settings.

A drawback of some functional approaches to religion, however, is their assumption that society has certain functional requisites. This assumption implies that society requires certain social functions, some of which are uniquely fulfilled by religion. Some theories, for example, assume that society requires cultural integration (i.e., a common core of beliefs, values, and commitments). If religion is then defined as that which provides cultural integration, the theory implies that religion is a requisite for society's existence (Parsons, 1944:86). Such an argument is circular, describing religion as that which provides that which is defined as religion. The assumption that religion is necessary is unproven.

The breadth of functional definitions is a mixed blessing. While functional definitions are less culturally and historically bound, this inclusiveness makes it difficult to use them for empirical studies requiring neat, quantifiable categories. Some critics say that functional strategies result in all-inclusive categories, defining virtually everything human as religion.

---

2. Geertz, 1966:13. Geertz was probably not surprised by the actual development and international expansion of Japanese "golf religion," the Church of Perfect Liberty. One minister of this church explained that golf is "a little miniature of life. We can learn about ourselves through golf. Our teachings say that in life, anger, worry, and sorrow destroy success, just as in golf. Like golfing, life is an individual game, where you seek individuality; in golf, as in life, it's always the next shot, another chance for perfection" (quoted in the *New York Times*, May 30, 1975).

Sociologists using functional definitions need to show why they include phenomena that participants themselves do not consider to be religious. From a functionalist standpoint, a good case could be made for considering psychotherapeutic groups as essentially religious phenomena. They give their members a sense of purpose and meaning, and they use symbols (e.g., words and gestures) to establish moods and motivations that members believe will help them cope better with "real life." Members of these groups, however, often do not recognize their own beliefs and behavior as religious; they may even intensely disavow the religious label. On the one hand, it seems fair to accept the participants' notion of what they are doing; on the other, it is an honorable sociological tradition to point out the facades behind which people mask their activities. Few sociologists would accept at face value prison staff members' assertions that prisons are mainly for the purpose of rehabilitating prisoners. Similarly, there may be good sociological reasons for questioning whether vehemently nonreligious groups are actually nonreligious. Representatives of Transcendental Meditation (TM) have argued strongly in court that TM is not religion and therefore should be allowed in the public schools. The religious aspects of TM's beliefs and practices are downplayed because the desired image of TM is as technique rather than religion. The fact that a movement's desired image may be promoted by denying its religious qualities is worth studying, but sociologists need not accept a group's self-definition (for essays promoting functional definitional approaches, see Geertz, 1966; Lemert, 1975; Luckmann, 1967, 1977).

The choice between substantive and functional definitions is finally a matter of strategy. Each approach has advantages that may recommend it to certain sociological tasks. The two strategies, however, result in very different interpretations of various issues such as social change, secularization, the relationship between religion and other institutions in society, and new forms of religion. An awareness of the limitations and scope of each definitional approach will enable us to evaluate these other issues more critically. This book draws largely on examples that would be defined as religion by both substantive and functional definitions. At the same time, however, because of specific attention to what *individuals* hold as religiously meaningful, this book also explores phenomena that only functional approaches would consider to be religion. This approach seems fruitful for two reasons: It allows us to apply a sociology of religion perspective to an interesting range of phenomena, and it raises broader theoretical issues.

ASPECTS OF RELIGION

There are many facets of religion that are important for the sociologist to recognize. In Western societies, much emphasis is placed on formal beliefs. Religious education consists of informing children of what one's group believes; children read religious textbooks of explanations, learn catechisms (i.e., statements of propositions to which a believer should assent), study scripture, hear sermons. Christianity and Judaism place relatively great emphasis on intellectual and formal belief. In other cultures, however, formal beliefs are relatively unimportant. Other important aspects of religion include ritual, religious experience, and community. Children learn all that is relevant to their faith by participating in group ceremonies and imitating their elders' experiences. The following aspects of religion refer not only to the individual's religion but more generally to the ways in which the religious group organizes itself to focus its shared meanings.

Religious Belief

Every religion has an essential **cognitive aspect.** The religion shapes what the adherent *knows* about the world. This cosmic knowledge organizes the individual's perceptions of the world and serves as a basis for action. If I believe, for example, that active, powerful evil spirits surround me, I will perceive "evidence" of their activity, and I will take actions to protect myself from them. My belief in evil spirits helps me explain other aspects of my life, such as why I cannot get a job and why I feel depressed and anxious. The same belief suggests appropriate actions, such as necessary prayers to ally myself with good spirits or use of amulets to ward off the influence of these evil spirits.

There is a tendency in modern Western societies to treat religious beliefs as "mere opinion," as opposed to empirical beliefs, which are treated as "knowledge." This distinction hides the fact that both types of belief are "knowledge" to the individual who holds them. If a person considers evil spirits to be real, they *are* real in their consequences; they shape the person's experience and actions. The individual who believes that evil spirits cause illness and the individual who believes that germs cause illness are both acting according to their "knowledge."

Most of us are quite familiar with formal religious beliefs, the statements to which adherents of a given religion are supposed to assent, such as a catechism or a creed. The entire enterprise of theology out of which formal beliefs are developed represents a highly specialized and intellectualized approach to religion. But religion also includes less formal kinds of beliefs such as myths, images, norms, and values. Myths of creation and rebirth, for example, are told and enacted in dance and song. These other kinds of symbolization are often more potent influences on behavior than intellectual beliefs.

Religious beliefs are not mere abstractions that are irrelevant to everyday life. People use their beliefs to make choices, interpret events, and plan actions. Myths, one form of religious belief, are paradigms of human existence. There are myths about all major aspects of human life: birth and rebirth, creation and transformation, one's people and place, marriage, work, fertility, sterility, and death. Myths are stories that provide a rationale for a group's actions. They can be metaphors for concrete social structure and for real human events. Individuals draw on these interpretations to give meaning and direction to their own actions. Indeed, the very language in which beliefs are expressed structures believers' perceptions of the world.

Religious beliefs also inform the individual what action is good and desirable or bad and to be avoided. They may tell the individual that marriage is good and right because the holy marriage of the gods is to be copied by humans. Religious beliefs may inform the individual that eating other people is wrong because the gods value people and therefore do not define them as appropriate food. Thus, an entire range of values, norms, and attitudes derives from religious beliefs.

Religious Ritual

Ritual consists of symbolic actions that represent religious meanings. Whereas beliefs represent the cognitive aspect of religion, ritual is the enactment of religious meaning. The two are closely intertwined. Beliefs of the religious group give meaning and shape to ritual performances. Ritual enactments strengthen and reaffirm the group's beliefs. They are ways of symbolizing unity of the group and, at the same time, of contributing to it. Ritual helps generate religious conviction. By ritual action, the group collectively remembers its shared meanings and revitalizes its consciousness of itself. This has important consequences for both the group and the individual member. The group renews its fervor and sense of unity, and individual members come to identify with the group and its goals (Durkheim, 1965:420; Geertz, 1966:28).

Ritual is an effective way of transforming space and time. Ritual places (such as a mountain or a shrine) can be transformed into the locus of power and awe. Time, too, can be changed, becoming a metaphor for sacred meanings and a catalyst for religious experience (Kertzer, 1988; Smith, 1987). For example, a pilgrimage is a ritual journey in which the participant enacts (both literally and metaphorically) a transition from one situation and self to another (Turner and Turner, 1978; see also Frankenberg, 1986; Pace, 1989).

Ritual practices (such as eating and drinking communion, kneeling, anointing, singing) are often ways of transforming body metaphors into physical, mental, and/or emotional realities (McGuire, 1996; see also Csordas, 1994). For example, an early Buddhist teacher (Kukai, 774–835 C.E.) taught "We become the buddha through our bodies" (quoted in Kasulis, 1993:310). He emphasized use of specific spiritual practices, including disciplined bodily postures, harmonizing sounds, and meditation practices, the end of which was to embody buddhahood.

Various religious groups place different emphasis on ritual. Eastern Orthodox, Roman Catholic, and Episcopalian Christians emphasize overt ritual more than Baptists or Methodists. The use of symbolism, such as processions, sacraments, candles, icons, and chanting, aids the collective remembering of the group's shared meanings. Even groups that do not consciously use rituals repeatedly symbolize their unifying beliefs. "Revival" meetings often emphasize spontaneity in worship and downplay formal ritual. Nevertheless, though the words of their prayers may not be formally set, members respond especially fervently to prayers that fit a familiar formula, and their responses are often equally stereotyped and expected. These periodic reenactments are just as much ritual as the "high church" ceremonies of Episcopalians.

As these examples suggest, the *content* of an act is not what makes it a ritual act; rather it is the symbolic meaning attached to the act by participants. This symbolic value is what distinguishes, for example, lighting a candle to beautify one's dinner table and lighting a candle of the menorah to commemorate the Feast of Lights (the Jewish holiday of Hanukkah). There are also ritual acts that have, over time, become remote from their symbolic significance. For example, Roman Catholic priests wear different colored vestments to celebrate Mass on various holidays, but many people in the congregation may not even know what the symbolism of pink or green vestments means, and thus the ritual significance of the vestments is empty for them. Ritual performed for its own sake, empty of meaning for participants, has led to the notion that ritual itself is deadening.

It is not that ritual per se is dead but rather that the relationship between the symbolism and the group's shared meanings is weak or severed. Events or beliefs that a given ritual has symbolized may, over time, become less important or even forgotten. Perhaps the group may have moved on to new ways of symbolizing itself and its beliefs. Or perhaps the empty ritual may be symptomatic of weak ties in the group itself, that it has no essential unity to celebrate. By contrast, ritual that is a vital symbolic expression of important meanings for the group can be a sustaining, strengthening, and enlivening experience of unity.

The dynamic potential of religious ritual suggests its link not only with religious belief but also with religious experience. Religious symbols, expressed in beliefs and rituals, have *real* power, which can be experienced personally by the individual. Ritual words and ceremonies can evoke experience of awe, mystery, wonder, and delight. Religions often emphasize the power of ritual words, as exemplified by the seriousness surrounding the pronouncement of the words "This is my body" in Christian communion services or by the expectation surrounding the exclamation "Heal!" in a faith-healing service. Ritual has the potential to produce special religious experiences for the group and its individual members.

Religious Experience

Religious experience refers to all of the individual's **subjective involvement** with the sacred. Although such experience is essentially private, people try to communicate it through expression of beliefs and in rituals. A communal ritual may be the setting for a personal religious experience. Thus, a person receiving communion (i.e., a communal ritual) in a Christian worship service may also experience an intensely subjective awareness of God. Prayer, meditation, dancing, and singing are other common settings for personal religious experience. Similarly, even a private experience has a social element because socially acquired beliefs shape the individual's interpretation of religious experience. The symbolism of various religious traditions shapes the interpretations of even highly mystical experiences through such images as the pilgrimage, perfect love and marriage, and rebirth or transformation (Underhill, 1961:125–148).

Individual religious experiences vary considerably in intensity. They range from momentary senses of peace and awe to extraordinary mystical experiences. Different religions place different emphases on religious experience. Most Christian denominations do not actively encourage highly emotive religious experiences, whereas in some Pentecostal groups, these experiences are central and eagerly sought. In many religions, extraordinary and intense experiences are segregated, appropriate only for certain members or on certain occasions. Thus, among the peoples of the northern Asian arctic region (e.g., Eskimos), extraordinary religious experiences are expected of certain members called shamans. *Shamans* are religious specialists who have undergone an intense encounter with sacred forces and emerged with special powers to effect good or evil on behalf of the rest of the group (see Eliade, 1964). Another example is the segregation of occasions for special religious experiences (e.g., initiation rites).

The content of religious experience varies. It may include pleasurable aspects such as a sense of peace, harmony, joy, well-being, and security. Religious experience may also produce terror, anxiety, and fear. While the content of the experience partly depends on the group's beliefs about what is being encountered, both the pleasurable and frightening experiences are related to the sense of power or force with which the sacred is believed to be endowed. The individual who experiences a sense of security does so because of the power of the sacred to protect from harm; the individual who experiences great fear does so because of the power of the sacred to cause grave harm. The notion of the sacred thus entails both harmful and helpful aspects. Personal experiences of this power can be overwhelming.

William James (1958:67) quotes an account of a religious experience that illustrates the intensity of some of these aspects:

> I remember the night, and almost the very spot on the hilltop, where my soul opened out, as it were, into the Infinite, and there was a rushing together of the two worlds, the inner and the outer. It was deep calling unto deep,—the deep that my own struggle had opened up within being answered by the unfathomable deep without, reaching beyond the stars. I stood alone with Him who had made me, and all the beauty of the world, and love, and sorrow, and even temptation. I did not seek Him, but felt the perfect unison of my spirit with His. The ordinary sense of things around me faded. For the moment nothing but an ineffable joy and exaltation remained. It is impossible fully to describe the experience. It was like the effect of some great orchestra when all the separate notes have melted into one swelling harmony that leaves the listener conscious of nothing save that his soul is being wafted upwards, and almost bursting with its own emotion. The perfect stillness of the night was thrilled by a more solemn silence. The darkness held a presence that was all the more felt because it was not seen. I could not anymore have doubted that *He* was there than that I was. Indeed, I felt myself to be, if possible, the less real of the two.

This kind of religious experience represents an "alternate state of consciousness"—a situation in which the individual's consciousness is relatively remote from the sphere of everyday reality. The person may experience being out of one's body or being one with something or someone else. For example, one person described such an experience as follows: "I came to a point where time and motion ceased. . . . I am absorbed in the light of the Universe, in Reality, glowing like fire with the knowledge of itself, without ceasing to be one and myself, merged like a drop of quick-silver in the Whole, yet still separate as a grain of sand in the desert" (quoted in Happold, 1970:133).

Alternate states of consciousness are not necessarily religious. Substances such as peyote, for example, can produce alternate states of consciousness, but the context in which they are taken (e.g., ritual) and the meanings attached to taking them (i.e., symbolism) determine whether they produce religious experiences. "Alternate states of consciousness" or "peak experiences" are only one extreme of a continuum of religious experiences (see, e.g., Maslow, 1964). Less dramatic religious experiences include a sense of the presence of God, a moving conversion experience, or a deeply absorbing ritual experience.

Contemporary society often ignores or discourages such experiences. Peak experiences such as those described, however, may be more common than generally acknowledged. Approximately 35 percent of a U.S. sample (N = 1,467) had at some time experienced feeling "very close to a powerful, spiritual force that seemed to lift you out of yourself" (McCready and Greeley, 1976:132; see also Greeley, 1987; Hood, 1985; Hood et al., 1989; Spilka et al., 1985). Perhaps a self-fulfilling prophecy may operate in a culture's encouragement or discouragement of such experiences. In those cultures that value and encourage special religious experiences, members have felt and acknowledged these experiences; in cultures that devalue or discourage these experiences, members neither seek nor recognize them.

While a sizable proportion of the U.S. populace has had some special spiritual experience, the culture in general does not particularly value such experiences. This culture places great emphasis on rational, intellectual, "objective" ways of knowing. Religious experience, by contrast, is a way of knowing by subjectively apprehending a reality. One woman described her own experience as follows: "I felt a deep sense of warmth and security and a startling awareness of how much the Lord cares for me personally." This kind of knowledge, although it may correspond with the official belief system of a religious group, is not simply learned, deduced, theologically debated, or received from church authorities.

Religious Community

Religious experience may also include the awareness of belonging to a group of believers. Rituals often remind the individual of this belonging, creating an intense sense of togetherness. Communal religious settings can produce a resonance of several individuals' experiences and thus an even deeper sense of sharing "inner time" (see Schutz, 1964). Like making music together, such

shared religious experience can create a sense of sharing vivid present time—
an experiential communion (Neitz and Spickard, 1990). For example, participants in the several days and nights of a Navajo "chant" can, through ritual
acts, attune their individual experiences toward the culturally valued sense of
harmony between the individual, social group, and environment (Spickard,
1991). The *ritual* creates a group religious *experience,* which parallels and rein-
forces *belief* and ties together the *community.* All four elements are, in this example, tightly intertwined (although they are not necessarily so close in all
religious settings).

The community of believers may be formally or informally organized.
Formal specialization of a group into an organization such as a church is a relatively recent historical development. The religious group—formal or informal—is essential for supporting the individual's beliefs and norms. Coming
together with fellow believers reminds members of what they collectively believe and value. It can also impart a sense of empowerment to accomplish their
religious and everyday goals. And the nature of the religious community illustrates the social context of religious meaning and experience. We will explore
this aspect further in the next two chapters.

THE DEVELOPMENT OF THE SOCIOLOGY
OF RELIGION

The very development of sociology and the sociology of religion in particular
is rooted in some of the social processes we will examine (especially the process
of secularization). Only as beliefs and social practices were perceived as something other than taken-for-granted realities could a scientific study of these
"social facts" become possible. The development of Enlightenment rationality
and paradigms of scientific thought encouraged examination of religion as
human action (see Robertson, 1970:7–33).

Ironically, these same developments also acted on religion, possibly changing it fundamentally. Thus, sociology has some complicity in this alteration; to
a certain extent, popularization of the sociological perspective on religion has
the effect of altering the nature of religion itself. The social sciences, as heirs
to Enlightenment humanism, have interpreted religion as a human construction, undermining its claims to transcendence. Some recent sociologists have
called attention to this perspective and questioned whether the social sciences
can validly claim to be value free if their very choice of an interpretive stance
implies certain value judgments. They ask whether it is possible to have a social science of religion that allows for the truth of religion (see Anthony and
Robbins, 1975; Bellah, 1970; Greeley, 1972b; and especially Johnson, 1977).

An important strand of classical sociology of religion has a prophetic ring.
Max Weber's doubts about the future of a fully "rationalized" society, Emile
Durkheim's concern about the impact of social change upon the cohesion and

moral unity of society, Georg Simmel's insights into the threat to the individual of increasing societal controls, and Karl Marx's questioning the continued legitimacy of capitalistic modes of organization all contain strong critical themes (see Durkheim, 1965; Marx, 1963; Simmel, 1971; Weber, 1958a). These theorists saw their contribution to a sociology of religion as part of a larger examination of the nature of modern society. They asked questions such as: In what direction is modern society going? How did it develop to this point, and what factors influence its further development? What is the place of the individual in modern society, and what is the impact of these broader social changes on the individual? These classical contributions are emphasized in several chapters of this book.

Later developments in sociology, especially in the United States, did not continue these themes of concern. The sociology of religion was largely ignored until the late 1940s and 1950s, when an upswing in church membership and participation, combined with increasing interest of religious leaders in institutional research, provided an impetus for sociologists to focus again on religion. The dominant theoretical perspectives on religion then current in sociology were, however, very narrow (see Berger and Luckmann, 1963). Religion was defined strictly in terms of its formal organizational setting: the institutional churches. Much of the research was circumscribed by the needs and interests of these organizations. Demographic features of parishes were studied, for example, in order to predict future building and staffing needs. The standard methodological approaches of that period were similarly restricting; survey research techniques were best suited for measuring clearly definable institutional behavior or opinions. A survey interviewer could more readily ask a respondent to name a denominational preference than to describe fully an entire personal meaning system.

This stage in the development of sociology of religion was characterized by several assumptions, notably the identification of "church" as "religion." Another general assumption of this phase was that individual religiosity could be equated with individual participation in religious institutions. A corollary assumption was that the subjective aspects of religion could be tapped by identifying the individual's religious opinions or attitudes (see Luckmann, 1967:20–23; some criticisms of these assumptions are developed further in Chapter 4). Perhaps as a by-product of this narrow definition of the field's scope, the sociology of religion has been rather isolated from other sociological knowledge and subdisciplines. It was not until the 1980s that sociology of religion seriously expanded its horizons to explore the important connections with other subfields such as women's studies, sociology of health, and mass media (Beckford, 1990).

Beginning in the mid-1960s, the sociology of religion greatly broadened its theoretical focus and research interests. New developments in Western religion spurred this expansion. Following publication of Anglican Bishop John Robinson's provocative book *Honest to God* (1963), theologians debated the viability of culturally bound Christian theology, and a school of theology proclaiming the "Death of God" developed. Simultaneously, Catholic theologians explored new interpretations of their tradition, and the Second Ecumenical Council (Vatican II) established far-reaching structural changes in that church. Religious participation in various human rights movements, the articulation of various national "civil religions," and the emergence of numerous new religious movements raised important sociological questions that pressed the discipline to expand its interpretive scope. Political events have brought religious issues and groups to the fore, stimulating research and analysis on such themes as the prominence of religion in conflicts in Iran, Lebanon, Sri Lanka, and Northern Ireland.

Recent sociology of religion is characterized by its broader emphasis. It encompasses not only Western but also non-Western religious expression; it sees parallels between religious behavior in this society and simpler societies; it is interested in both institutional and noninstitutional religious behavior. Its theoretical concerns reflect both a thoughtful return to classical themes and an ability to go beyond the classics. This development is part of a generally larger interest within sociology in building critically on classical theory. Recent sociology of religion is more central to the larger sociological enterprise than before, asking what can we understand about the nature of society through examination of religion? This focus leads current sociology of religion to themes that are central in contemporary theory, such as the legitimation of society, the individual-to-society link, and the impact of social change (Robertson, 1977).

SUMMARY

This chapter has examined key features of the sociological perspective on religion: its objectivity and empirical focus. Definitions of religion are matters of serious academic debate, and the choice of definitional strategies (i.e., substantive or functional) influences the observer's conclusions about the place of religion in contemporary society. To understand how religious groups organize themselves around shared meanings, it is necessary to examine several aspects of religion: religious belief, ritual, experience, and community.

The development of the sociology of religion is itself rooted in some of the social processes that have influenced religion's role in society. Only as beliefs and social practices lost their taken-for-granted quality could a scientific study of them proceed. Sociology of religion was central to the focus of classical sociology—its concern about the emerging forms of societal legitimacy, social controls, and the place of the individual in modern society. Recent sociological thought examines these issues with renewed fervor.

THE FOCUS OF THIS BOOK

These theoretical themes are the basis of the organization of this book. Chapter 1 has described the scope of the field and illustrates how different strategies for defining religion result in different interpretations of the location of religion in contemporary society. Chapter 2 discusses the ways in which religion provides meaning and belonging for the organization of individual lives and social groups. The capacity of religion to provide meaning and belonging is further illustrated in Chapter 3, which explores how the individual's religion is socially acquired, maintained, or changed. The individual's religion cannot be narrowly equated with participation in specifically religious organizations, as Chapter 4 shows; rather, it comprises a complex array of beliefs and practices organized (sometimes not very coherently) into a personally meaningful and useful way of thinking and doing—a worldview. People do, however, affiliate themselves with groups to express and maintain their religions, and specifically religious organizations have identifiable characteristic patterns, as described in Chapter 5 on religious collectivities. These patterns are dynamic, resulting from tensions within the religious group and between the religious collectivity and the larger society. Chapter 6 explains the dual potential of religion to promote social conflict and to enhance social cohesion; these aspects of religion are also related to religion's role in social change. Chapter 7 examines the conditions under which religion is likely to promote or inhibit social change. Various social changes also influence religion and its social location. Chapter 8 discusses several interpretations of such changes connected with modernization and the structure of the modern world. These theories are directly related to some of the broader issues of contemporary sociology.

Because of this focus, this book de-emphasizes much of the respectable but narrow material from the sociology of institutionally specialized religion. Although questions about the organization of churches, synagogues, and other religious bodies are important, they neither address the broader issues raised by contemporary sociology nor lend many new insights to sociology as a whole. Thus, this book will include relatively little discussion of the clergy, congregational polity, or membership growth or decline. For the same reason, methodological issues discussed in this volume suggest ways of exploring individual religiosity, which may often be only loosely related to any religious organization such as a denomination. The impact of globalization and later modernity on religious life has made these questions much more significant than they once were, and certainly more significant than any analysis that limits itself to the study of religious organizations. Yet much of this emphasis is relatively new, so not many research results exemplify this broader focus, and whole areas of relevant information remain untapped. The sociology of religion has produced much excellent work, but some of the weaknesses of data, methodology, and theory pointed out in this book suggest areas for further development. The implications of contemporary sociology are far reaching, and I hope to share with the reader the intellectual excitement of pursuing some of these issues.

RECOMMENDED READINGS

Robert N. Bellah. "Christianity and Symbolic Realism." *Journal for the Scientific Study of Religion* 9(2), 1970:89–96.

Clifford Geertz. "Religion as a Cultural System." *Anthropological Approaches to the Study of Religion,* Michael Banton, ed. London: Tavistock, 1966:1–46.

Melford Spiro. "Religion: Problems of Definition and Explanation." *Anthropological Approaches to the Study of Religion,* Michael Banton, ed. London: Tavistock, 1966:85–126.

2

The Provision of
Meaning and Belonging

R eligion represents an important tie between the individual and the larger social group, both as a **basis of association** and as an **expression of shared meanings** (cf. Greeley, 1972a). In this chapter, we examine the provision of meaning for the individual and the larger social group. This provision of meaning is linked with the communal aspect of religion; a community of believers maintains a meaning system and mediates it to the individual. The religious processes described in this chapter are interesting in themselves, but we must ask the broader question: What does our understanding of religion tell us about society itself? Comprehending how the individual and the larger social group are linked in a community of shared meanings is important for the larger issue of how society is possible.

RELIGION AND SYSTEMS OF MEANING

The capacity of religion to provide meaning for human experience has been a major theme in the sociology of religion since its early emphasis in the works of Weber (1958a). **Meaning** refers to the interpretation of situations and events in terms of some broader frame of reference. For example, the experience of losing one's job is given meaning when it is interpreted as "bad luck," "market forces," "the boss is trying to get rid of pro-union workers," "ethnic discrimination," or "God's will." As this example shows, meaning here refers to ordinary, everyday interpretations of experience. Sometimes meaning is expressed in grand, theoretical terms, such as elaborate theories of meaning formulated by philosophers and theologians. Most of what is meaningful to people in their everyday lives, however, is less complex.

Have you ever tried to write your autobiography? The ways in which you interpret your experiences and the events, persons, and experiences you choose to remember, as well as how you explain them, reflect the meanings you apply to events in your own life. This chapter discusses how that kind of meaning is attached to individual and social life.

People typically choose their personal meaning from a larger, socially available meaning system. Although the parts that constitute this personal meaning system are all interrelated, they are not necessarily coherent or internally consistent. People may believe in the idea of germs and in the seemingly contradictory idea of evil spirits as explanations for illness. They may simply use the concept of germs to interpret some illnesses and evil spirits to explain others.

Most historical religions are comprehensive meaning systems that locate all experiences of the individual and social group in a single general explanatory arrangement. A comprehensive meaning system such as these is called a **worldview** (Berger and Luckmann, 1966). We will use the concepts of "religious worldview" and "religious meaning system" more or less interchangeably because most historical religious meaning systems have been comprehensive. Nevertheless in modern society, religious meaning systems compete with many other worldviews. Individuals are less likely to use any single comprehensive meaning system but may apply religious meanings to segments of their lives. For example, some meaning systems (e.g., astrology) explain only certain spheres of their adherents' lives, and believers find other meanings elsewhere.

According to some sociological definitions of religion (especially functional definitions, described in Chapter 1), any comprehensive meaning system is fundamentally religious, regardless of its content. For the purposes of this discussion, however, it is sufficient to say that comprehensive meaning systems that are not overtly religious (e.g., psychotherapeutic or political meaning systems) share most characteristics of specifically religious meaning systems. The process of conversion from one comprehensive meaning system to another is comparable, regardless of the belief content of either meaning system. We must also keep in mind that meaning systems are not abstract but are created and held by people.

According to Peter Berger (1967), the provision of meaning is particularly important for an understanding of religion because of the ways that meaning links the individual with the larger social group. *Meaning is not inherent in a situation, but is bestowed*. A person could interpret an event, such as losing one's job, by applying a wide variety of possible meanings. Although the individual may examine the event itself for clues about which meaning "fits" best, the final choice of meaning is *applied to* the event. The person fired might conclude that, although the boss cited uncooperative attitudes, the "real" meaning of the event points to a greater mission as an organizer of workers elsewhere. The experience is given meaning by this choice of interpretations. Attaching meaning to events is a *human* process (Berger, 1967:19).

Geertz (1966:40) suggests that religion serves as a template in establishing meaning. It not only interprets reality but also shapes it. The template of religion "fits" experiences of everyday life and "makes sense" of them; in turn, this meaning shapes the experiences themselves and orients the individual's actions. A meaning system in which Satan is prominent can explain past events and experiences (e.g., temptation or illness) as evidence of Satan's effects and can shape future experiences (e.g., by warning believers to avoid certain places or activities where Satan is likely to ensnare them).

By examining how this meaning-giving process occurs, we can understand some of the ways religion links the individual with the larger social group.

Meaning for the Social Group

Meaning systems interpret an entire group's existence. Thus, the group's ways of doing things and its very existence are assigned meaning. The group, for example, may explain its moral norms as instituted by its god; its pattern of family life may be interpreted as copying the family of the gods; and its history may become meaningful as the story of the gods' relationship with their people. As Berger (1967:29–33) points out, the meaning system is both *explanatory* and *normative;* that is, it explains why things are the way they are and prescribes how things should be. The meaning system of the group "makes sense" of its *social order*—the present/existing and future/desired social arrangements of a group, such as its form of authority and power, its stratification system and allocation of roles, its distribution of resources and rewards, and so on. These qualities make the meaning system a strong legitimation for the social order of the group.

A **legitimation** is any form of socially established explanation that is given to justify a course of action. Legitimations include any explanation of social practices: Why do we do things this way? Why should we behave according to that norm? Why do we have this position in society? Particularly important are those legitimations that establish authority in the group (Berger, 1967:29ff.).

Legitimations are expressed in a wide variety of forms. Myths, legends, proverbs, folktales, and history are all invoked to justify certain social arrangements. Taken together, these legitimations may be viewed as the "story" out of which a group lives. These examples show that legitimations are seldom intellectual in form. Intellectual legitimations, such as those developed in philosophy, theology, and political and economic theory are a highly specialized form of legitimation. Nonintellectual forms are far more common.

In India, the ancient Hindu myth of four *varnas* legitimated the division of labor and subsequent rigid social class system developed by the thirteenth century (and continued well into the twentieth century). According to this myth, the highest social group, the Brahmins, issued from the mouth of the Creator. Duties assigned them by the Creator included various religious functions, study, teaching, sacrifice, almsgiving, and the perpetuation of the sacred literature. The next group were Kshatryas, who were born from the arms of the Creator. Their special gift was power, and their duties included fighting and governing. From the thighs of the Creator came the Vaishyas, who were given the gift of strength; they were expected to cultivate the land and engage in other productive labor. The lowest social group, the Sudras, came from the Creator's feet. Theirs was the duty of serving the other three varnas (Lemercinier, 1981:164). This creation myth used potent imagery of the functions and value of parts of the Creator's body to explain the human social structure, to justify the privileges of the higher classes, and to prescribe appropriate behavior of all social groups. The myth itself exemplifies a nonintellectual form of legitimation, but complex intellectual ideologies and legal codes were also eventually developed to justify the Hindu caste system, as further described in Chapters 4 and 7.

Legitimations explain the ways in which the social group has behaved, but they also shape future action by justifying the norms of appropriate or desirable action. Religious legitimations make especially strong claims for the bases of order and authority and for the specific arrangements of the social order by their references to a higher authority. Thus, the social order is represented as more than human convention. The European idea of the "divine right of kings," for example, implied that the king ruled not merely as a political arrangement but as a permanent, God-given right. Contemporary examples of social arrangements that are religiously legitimated include the sanctity of marriage, the right to own and defend property, and the moral rightness of waging certain kinds of war.

The effectiveness of these legitimations often involves a certain amount of *mystification* (Berger, 1967:90). Anything suggesting that the social arrangements are purely human convention is allowed to be forgotten or is deliberately masked. The recent way of doing things is promoted as the "authentic tradition," and contrary evidence is not emphasized in the group's tale of its own history. For example, the Christian ideal of the institution of marriage is often represented as being exactly the ideal held during the whole history of Christianity. Actually, however, both the ideals and practices of marriage have changed considerably, and current religiously legitimated arrangements, such as the practice of church weddings, are relatively recent.

The early Christian churches were ambivalent about marriage; some early church leaders thought that the ideal for committed Christians was celibacy or even castration, with a chaste marriage (i.e., without any sexual activity) as an accommodation for those who could not meet the difficult ideal (Bullough and Brundage, 1982; Elliott, 1993). Until the sixteenth and seventeenth centuries, ordinary people typically wed by private self-marriage (later formalized as "common-law" marriage). Not until the sixteenth century was there a church prohibition of the practice of concubinage (Brundage, 1982). For most people, self-marriage was an acceptable pattern through the Middle Ages, though couples sometimes went to the church steps for a blessing. Only in the sixteenth century was church marriage declared a sacrament, but during the subsequent Reformation, many Protestant groups denied that marriage was a sacrament. Most Protestant groups did, however, keep the practice of church weddings, and they developed their own rituals for these ceremonies. (Noonan's 1965 analysis of Roman Catholic marriage and family norms and Nelson, 1949, provide further examples of how Christian ideals of marriage and other social institutions have changed considerably over the years.) Thus, we see how the effectiveness of legitimations of present day marriage practices depends on treating these practices as traditional and on forgetting or de-emphasizing those parts of history that are inconsistent with present norms and practices.

Although religious legitimations are generally used to justify existing social arrangements, they are also potent forms of criticism of the existing social order, as Chapter 7 shows in some detail. The Old Testament prophet Amos legitimated his denunciation of people's ways by reference to their God. Religious legitimations may be invoked to justify even revolutionary action. Thus, religious legitimations are not solely a tool of the dominant group but may justify actions of subordinate and dissenting groups as well.

The Individual's Meaning System

The individual does not construct a personal meaning system from nothing. An individual's meaning system is learned, for the most part, during the process of socialization. The interpretations that seem most plausible to a person are likely to be those that are familiar and held by others who are important to that individual. So although each individual operates with a highly personal meaning system, that set of meanings is greatly influenced by family, friends, institutions (e.g., education), and the larger society. While the individual comes to accept some of the meanings communicated during socialization, the resulting meaning system is not, however, an inevitable product of socialization. The individual can reject or modify meanings communicated by others. When several competing meaning systems are presented as alternatives (which happens especially in modern society), the individual can choose which meanings to accept. The individual may accept the meaning system presented by a subgroup in society and not that of the larger society. Yet all personal meaning systems gain effectiveness by their link with some community in which they are shared.

Meaning and self-identity are intertwined. People locate themselves and their personal actions in a larger social order by means of their meaning systems. The individual selects subjectively meaningful interpretations of events and experiences from the larger interpretive scheme provided by the meaning system, and that personally held meaning system informs the individual's sense of self. The applied meaning system tells the individual what "kind" of person one is, the importance of the roles that one performs, the purpose of the events in which one participates, and the significance of being who one is. A meaning system, in other words, *makes sense* of one's identity and social being. Some societies emphasize subjective choice of meanings; others take for granted group meanings, but all meaning systems are based on an intersection of self and social group.

People can bestow meaning on a situation in various ways. Beliefs are important in this process. Ideas can help locate an experience in a meaning system, but a meaning system cannot be reduced to its formal belief content. Miracles, magic, ritual, and symbols also contribute to a pervasive sense of meaning. The individual can apply meaning to a situation by performing an appropriate ritual for the event. Ritual practices and other religious actions, such as sharing a ritual meal, are ways by which the individual may subjectively—often bodily, as well as emotionally and mentally—experience social meanings. Similarly, by interpreting an event as miraculous or magical, the individual places special meaning on it. The application of meaning to human experience entails social processes. Through everyday conversation, the individual tries out interpretations of experiences. Interaction with others—especially with persons whose response "counts" to the individual—is a significant part of the process of bestowing meaning on a situation.

At the same time, the meaning system informs the individual of the values and norms of the larger group. The following interpretive statements illustrate the variety of ways in which norms and values can be embodied in a meaning system:

"Cleanliness is next to godliness."
"Masturbation is likely to lead to insanity."
"Most of the problems with youth nowadays are due to bad mothers who go out to work instead of taking care of their children."
"Doctors should make a lot more money than others because they had to

work so hard and long to become qualified, and their skills are more important to society."

"People who commit adultery will be punished by God."

"He brought his illness on himself by his bitter and hateful outlook on life."

"Hard work and patience are always rewarded."

Even though you may personally disagree with some or all of these statements, you can see how each statement can be used to *interpret* situations or experiences and to *evaluate* general kinds of behavior. Meaning systems embody norms—social evaluations of behavior. To the extent that one applies these normative interpretations to one's own behavior, the meaning system brings an evaluative element to one's identity. Thus, in terms of the interpretive framework, one may conclude: "I am a sinner," "I am a good mother," "I am a successful warrior," "I am ill-fated," or "I am a virtuous person."

The individual's meaning system makes possible the evaluation of past actions and the motivation of future actions. The ability to perceive events as ordered in some way enables one to plan and orient one's actions. If things are experienced as "just happening" in a chaotic, meaningless string of events, the person literally does not know what to do. If, however, these events are given meaning, their interpretation implies an appropriate course of action (even if that action consists in seemingly passive responses such as "suffering through it" or "praying over it"). By relating mundane social life to the realm of the transcendent, religion is particularly effective in motivating the individual's participation in the larger social group. The classic example of such motivation is the concept of *vocation,* the idea that one is "called" to an occupational or economic status by God. This idea gives special meaning to everyday work. By understanding such work in terms of vocation, individuals gain a sense of purpose and value to their labor, and society gains the willing contribution of its members (see Weber, 1958a:Chapter 3).

Both the individual and the social group draw on religion to give meaning to their existences. The meaning system provides interpretation for their experiences, locating human lives and events in terms of a larger framework. Religion also serves as an important form of legitimation, or justification, for both the individual and the social order. Religion interprets and evaluates the "way things are to be done" in the social group. These legitimations provide meaning for individual members of the group who accept these explanations and incorporate them into their ways of thinking about the world.

CRISIS OF MEANING

The meaning system of the individual or group can integrate most routine events into an understandable pattern, a meaningful whole. Some events and experiences, however, are not so easily interpreted within the existing meaning system. The individual who experiences the death of a loved one, a painful illness, or serious economic misfortune may not be helped by the existing personal meaning system. An entire group can undergo similar meaning-threatening experiences: oppression by an enemy, famine, earthquake, or economic depression. Such events are particularly meaning-threatening if they appear to contradict important aspects of the existing meaning system. Groups who believe in a loving, personal god have more difficulty reconciling disastrous events with their meaning system than do groups whose god is remote and capricious.

Another situation that creates a problem of meaning is serious discrepancy between the ideal that a group promulgates and the actual practice. Inequality and injustice can be especially meaning-threatening when they are inconsistent with the group's ideals. Because the threat of the meaninglessness of such events is so great, individuals and groups try to build special legitimations into their meaning systems to justify these apparent contradictions or discrepancies. In other words, part of the meaning system itself interprets events and experiences that would seem to contradict the meaning system. These meanings are responses to the *problem of theodicy* (Weber, 1963: Chapter 9).

Theodicies

Theodicies are religious explanations that provide meaning for meaning-threatening experiences. Most religions, for example, offer theodicies of suffering and death. The content of these explanations differs among the various religions, but the desire to find meaning for such experiences appears virtually universal. Disaster and death create a problem of theodicy not because they are unpleasant, but because they threaten the fundamental assumptions of order underlying society itself (Berger, 1967:24). Theodicies tell the individual or group that the experience is not meaningless but is rather part of a larger system of order. Some successful theodicies are, in fact, nothing but assertions of order. A woman discussing her personal meaning crisis after her husband's premature death said: "I finally came to understand that it didn't matter whether *I* understood why he died when he did, but that God had a reason for it, and that was all that mattered." For this believer, knowing that an order exists behind events was more important than knowing *what* that order is.

Similarly, theodicies do not necessarily make the believer happy or even promise future happiness. A person suffering poverty and disease may be satisfied with the explanation that the situation has resulted from sins committed by the ancestors. Such an interpretation offers little hope for overcoming the poverty or disease, but it does offer meaning. It answers the question, "Why do I suffer?"

The capability of religion to provide meaning can be illustrated by how a society handles the especially meaning-threatening situation of dying and death. There is an inherent problem of meaning in the prospect of one's own or a loved one's death. Death seems to negate the individual's or group's sense of order. For this reason, the way in which a society handles the process of dying is revealing of its larger meaning system. Long before most individuals face death, their religion has been providing meaning for their various life stages (e.g., by rites of passage into adulthood or by legitimating the norms and prerogatives of old age). In many societies, the meaning system somewhat normalizes dying as a further stage of human development. Some meaning systems affirm an afterlife or a rebirth; others emphasize that people live on through their offspring or tribe.

In American society, dying and death have become particularly meaning-threatening. This problem of meaning partly results from the undermining of traditional theodicies; formerly used explanations have become less satisfying to many persons. The problem of the meaning of death is especially acute in this society because its value system assigns comparatively great worth to individual lives. Numerous social arrangements contribute to the problem. The dying person is typically segregated from one of the foremost meaning-providing social supports—the family. Furthermore, the specialized roles of those tending the dying seldom include the provision of meaning. Unlike their counterparts in traditional medical systems, modern doctors do not generally consider as part of their role helping the patient assign meaning to illness or death (Kleinman, 1988). Death and dying have become generally secularized processes, and the problem of their meaning has become acute.

Anomie

Sometimes a meaning system is completely unable to absorb a crisis experience. The theodicies applied may not be effective in reintegrating the experience into the meaning system, or the group supports of the entire meaning system may be so weakened by the crisis that people are unable to restore a sense of order and meaning. When an individual or group has lost its fundamental sense of meaning and order, ongoing social life becomes virtually impossible. Why should one want to do anything if everything seems senseless? When a meaning system is so weakened, the moral norms it supports are also undermined. And with no underlying order to existence, there is no apparent basis for distinguishing between "right" and "wrong" ways of behaving.

It becomes necessary, if social life is to continue, to establish a new basis of order and a new meaning system. If a group's way of life has been thoroughly disrupted by military conquest, for example, the group might reorganize itself around a social movement (e.g., followers of a new prophet) that offers a new basis of order and meaning for group members. The new meaning system sometimes consists of embracing new meanings (e.g., those of the conquerors), a rearrangement and reaffirmation of old meanings (e.g., in nativistic movements), or *syncretism*—the interweaving of new meanings with the older meaning system. In colonial Brazil, for example, uprooted, enslaved Africans were allowed to celebrate the *congada*—a symbolic coronation of the King of the Congo, absorbed into the annual Catholic festival of Our Lady of the Rosary, who became patroness of blacks. Several symbolic homologies between Western African religion and Brazilian Catholicism made fertile ground for syncretism (Camara, 1988). Regardless of the source of elements for the new meaning system, its importance lies in its ability to reorganize the basis of order for the group. Social life is given new meaning.

The situation of crisis in the group's meaning and order is described in Durkheim's concept of *anomie* (or *anomy*). The word *anomie* means literally "without order." Durkheim applied this concept to social situations in which there was a deregulation of the public conscience. *Anomie* means a crisis in the moral order of a social group (Durkheim, 1951:246–257; see also Berger, 1967:22ff.). Although anomie has serious consequences for the individual (e.g., it might lead one to suicide), it is primarily a social situation referring to the inability of the social group to provide order and normative regulation for individual members. Religion's capability of providing meaning and order suggests that religion functions as a protection from anomie in two ways: A firm religious basis of order is a buffer against the occurrence of anomie in the first place; and if the group does experience an anomic situation, religion can potently respond to the crisis of moral meanings. Meaning is fundamental to a sense of order—without it, there is chaos. Indeed, Berger suggests that the opposite of "sacred" *is* "chaos." He states, "The sacred cosmos, which transcends and includes man in its ordering of reality, thus provides man's ultimate shield against the terror of anomy" (Berger, 1967:27; see also Geertz, 1966:14).

Mazeway Disintegration

Another way of understanding social life in such crises is by examining the parallels between people's reactions to major physical disasters and their responses to cultural crises. Wallace (1956, 1957) calls both of these responses **mazeway disintegration.** The cultural mazeway is the socially constructed and learned patterns and rules for interaction that a given group has developed over many generations as "the way" to achieve their wants. Wallace suggests that people in such crises are initially unable to act because their cultural mazeway has fallen apart. The crisis of mazeway disintegration is resolved by reaffirming their identification with some cultural system—traditional, new, or syncretic. For this reason, religious movements that arise to accomplish this reaffirmation may be considered "revitalization movements"; that is, they enable the social group to come alive again.

Although massive disruptions in a group's way of life (e.g., conquest and famine) are obvious starting points of such crises, less dramatic change is more common, and the sense of crisis is less profound. Members of the group may feel only a vague sense of malaise, a discomfort with the way things are going and a general ambiguity about what they as individuals should do. Rapid social change often leaves people unsure about where they "stand." Norms become open; although the old regulations no longer hold, new ones have not been crystallized. Examples of such ambiguities are found in the whole area of gender role expectations in contemporary societies.

An important reason why rapid social change is so threatening to a group's meaning system is that the very *fact* of change undermines many legitimations of the existing social order. It temporarily humanizes the established rules and patterns of interaction by showing them to be changeable products of human convention. Before such change, the group's ways of doing things may have appeared immutable, but the very fact of change belies this perception. In this context, it is possible to understand, for example, the exasperation of a Roman Catholic who, during the period immediately after Vatican II, exclaimed, "How can it be that last year eating meat on Friday was a terrible and grave sin, and this year it is merely an optional and private sacrifice?" Even more threatening are the perceived changes in a social group's fundamental norms and values, such as patterns of family life. Periods of dramatic social change seriously challenge the existing meaning system of a group. Often these periods are marked by the rise of new religious movements that affirm an "improved" meaning system, typically claiming that its own ways are not changeable because they stem from a higher authority than that of the old meaning system.

To summarize then, crises of meaning may occur on the personal level or for an entire social group. The impact of crises of meaning on groups and individuals highlights the significance of the meaning-providing capabilities of religion. Nevertheless, we should not assume an overly intellectualized image of humans. Although a strong case can be made that meaning is a fundamental social requirement (Luckmann, 1967; see also Berger, 1967), we must keep in mind that all people do not necessarily equally desire to find meaning for all aspects of their lives. Similarly, some people are more tolerant than others of inconsistencies in their meaning systems. Some people desire more intellectually elaborate systems of meaning; others are satisfied with quite simple views. Furthermore, meaning does not operate only at a cognitive level; for many people, ritual practice and religious experience effectively sustain the meaningfulness of their lives. Yet historical religions and quasi-religions address similar problems of meaning: discrepancies between the ideal and actual practices in the society; personal suffering and death; and group crises such as war, conquest, and natural disasters. By providing meaning in the face of meaning-threatening situations, religion enables the individual or group to cope with the situation. If the meaning-threatening situation can be successfully reintegrated into the group's sense of order, social life can continue.

COMMUNITY AND RELIGIOUS BELONGING

A direct relationship exists between a community of believers and the strength of its shared meaning system. The unity of the group is expressed and enhanced by its shared meanings. The group's meaning system, in turn, depends on the group as its social base for its continued existence and importance. The idea of the "church" (i.e., community of believers) is not merely an organizational feature of religions but expresses a fundamental link between the meaning system and the community that holds it.

Religion as the Expression of Social Unity

Durkheim treated the essentially communal quality of religion as definitive. The collective nature of religion was central to Durkheim's analysis. He concluded that religious rituals and symbols are, at root, representations of the social group. These *collective representations* are the ways by which the group expresses something important about itself to its own members. Thus, by participating in group rituals, individual members renew their link with the group, and they learn and reaffirm shared meanings (Durkheim, 1965:257). We will explore Durkheim's theories of religious belonging further in Chapter 6.

For Durkheim, however, collective representations were not merely mental representations but were also collectively experienced. Thus, shared religious meanings expressed the group's unity, while shared experience produced that unity. Sociology has not developed an adequate understanding of religious experience, in part because the experiential aspect of religion has too often been treated as purely private (Spickard, 1993). An alternative approach, by contrast, suggests that some religious experience may reflect intense *intersubjectivity*. If people can share experiences intersubjectively, valued qualities, such as *communion* (literally, "union with") and *compassion* (literally, "co-feeling") can be experienced as actually—not just figuratively—real.

Plausibility Structures of Meaning

Berger emphasizes that meaning systems require social "bases" for their continued existence. These social bases are called **plausibility structures**—specific social processes or interactions within a network of persons sharing a meaning system. As the term implies, the meaning system continues to be plausible (i.e., believable) within these social structures. Berger asserts that *all* religious traditions require specific communities of believers for their continuing plausibility (1967:46). A sound plausibility structure allows the meaning system to be held as a common, taken-for-granted entity. Similarly, it strengthens the ability of individuals to believe, for the group gives them social support and reinforcement in their worldview.

Historically, most religious worldviews were coextensive with the societies that held them. Relatively simple societies exemplify especially well the link between the group's meaning system and networks of fellow believers. If one were born and socialized into a remote mountain tribe, for example, the religious meanings and rituals of the group would be taken for granted. Simply belonging to the tribe would make one a part of the religion, and the whole tribe would serve as social support for one's belief and practice. Indeed, it would be difficult, if not impossible, for a member isolated in such a tribe to conceive of any other meaning system.

A societywide plausibility structure still exists in many relatively complex societies. In some parts of Latin America, religion is pervasive in all aspects of life, and a single, particular worldview dominates the society. Most people spend their daily lives solely in the company of people who share major elements of their worldview. Such massive social support for shared religious beliefs and practices makes it relatively easy for believers to take them for granted.

The significance of plausibility structures can also be seen in situations where believers of a worldview are separated from the social group that serves as its basis of plausibility. Missionaries typically went out in closely knit bands among the "unbelievers" (i.e., people with a different worldview from their own). This arrangement was not merely for mutual physical protection but—more importantly—for mutual protection of their view of reality (see Cohen, 1990, on how native worldviews influence missionaries). A meaning system shared even with only a few significant others is stronger than one held alone and surrounded by persons who do not share that meaning. In this context, we can appreciate how grave a punishment it was for some societies to exile a member. Exiles were cut off both from other members of the society and from the social support for their meaning system.

Pluralistic societies, by contrast, are characterized by the relatively unrestrained competition of several different meaning systems. Pluralism itself makes the maintenance of a meaning system problematic because it undermines the taken-for-granted quality of any single worldview. In pluralistic situations, each religious group must organize itself to serve as its own plausibility base because the society as a whole does not support its meaning system. Thus, in pluralistic societies such as the United States and Canada, religious groups often emphasize their communal ties, for example by identifying strongly with an ethnic cultural heritage that sustains both the meaning and belonging aspects of the religious group. In the face of societal pluralism, many religious groups try to create a cultural enclave, characterized by frequent interaction with fellow believers and physical or symbolic withdrawal from the "world" of competing beliefs and values.

By contrast, religious groups that cannot sustain the experience of "belonging" in a pluralistic context may find their members less committed (see Chapter 3). One U.S. study found that "declining community attachment, *more so than the erosion of traditional belief per se,* is the critical factor in accounting for the declines in church support and participation" (Roof, 1978:136). The pervasiveness of community cultural reality is illustrated by the traditional rural Baptist church in the South, where kinship networks, shared sense of continuity of tradition, and everyday social practices and roles (such as gender roles) serve as plausibility structures for evangelical Christian beliefs and experience (Heriot, 1994). Persons with strong local community identification have a structure of social support to maintain the plausibility of their shared meaning system.

Modern societies, however, are characterized by weakened ties to a local community; urbanization and high levels of geographical mobility disperse families and communities. Increasingly, through mass media, education, and other wider social influences, even nonurban communities are exposed to competing worldviews. When a religious group holds a worldview not supported by the larger society, it must construct its own plausibility structures. Other processes that illustrate the way in which religious belonging supports a meaning system include socialization, conversion, and commitment, discussed further in Chapter 3.

EXTENDED APPLICATION: MILLENARIANISM AND DUALISM IN CONTEMPORARY SOCIAL MOVEMENTS

Many of the concepts presented in this chapter are particularly relevant to an understanding of contemporary meaning systems. The last part of this chapter is an extended example that applies some of these concepts to modern social movements. We begin by noticing certain themes common to many contemporary movements: millenarianism and dualism. What do these themes imply about the movements' responses to problems of meaning? More important, what does the attraction of such movements imply about recruits' relationship to the existing social order? This essay is one interpretation of the significance of millenarianism and dualism in contemporary social movements.[1]

These two themes appear regularly in the worldviews promulgated by many developing religious movements. A **dualistic worldview** holds that reality consists of two irreducible modes—Good versus Evil. **Millenarianism** is the expectation of an imminent collapse of the entire social order and its replacement with a perfect new order. The term is derived from occult Christian predictions that the world would end one thousand years after Jesus' birth. The occurrence and appeal of movements that emphasize these two elements reveal some of the problems of meaning in contemporary society.[2]

These two elements of belief are not merely quaint characteristics of the movement but are central to their appeal. The appeal of the dualistic interpretation of reality and of millennial expectations results from a need for a new, firmer order in a time when the old basis of order appears to be collapsing. The elements of dualism and millenarianism represent the assertion of order in the face of felt disorder.

Two Opposing Principles: Good and Evil

A dualistic worldview depicts all reality as consisting of two fundamental modes or opposing principles—one Good and the other Evil. This dualistic perspective is especially important in the belief systems of groups growing from Western religious perspectives (e.g., Islam, Judaism, and especially Chris-

---

1. These themes are important indicators about the significance of contemporary social movements, but they are not sufficient explanations; important economic, political, and social structural factors are also involved.

2. Because this essay is presented as an example, it is necessarily somewhat oversimplified. See referenced sources for some of the complexities of interpreting social movements. The essay suggests only that dualistic and millenarian theodicies are effective responses to problems of meaning posed by anomie. That does not mean that these movements are "caused" by anomie. A full explanation of these phenomena would need to show that recruits had experienced anomie. It would also need to analyze the sources of problems of meaning in contemporary society.

tianity). The idea of warring forces of Good and Evil has been recurrent in Christian ideologies, and contemporary expressions of this idea can be traced to enthusiastic religious movements of the sixteenth and seventeenth centuries. Some form of dualism is used by many Christian groups to explain the nature of humankind, the presence of evil, temptation, sin, suffering, and the need for God's help.

Many emerging religious movements particularly emphasize dualistic interpretations of the world (see, for example, Aho, 1990, 1994; Barker, 1984; Bird, 1979; Richardson et al., 1979; Tipton, 1982a; Westley, 1983; Wilcox et al., 1991). Through a new emphasis on mystery, magic, and miracle, believers perceive their world as full of evidence of the action of good and evil forces. Believers experience the world as remystified by the presence of God's spirit; it is also remystified by the presence of Satan in numerous forms. Just as God is influential in their world, so too is Satan immediate and active. Believers see the direct influence of God in the beneficial and pleasant events of their lives; at the same time, they see the Evil One as the source of doubts, temptations, dissension, sickness, and other troubles. This remystification of everyday life is in marked contrast to the usual, more secular interpretation of events.

Some dualistic belief systems identify and personify the forces of Evil in movements, issues, or people that threaten the values of the group. A member of one youth-oriented Pentecostal sect stated:

> This conspiracy is what the Bible calls the false church. There's a false church, the Antichrist, and a true church. This elite is the Antichrist. It has the highest religious leaders, too—the World Council of Churches. They've got definite people, like . . . , who are in with this conspiracy that sit on the board of the Council (quoted in Tipton, 1982a:91).

Dualism in the ideology of some Black Muslims identifies the forces of Evil with the white oppressors (Lincoln, 1989). Similarly, various fundamentalist Christian groups identify communism with the forces of Evil (cf. Ammerman, 1991; Himmelstein, 1983:17). Another dualistic religious group is the Unification Church of the Reverend Sun Myung Moon, whose belief system holds that the fundamental struggle is between communism and the "God-fearing world" of which the United States is the center (Anthony and Robbins, 1992a).

Dualism is also a prominent feature of many religious nationalisms: They interpret "our" people as God's people and "their" people as identified with the forces of Evil. In this way, dualism serves as a political ideology. In 1977 one militant Islamic fundamentalist wrote:

> Some people today divide the world into democratic or fascist [systems], labor or liberal parties. Islam does not recognize any parties except two: the Party of God and the Party of Satan. The Party of God are those who act as His agents on earth and govern by His laws. The others are the Party of Satan—whatever variation there may be in their system of

government and regardless of the conflict among them. In the end, they are one coalition formed to oppose God the Mighty, the Omnipotent (quoted in Arjomand, 1989:119–120).

The cosmic opposition implied in dualism provides a ready explanation for any felt opposition to one's own belief system. Numerous religious movements identify their own beliefs and practices as on the side of Good and decry competing religious movements as on the side of Evil. Often, however, Evil is identified with vague general forces that seem to work against the commitments of the religious group. Thus, one neo-Pentecostal member stated:

> I think Satan does his best. I think it's time we recognized that the power of evil lives in the world and that we as individuals have to cope with the evil that's in the world today. And again that's another thing, another advantage of having a prayer community. It's because we gather together for strength and you sort of get to be invincible in a group. If you try to hit it out on your own, I mean obviously, you just can't with those devils. You just don't stand a ghost of a chance. You know, I know a lot of people don't believe the devil exists, but I really don't think things would be as bad as they are, the evil in us, the evil in man. The evil that's in man is the evil that's an extension of possession of the devil. The devil is using that person just like God can use us. . . . I do believe that a lot of evil, a lot of sadness and sickness in this world is brought out by Satan.
>   I think it's in his scheme of things to turn us against God (quoted in McGuire, 1982:150).

The persistence and strength of belief in dualistic beliefs suggest that dualism is not merely one characteristic of these movements but may indeed be central to their appeal.

Dualism as a Response to Anomie. Dualistic interpretations are useful to both the group and the individual, especially as a response to anomie, which results in a sense of normlessness and powerlessness. Being out of touch with the source of moral power means being powerless in the face of the nameless terrors of the disordered universe. Only a firm reordering of the social order can provide an effective protection against such terror (Berger, 1967:27). A new order must be established and maintained continually against the occurrence of further order-threatening phenomena. Even a firmly established new order is threatened by occurrences such as suffering and death, which perpetually raise the problems of "meaning" and "order."

A dualistic worldview both provides a framework for the *construction* of a new order and legitimates its *maintenance*. Religious dualism is particularly effective in this reordering because it holds that both the sources of all problems and the "real" solutions to these problems derive from a transcendent realm (see Lofland, 1966:36). Dualism enables believers to name the sources of their anxieties, fears, and problems. This identification is in itself an important source of believers' sense of order and control. Identifying certain difficulties as caused by evil forces also implies a clear-cut course of action. Believers locate their personal courses of action within a cosmic struggle, a continual battle between the forces of Good and the forces of Evil. This identification provides explanations of all events—good or bad—that occur in their lives, and it gives meaning to everyday existence. Even trivial aspects of daily life become part of the order implied in the dualistic worldview.

This order-giving potential of dualism is partly responsible for believers' sense of spiritual opposition. The believer perceives frequent, regular evidence of the immediacy and influence of both good and evil spirits. This evidence, it should be emphasized, is usually in the form of actual events in everyday life. A dualistic worldview contains a built-in tendency to ascribe all good events to God and all bad ones to Satan. The tidiness and order of the dualistic interpretation of the world are part of the basic appeal of such movements. Dualistic figures of ultimate Good and Evil simplify the world to people who are overwhelmed by the ambiguity and complexity of modern life.

In facing the perceived opposition, the group gains cohesion and strength. This sense of opposition is also a useful legitimation for maintaining the new order because it enables believers to see significance in subsequent events, even those that would otherwise threaten their new beliefs. The dualistic worldview heightens the sense that the opposition is a conspiracy against believers and their group (cf. Barkun, 1974; Beckford, 1975b; Westin, 1964).

There are two sides to the believer's sense of order. On the "negative" side is a sense of fearsome powers bent on a conspiratorial attack on believers and their values. On the "positive" side, the same sense of order provides the believer with a feeling of harmony and symmetry. Both positive and negative aspects involve the forming of patterns and imposing of a sense of order (Weil, 1973.178–179). The believer's worldview is founded on the expectation that everything *is* ordered; therefore, order and patterns in the world are subsequently perceived. The "positive" side of this spiritual order can be seen in the believer's constant discovery of "holy coincidence," "providence," and other evidences of the pattern of God's work in everyday life. The dualistic worldview gives form and a "cast of characters" to the patterns that believers perceive in the world around them.

Dualism as Theodicy. Dualism also offers the believer an explanatory system for reinterpreting order-threatening events. As such, it functions as a theodicy. The theodicy provides meaning but not necessarily happiness, informing the believer, "You may be miserable, but be comforted that there is, at least, meaning to your suffering."

Dualism is a particularly effective form of theodicy because it is a closed system of legitimation; that is, built into the legitimating system itself is an "explanation" of every argument against the system. Thus, the same theodicy can explain both good and bad events, opposition and confirmation. It also provides a basis for self-justification and for the moral condemnation of others (cf. Rokeach, 1960:69).

Religious groups often use the dualistic theodicy to explain the doubts and uncertainties experienced by new believers: "Your doubts and anxieties are because the devil is trying to confuse and tempt you back to your old way of life." This theodicy gives meaning to the natural doubts and sense of uncertainty encountered by the new believer in the resocialization process. Armed with a concrete identification of the source of these difficulties, the new believer has a greater sense of power in facing such problems of belief. The theodicy that fears and doubts are planted by Satan is a welcome explanation.

The foremost function of dualism is the reshaping of believers' interpretations of everyday events. Believers reinterpret even their own roles. Dualism bestows significance on events that formerly seemed meaningless. Human failure, suffering, social problems, personal difficulties, and death are all given meaning. Events that formerly appeared to be random, haphazard, and disorderly come to be perceived as part of a clear pattern. One's sense of ambiguity and insecurity is resolved by perceiving one's personal role as part of "something big"—a great cosmic struggle in which one gains power and purpose by siding with the good forces.

Dualism enables the group to create order out of chaos. The source of threat is not only the external chaos of the anomic social world; intragroup problems constitute a threat, too. Internal conflict is especially disruptive because it undermines those interpersonal relationships that support the believers' worldview. Problems of "disorder" are inherent in the establishment of any new group, and much ambiguity arises during the developing stages of a movement. Also, if members perceive their new belief system to be a departure from their previous one, the fact of change may cause some uncertainty. Emphasizing the group's role in a cosmic dualistic confrontation enables the group to resolve its problems of uncertainty and ambiguity (Slater, 1966). Dualistic beliefs affirm the boundaries between the group and "the world," and they provide mechanisms (e.g., exorcism, confession, and healing) for dealing with internal conflict.

The group constructs and perceives its world in dualistic terms: Good versus Evil, truth versus deceptions, light versus darkness, purity versus pollution. For example, a recent dualistic religious movement in the United States symbolizes the forces of evil as all "impurity": germs, viruses, bacteria, dirt, sludge, filth, garbage, diseases (especially addictions, cancer, herpes, and AIDS), human "pests" and "rodents," "unclean" persons (especially immigrants and "impure" racial groups) (Aho, 1990:85). Dualistic boundaries serve to enhance member commitment, even commitment to sacrifice for the group's goals (cf. Douglas, 1986).

The dualistic theodicy enhances the individual believer's sense of security. Recruits to these movements often express a strong desire to know for sure where they "stand" before God. This desire for certainty is fulfilled by a secure sense of order, which affirms a clear duality between the forces of Good and Evil and shows a distinct path for the believer who would side with Good. This simplified course of action informs the member that, having sided with the Good, one is safe and secure.

The Imminence of the Millennium

A dualistic theodicy is particularly forceful if combined with a millenarian theodicy. **Millenarianism,** the expectation of an imminent disintegration of the entire social order and its replacement with a perfect new order, is a recurring theme in religious movements. The millenarian theodicy informs the believer that the present chaotic social order, with its misery and deprivations, is transient and that in the coming new order the believer will have a better life and no longer experience the turmoil and malaise of the present. Millenarianism offers both an explanation of the wrongness of the present social order and a hope for future change; as such, it is a potent source of change-oriented behavior.

The millennial dream that the perfect new order is imminent is also a response to anomie. When the old order is seen as nearly defunct, the new order must be near (see Barkun, 1974). The perceived disorder of the United States is especially upsetting to some groups who believe that the nation has a special purpose in God's plan. One television preacher said:

> America's star is sinking fast. If Christians don't begin immediately to assert their influence, it may be too late to save America from the destruction toward which it is plunging. And, since America now stands as the key base camp for missions around the globe, to fail to save America now would almost certainly be to miss its last opportunity to save the world (James Robison, from *Save America,* as quoted in Hadden and Swann, 1981:97).

Similarly, the Reverend Moon warned:

> But today America is retreating. It's not just an accident that great tragedy is constantly striking America and the world, such as the assassination of President Kennedy and the sudden death of Secretary-General Hammarskjöld of the United Nations, both in the same decade. The spirit of America has declined since then. Unless this nation, unless the leadership of the nation lives up to the mission ordained by God, many troubles will plague you. God is beginning to leave America. This is God's warning (quoted in Anthony and Robbins, 1978:84).

People who view this society as badly out of order and consider the established religions impotent to do anything about the "problems" are likely to hope for dramatic change. The ultimate religious "solution" is the end of the world. The coming of the millennium will not only destroy the old problematic order but will also vindicate the faith of the believers.

Religious Millenarianism. The idea of an apocalypse—a dramatic end of the world—is not a new religious theme. Many religious groups have anticipated the imminent end of the social order they define as ungodly. Some have expected a total physical cataclysm in which the world would end; others have awaited a God-sent revolutionary end in which the existing social order would cease and be replaced by a perfect new order. This latter hope was frequently combined with the anticipation of a messiah who would bring about the new order. Such themes are particularly prominent in Western religions, especially Christianity (see Barkun, 1974; Bettis and Johannesen, 1984; Cohn, 1970, 1993; Festinger et al., 1956; Moses, 1993; Talmon, 1966; Wilson, 1970).

Likewise, several Shi'ite (Muslim) sects combine a messianic and millenarian expectation in their belief that the Twelfth (Hidden) Imam would reappear as the Mahdi (bringer of justice to the world) and the Riser of the Resurrection (who would preside over the end of the world and the last judgment). Khomeini did not need to make explicit millenarian claims to tap the apocalyptic religious and political mood that prevailed in 1978 among many adherents of these Shi'ite sects in Iran in order to mobilize the revolutionary potential of the religion (Arjomand, 1993). While messianic hopes are linked with millenarianism in Christianity and Islam, other religious groups such as Sikh "fundamentalists" in India have used millenarian visions of alternative social systems in which their oppression by a more powerful religious group would end and utopias of harmony, purity, and justice would prevail (Oberoi, 1993).

Millenarian movements were widespread in the United States in the nineteenth century, sometimes combined with a messianic vision of a special destiny for the nation. The Mormons, Millerites (and their offshoot, Seventh Day Adventists), Christadelphians, and Jehovah's Witnesses are examples of groups in which the millenarian expectation was or is central (see Wilson, 1970). Nativistic religious movements, which envisioned the millennium as the dramatic restoration of a traditional "native" way of life, arose in the nineteenth century among several Native American tribes, whose social and economic plight was desperate. The Ghost Dance,[3] Handsome Lake, Indian Shakers, and the peyote cult exemplify these movements by which Native Americans responded to the profound disruption of their traditional way of life wrought by white settlers, government policy, and the army (see Aberle, 1966; Barnett, 1957; Mooney, 1965; Myerhoff, 1974). A black syncretic movement, the Lost Found Nation of Islam (i.e., Black Muslims) appealed to both the millennial possibilities in Islam and U.S. blacks' familiarity with Christian millenarian themes. Its early teaching was that 1914 ended the 6,000 years of the "white devils'" rule, after which the lost Nation of Islam would finally realize its destiny (Lincoln, 1989). Later, however, the movement emphasized orthodox Sunni versions of Islam and abandoned this prophecy, while a splinter group headed by Louis Farrakhan reasserted it (Mamiya, 1982; Mamiya and Lincoln, 1988).

Belief in the imminence of the millennium is widespread among new religious movements as well as in Christian fundamentalist denominations and sects. Peculiarly modern forms of millenarianism include groups that base

---

3. The Ghost Dance was a highly traditionalistic movement that foresaw the destruction of the white man. Although the movement itself was pacifist, its appeal and potential for uniting diverse tribes frightened the frontier army, culminating in the Battle of Wounded Knee, in which over two hundred men, women, and children were killed.

their millennial expectations on UFOs and extraterrestrial communications (see Balch and Taylor, 1976, 1977; Festinger et al., 1956; Kerr and Crow, 1983). The Order of the Solar Temple, whose members in Canada and Switzerland may have committed mass suicide in 1994, based its millenarian vision on the Western occult tradition (e.g., Gnostic Christianity, Kaballism, Rosicrucianism, and hermetic freemasonry)(*Millennial Prophesy Report,* November, 1994). Dispensational premillennialism among conservative Christians has been used for political ends. Building on the long history of millenarian thought, dispensational premillennialism proposes that the world's history is divided into periods of "dispensations," each ended by a wrathful judgment by God. Accordingly, the world is now in the final dispensation, which will end in a "great tribulation" marked by horrible wars and the destruction of much of the world's population; subsequently, Christ will return to reign. Many dispensationalists also believe that the "righteous" (i.e., true Christians like themselves) would be spared this tribulation by the "rapture" in which Christ would "catch" them up before the last seven years of the world. Unlike other millenarian movements that held similar beliefs, this movement actively promotes certain political and military actions that are believed to hasten the end time (Harrell, 1984; Robertson, 1988). For example, despite their widespread anti-Semitism, dispensationalists have raised much money to help right-wing Jewish extremists reclaim the Temple Mount in Jerusalem from the Muslims who have one of their holiest, centuries-old shrines on the site. Unlike other millenarians who merely decry the crises and chaos as signs of impending doom, some Christian dispensationalists actively promote crises and strife.

Social and Political Millenarianism. Not only specifically religious movements, but also many other social and political movements, have dominant millenarian themes. Several "isms" of the nineteenth and twentieth centuries project a secular millennium. Thus, many forms of nationalism are based on a vision of a people realizing their destiny in the destruction of the old social order.

The *Sendero Luminoso* (Shining Path) movement, engaged in armed guerilla struggle in Peru, is a contemporary political expression that uses highly millenarian rhetoric. The articulator of its ideology, referred to by his *nom de guerre* as "President Gonzalo," said in an early speech:

> The trumpets start to sound, the rumor of the masses grows and it will grow more, it will deafen us, it will attract us to a powerful vortex . . . and in this way it will create the great rupture and we will be the makers of the definite dawn. We will transform the black fire into red and the red into light. That we are, this is the reconstitution. Comrades, we are reconstituted! (quoted in Diaz–Albertini, 1993:172).

The millenarian and messianic appeal of the Russian Revolution is another example (Murvar, 1971). Indeed, a persuasive case can be made that the major revolutions of this century (Nazism, Soviet and Chinese communism) were millenarian movements.[4] The potential for modern movements to create and sustain a sense of urgency and imminent disaster in a larger mass population can be partially attributed to the impact of mass media (see Barkun, 1974; Cohn, 1970). Even without reference to a supernatural realm, these belief systems are millenarian responses to problems of order and meaning.

Some social historians observe that there appears to be a cyclical pattern to the rise of millenarianism. In U.S. history, for example, the 1840s, 1890s, 1930s, and 1960s have been identified as periods of intense millenarianism in both religious and other social political forms. If there are recurring patterns, do these periods have certain factors (such as economic conditions, regional growth or decline, moral or political crises) in common (Barkun, 1985; Hammond, 1983; McLoughlin, 1983; Smith, 1983)?

Millenarian movements, even explicitly religious millenarianism, may have an inherent political quality to them, because members can draw the conclusion that the enormity of the end of "the World's" time invalidates previously binding rules and laws, including established governments' laws. Thus, millenarian movements have a built-in potential for **antinomian** behavior, because their key beliefs undermine the temporal basis of the political order (Fields, 1993). Often government authorities correctly sense these movements' challenge to law and the existing order of the nation, but they fail to appreciate that the antinomian threat is deeply embedded in an apocalyptic vision that completely transforms the meanings of time, order, and behavior. Had U.S. authorities taken seriously the Branch Davidians' highly apocalyptic and millenarian beliefs as a basis for their behavior before and during the 1993 siege of their compound (resulting in the fiery deaths of eighty-six people), they would have realized that their approach to dealing with the group's antinomian conduct was counterproductive (O'Leary, 1994; see also, Robbins and Anthony, 1995).

Dualism and Millenarianism in Social Movements

Expectation of the millennium is not merely an incidental item of belief; rather, it is a significant part of the group's response to the problem of anomie. If people have experienced anomie, the solution *par excellence* is not only to affirm a new order for the present but also to believe in a future reordering that will reintegrate all problematic aspects of the present situation into an overall, meaningful perfect order.

The imminence of this dramatic reordering creates a sense of urgency and makes the mundane business of daily life "in the world" seem trivial. Millen-

---

4. The messianic tone of Maoism had a precursor in the mid-nineteenth-century Chinese Taiping rebellion, a civil war in which some 20 million persons were killed in combat, massacres, and purges. The rebellion was based on a syncretic religious message that combined elements of Christianity—especially the idea of heavenly reign after the return of the Messiah—with popular Confucianism and Buddhism to justify both the authority of Hong Xiuquan and his platform calling for the destruction of old economic hierarchies to be replaced with egalitarian sharing of land and resources (Spence, 1996).

nial expectations therefore enhance members' commitment to the group through which they will help usher in the new order. The idea that the end of this world is near makes it much easier for members to give up possessions, jobs, education, and other things that have value only in the nearly defunct world. Like the dualistic theodicy, millenarianism encourages pattern forming. Believers interpret seemingly random events as evidence for the imminence of the millennium. For example, in the 1990s some Christian, occult, and psychic groups saw signs of the impending end in current events: earthquakes and volcanoes, ecodisasters, wars in the Mideast, UFO and comet sightings, economic crises, and so on. The approach of the year 2000 spurs some millennial expectations, while other believers attempt elaborate calculations based on symbolic numbers in the Bible, the occult prophecies of Nostradamus, and so on.

The rhetoric of millenarianism combined with dualism provides a persuasive motivation for social movements, because it undermines the authority of the existing order, reorders the collective notion of time, and suggests a line of action in the face of that which the movement defines as evil (O'Leary, 1994).

The millenarian vision is likewise a theodicy for the discrepancies within the present order. Even after conversion, members are likely to have difficulties—friends and relatives "not understanding," prayers seemingly not answered, doubts and uncertainties, and the society still not on the "right track." The belief in the coming of the millennium relativizes the problems and opposition of the present by the knowledge that all of these will be overcome in a glorious future. Members can feel that although "things" are really bad now and will probably get worse, they are not personally threatened by the disorder and ambiguity because they know that they are allied in the present with the source of perfect order and will have a privileged position in the unknown glorious future.

Although infinite possible meanings could be attached to any given event, the themes of millenarianism and dualism illustrate how meaning systems interpret events and locate them in some larger framework. We have seen how individuals and groups use their belief systems to bestow meaning on the social world; and how elements of belief such as imminence of the millennium, messianic expectations, and dualism are responses to a social situation that appears chaotic, ambiguous, or hopeless. Individuals who hold dualistic or millenarian worldviews relate to their social world accordingly. Their worldview organizes their experiences and influences their associations and their very identities.

SUMMARY

This chapter has outlined some of the ways in which the meaning-providing and belonging components of religion are linked. Religion is both a basis of association and an expression of shared meanings, the importance of which depends largely on the social support of a community of believers.

Religious meaning systems are ways of interpreting events and experiences. They assign meaning to a group's existence and to an individual's identity. The meaning system of a group interprets its social order, serving as a legitimation of social arrangements. Meaning systems also locate human lives and events in terms of a larger framework, identifying the individual with an overarching order. The significance of meaning in individual and social life is illustrated by situations in which meaning is threatened. Theodicies are religious explanations that offer meaning for meaning-threatening experiences such as suffering and death. Some crises of meaning profoundly disrupt the social order. The concepts of *anomie* and *mazeway failure* describe the impact of such crises on the individual and group.

Analysis of the themes of millenarianism and dualism in contemporary social movements illustrates how groups bestow meaning on the social situation. These two themes can be interpreted as responses to problems of meaning and order. They represent an effort to posit a new order in response to the ambiguity and confusion of meanings produced by social change. Millenarianism and dualism are theodicies that locate believers' experiences in terms of a larger framework of order and meaning.

Religious belonging is a natural outcome of shared meaning systems. At the same time, a community of believers (however small) is important for maintaining the plausibility of a meaning system. Where a particular worldview has monolithic status in a society, these plausibility structures pervade the entire society. In pluralistic societies, by contrast, worldviews are in competition. The interrelationship of the meaning-providing and belonging aspects of religion is well illustrated in the social processes of religious socialization, conversion, and commitment, to be explored in Chapter 3.

RECOMMENDED READINGS

Nonfiction

Peter L. Berger. *The Sacred Canopy: Elements of a Sociological Theory of Religion.* Garden City, N.Y.: Doubleday, 1967. Berger's interpretation of the contemporary religious situation is based upon a carefully developed theory of identity.

Thomas Luckmann. *The Invisible Religion: The Problem of Religion in Modern Society.* New York: Macmillan, 1967. Luckmann develops a theory of modern religion, interpreting nonofficial religion, privatization, and individual religious forms.

Steven M. Tipton. *Getting Saved from the Sixties: Moral Meaning in Conversion and Cultural Change.* Berkeley: University of California Press, 1982. Tipton explores the meaning systems and ethical styles of 1960s youth participating in a Pentecostal Christian sect, a Zen meditation center, and a human potential organization (est). The book illustrates well how culturally available meaning systems work or fail to work as individuals develop personal meaning systems in a time of rapid social change.

Max Weber. *The Sociology of Religion.* Trans. E. Fischoff. Boston: Beacon, 1963. This book contains Weber's mature thinking on several important aspects of religion, including theodicy, the idea of salvation, asceticism and mysticism, the role of the prophet, and the religion of nonprivileged classes.

Other Books

Issues of religious meaning and belonging are often well illustrated in fiction and literary accounts. The following sources are useful:

Walter M. Miller, Jr. *A Canticle for Leibowitz*. New York: Harold Matson, 1959. Describes the role of religion in regenerating civilization following nuclear war, how the Albertian Order of Leibowitz was founded, and how it struggled against the Dark Ages that followed the nuclear holocaust. The few artifacts upon which the new knowledge was built included sacred texts written by the Blessed Leibowitz, such as the following fragment: "Pound pastrami, can kraut, six bagels . . . ."

Sheri S. Tepper. *Raising the Stones*. New York: Doubleday, 1990. Science fictional depiction of cosmic clashes of religions. Clever portrayal of patriarchal and dualistic religions. One finds oneself empathizing with the people whose god appears to be—well, a fungus.

Kurt Vonnegut, Jr. *Cat's Cradle*. Baltimore: Penguin, 1963. A science fiction account of the end of the world and the ultimate religious movement, Bokonism. Serious humor about the construction of sacred texts, symbols, and meaning.

Elie Wiesel. *Night*. New York: Pyramid, 1960. A moving personal journal of the author's experience as a child in a Nazi concentration camp. The issue of meaning is implicit and powerful.

3

The Individual's Religion

S ociology of religion emphasizes religious groups and social expressions of religion. The individual members of religious groups are, however, social *actors*—that is, persons with motives and meanings of their own. Although the individual actor's attitudes, conceptions, and behavior may be strongly influenced by social groups, there is no neat or deterministic correlation between what the group believes and what the individual member personally holds central. To understand religious behavior, we must know how religion shapes and is expressed by the individual actor. Chapter 2 suggested some of these ways. Religion provides meaning for the individual's life, enabling one to interpret, evaluate, and project experiences in everyday life. Also, social groups are essential in shaping, maintaining, or changing the individual's worldview.

This chapter examines the processes by which the individual adopts and becomes committed to a religious worldview and community. First, we will explore the social forces shaping and maintaining the individual's worldview as it develops and changes over the life cycle. Second, we will analyze the process of conversion by which an individual may dramatically change that worldview. Finally, we will examine the process of commitment, by which old and new believers alike commit themselves to the group of fellow believers.

SHAPING THE INDIVIDUAL'S RELIGION

The meaning-providing and belonging features of religion, described in Chapter 2, are important to shaping the individual's religion. The individual's meaning system is socially acquired, socially maintained, or socially changed. The individual subjectively experiences religion in social contexts, influenced by socially shaped, learned meanings.

Meaning, Belonging, and Identity

These learned meanings are conveyed by specific people. The child does not (indeed cannot) internalize the group's meaning system without having a sense of belonging to the group in which that meaning system is grounded. Typically, in societies with relatively homogeneous meaning systems, belonging to the group means coming to take its worldview for granted.

In societies with competing worldviews, socialization into one's own group includes some awareness of the existence and differences of other groups around one's own. The child's identity typically includes an important sense of belonging to a specific religion, ethnic group, or nation, as well as a feeling of contrast with others not of that religion, ethnic group, or nation.

Religion is often an important part of how an entire *group* thinks of itself: what it means to be "one of us." For example, the religious identity of "Ulster Protestant" includes religion and other political and cultural meanings pertaining to belonging to a particular community. The connection of religion and group identity is explored in more depth in Chapter 6. Another link of religion and identity includes all the ways that religion informs the individual's self-identity.

Self-identity refers to each person's biographical arrangement of meanings and interpretations that form a somewhat coherent sense of "who am I." Often the question "Who am I?" is answered in terms of "This is where I belong." Thus, a woman might describe herself as "a mother, a wife, a Catholic, a Polish-American, a member of the town volunteer ambulance squad, in the church choir, and vice president of the PTA." These roles represent not merely formal memberships but—more important—social locations of her identity. Religion pervades all such social roles in relatively undifferentiated societies. In modern, highly differentiated societies, however, religion is only one significant source of the individual's sense of belonging and identity.

Self-identity is profoundly *social,* because it is constructed through ongoing interactions with others. Although those others—especially important persons like parents, spouse, close friends—are powerful influences on the individual's developing self-identity, their influence is never deterministic. Nor is identity fixed; it is continually changing as the individual takes on new roles, interacts with new people, and grows older. Even in very rigid, traditional societies, for example, a woman's self-identity as a new mother is very different from her later self-identity as an old woman.

The construction of self-identity in modern societies may, however, be qualitatively different from that in traditional societies. If, in modern societies, the individual's social situation is based less on ascribed (socially fixed) statuses and roles, it is probable that self-identity is likewise more malleable, selectively constructed, and changing (cf. Giddens, 1991:53–55). In late modernity, individual self-identity is more eclectic, more like bricolage—an edifice constructed from a wide range of culturally available options (Beckford, 1989; Luckmann, 1967).

Components of the individual's self-identity, including religion, family identification, ethnic identity, aesthetics, and even gender, appear to be more open to reflexive *choice* in the modern world. Religion thus becomes an optional resource for the ongoing project of constructing personal identity. If this thesis about modern self-identity is accurate, it means we need to think of such social processes as childhood socialization, conversion, and commitment as more tentative and changeable ongoing projects.

Like all identity projects, the individual's religion develops and changes over the life course. A person is not born with a full-blown set of religious beliefs and practices; religion is developed and nurtured (or ignored) in the socialization of the child. Although one may have reached an identifiable religious perspective or commitment by young adulthood, it would be a mistake to consider this the end of one's religious development. The individual's religion continues to develop and change, perhaps less dramatically, throughout the rest of life. Indeed, the very meaning of "being religious" changes in different periods of life, and the place of religion in the individual's life also changes. In keeping with this dynamic perspective on religious development, this section follows the life cycle of the individual believer. It focuses (as does religious practice itself) on certain critical periods: early childhood, adolescence, marriage and procreation, and old age and dying. One intermediate period, for which there are no special religious rituals or events, is included: middle age.

One of the difficulties in describing an adequately complex picture of religious development in modern societies is that much of the research in this area has focused on narrowly defined church-oriented religion. Not only does much modern religious life take place outside religious organizations, but also religious affiliation is not a very satisfactory indicator of each individual's actual religion. (See Chapter 4 for related methodological issues.) The following description, then, suggests a line of reasoning that future studies might pursue in exploring the relationship between religion and family, community, identity, life cycle, and human development.

Childhood, Family, and Community

Early childhood is a critical period in the development of the individual's religion. The child begins to learn what it means to be "one of us" (our society, our ethnic group, our religion, our family, our tribe, etc.). Specifically in socialization, the individual internalizes the social group's moral norms and basic values. Socialization also accounts for the development of the individual's attitudes and values, such as attitudes about authority or the relative value of acquiring material possessions. These attitudes and values are closely linked with religious belief, and they vary according to the type of religiosity that individuals acquire through socialization into their group.

In relatively simple societies, there is no distinction between socialization into the larger group and religious socialization. Becoming religious—however defined in that society—is simply part of gradually becoming an adult, responsible member of society. Religious roles in simple societies are not differentiated from other roles such as mother, healer, or chieftain. Similarly, the child's religious roles simply involve the child's participation in ongoing social activities. The child might assist the father in his work and observe the father praying over his tools. Or a child might learn, in helping the grandmother, that one should bless the hearth and fire each morning upon rekindling the fire. He or she might learn the central myths of the tribe from sitting beside a storyteller after a meal. The pervasiveness of religion in simple societies enables the religious socialization of children to occur informally and continuously. Complex societies, by contrast, place more emphasis on intellectual learning and formal religious knowledge. The society's differentiation of religious knowledge from other relevant formal knowledge complicates the child's acquisition of religious knowledge.

Learning the content of the meaning system of one's group is, however, not sufficient. Children need to internalize this meaning system, to make it their own. A computer could teach a child an entire catechism, but that knowledge of itself would probably have very little significance in the child's life. Internalization of the group's meaning system occurs through interaction of the child with specific other members of the group who mediate the group's way of thinking and doing to the child. In early socialization, the child's family is the foremost influence.

The Family. The first, perhaps most enduring source of the individual's sense of belonging is the family. In the child's socialization, the family is something of a "boundary structure," connecting the borderline between the foundations of personality and the child's beginning participation in society (Parsons, 1969:424). The very young child is not *in* society but is a part of the family, only indirectly participating in society. Later full participation in society is organized according to the values, motivations, and attitudes incorporated during socialization into the individual's personality.

Rituals and symbols allocate identity even in early childhood. One of the most significant and widespread rituals is naming the child. The choice of given names, the affirmation of kinship networks (e.g., naming the child after a grandparent), and the bestowing of the family name (or refusal to bestow it) are highly symbolic acts. Many groups have special ritual occasions for giving a new family member this initial identity. Other occasions of childhood (e.g., birthday celebrations) also affirm the child's identity. The status of the child is relatively clear in most societies; thus, major rituals and symbols of identity are often used to make important transitions, such as the change from childhood to adulthood, in identity.

Both in the family and in formal religious education, parents may try to reproduce a religious worldview for their children. Many parents value some religious socialization for their children, and children's religious training is a major consideration in adults' decisions to reestablish their church attendance (Gallup and Castelli, 1989:144; Roozen et al., 1990). One study (Wilson and Sandomirsky, 1991) found that marriage and parenthood were the strongest predictors of whether women who were previously religiously unaffiliated would join and become active in a religious group. The authors suggest that, for women at least, the decision to affiliate with a church or synagogue is an interdependent one, aimed at a collective good for the whole family. A related phenomenon, however, is that once the family "nest" is nearly empty, parents of the current middle-age generation (sometimes called "baby boomers") are less likely to participate in their religious group than parents of younger children (Roozen, 1993).

The urge to attend church for the sake of the children is hardly universal, however. A substantial drop has occurred in the proportion of U.S. children receiving specifically religious training. In 1952 only 6 percent of adults in the United States had received no religious training as a child; this figure had risen to 9 percent in 1965 and to 18 percent in 1988. According to a 1988 survey, about seven of every ten adults with children ages 4 to 18 were providing religious training (Gallup and Castelli, 1989:66–68).

One study found that family roles themselves—actual or fantastic—were an important factor shaping children's images of God. Children identified God with parental (or grandparental) roles and qualities: big, authoritative, nurturing, understanding, forgiving, controlling, wise, angry, and so on (Heller, 1986).

The family is more than an agent of religious socialization. It is often, indeed, a primary religious group, a fundamental unit of the institution of religion. The trite adage "the family that prays together stays together" is probably reversed in its reasoning. It is entirely likely that any cohesive family expresses, rather than originates, its unity in rituals and symbols that, in most cultures, are also fundamentally religious. At the same time, religion has historically legitimated the family, established rituals to celebrate family unity, and provided norms and social controls to protect the institution of the family. The scope of the family unit in modern society is greatly narrowed from the extended family and large kinship networks of earlier periods. As a result, the primary agents of socialization in the family are fewer, usually only the parents and older brothers and sisters.

The Community. Another major source of identity is based on the individual's sense of belonging to a distinctive group—tribe, nation, ethnic group. The U.S. and Canadian history of immigration makes ethnicity an important factor in many people's experience. In a culturally plural society, a close-knit ethnic community may be an important plausibility structure supporting the group's religious worldview (Berger, 1967:46). Ethnicity may be an important basis for a person's choices of neighborhood, friends, job, marriage partner, leisure time activities, and organizational memberships. For another person, ethnicity may be irrelevant to such choices.

The process by which the child is socialized into the family and ethnoreligious group is gradual. It begins with the child's simple experiencing of group membership as a taken-for-granted part of life. Later the child may learn terms for identification with the group (i.e., "I'm a Catholic," "My people are Korean"). This simple identification proceeds from a confused usage of these terms to a clearer conceptualization of their meaning and of belonging to one or another group. One study found that young children (ages 5 to 7) had a vague impression of their religious affiliation as a kind of family name. One 6-year-old responded that it was impossible to be a Protestant and an American at the same time because "you can't have two names," unless perhaps you moved. Somewhat older children (ages 7 to 9) understood religious belonging in terms of wearing particular symbols or of characteristic behaviors (e.g., wearing religious medals or having a Christmas tree). Only much later, usually in adolescence, did children have conceptions of religious belonging that included nonobservable qualities such as faith or belief. Also in later stages of development, some children explained their own identification in terms of contrast to other religious groups (Elkind, 1964; on religious development, see Spilka et al., 1985).

Such studies unfortunately focus on too narrow a definition of religious identity and belonging. Although the individual's gradual self-identification in terms of official religious affiliation is one component of religious identity, we need to know more about the development of other aspects, especially commitment and loyalty. The relationship between identity and values, attitudes, and behavior also needs further exploration without narrow restriction to church-oriented religion. The communities of family, neighborhood, friendship, and ethnoreligious group provide an initial sense of belonging and a foundation for identity; subsequent communities provide a base of support for maintaining that identity or for changing it (as our later discussion of conversion will show). They are also the social base for the continued plausibility of the group's entire meaning system (Berger, 1967:45–47). Thus, the meaning-providing and belonging aspects of religion are directly connected.

Social changes and developments such as pluralism (i.e., the coexistence of competing worldviews) seem to have undermined some of the cohesiveness and effectiveness of these communities. As children emerge from the taken-for-granted small communities of family and neighbors, they will probably encounter a bewildering variety of other groups with other belief systems and ways of life. Some individuals remain immersed in the initial network that supported the plausibility of early beliefs and identity; others detach themselves from these bonds and seek elsewhere a social base for their beliefs and sense of who they are. The very existence of this element of choice in identity and belief system characterizes only certain kinds of society. The possibility of detaching oneself from the taken-for-granted beliefs and social groups into which one has been socialized provides greater freedom—and simultaneously makes both belief and identity problematic (Berger, 1967:137, 151).

There is considerable evidence that both ethnicity and religion may have become less salient as sources of identity for perhaps the majority of Americans. For white Americans at least, ethnic identity, like religious identity, has become essentially voluntary (Hammond and Warner, 1993). Since race is still largely an ascribed characteristic in the United States, nonwhites experience more limited options for their ethnic self-identities (Waters, 1996). Nevertheless, they also construct ethnic self-identities creatively and eclectically by choosing *which* cultural elements to make personally meaningful (cf. Conzen et al., 1990; Sollors, 1989). For example, one individual might choose to make participation in an ethnic parish a central expression of her self-identity, while another might emphasize ethnic dances and festivals.

In a pluralistic context, it does not even matter whether the chosen ethnic identity is accurately part of the individual's ancestry. For example, one black woman who was an enthusiastic member of a Brazilian dance troupe in Chicago explained that she had no reason to believe that her ancestors came from Brazil or the parts of Africa from which Brazilian blacks may have come, but Brazilian music and dance were now the most valued part of her current identity as a black person in the United States. In modern, pluralistic societies, individuals are relatively freed of structurally given identities, relatively freed to pick and choose culturally available elements to construct a picture of who they are (see Conzen et al., 1990; Sollors, 1989). Ethnic and religious elements of self-identity thus function as culturally plausible myths—stories out of which one lives.

Problems of Family and Community in Contemporary Society.
Many processes in contemporary society have seriously weakened the ability
of family, community, and religion to offer the individual a stable source of
belonging and identity. Neighborhood communities are decreasingly characterized by face-to-face intimacy and personal, affective bonds. Furthermore,
family, neighborhood, and religion alike have been undermined by **privati-**
zation—the process by which certain institutional spheres become removed
from effective roles in the public sphere (especially in the arena of economic
production and power). Chapter 8 will examine the impact of privatization
on religion in greater detail, but it is important here to note the link between
privatization of the family and religion.

The power of religion to legitimate other institutions such as the family
has been undermined. There is little support for the family from public institutions. What support exists is largely due to the family's significance as an
economic unit of consumption and to the family's symbolic importance (e.g.,
in the rhetoric of politicians). These processes, together with the pluralism of
society, have weakened the norms governing family interaction. At the same
time, however, greater demands are made on the family. The weaknesses of
the neighborhood and community in providing support for members' identities result in greater pressure on the family to be a complete repository of
identity (Brittan, 1977:58, 59).

Economic and demographic changes have made the type of family ideal-
ized in the 1950s (i.e., two or three children living with both biological parents, the mother not working outside the home) today only a small minority
of U.S. families (Marler, 1995). Single-parent families, blended families, adop-
tive families, homosexual families, commuter families (where one spouse
works in a distant city and commutes "home" on weekends) and dual-career
families are also prevalent patterns for family life. In the last decades of the
twentieth century, perhaps the biggest single stress on the family's ability to
meet its members' needs has been workplace and related economic changes
that have required parents to spend greatly increased proportions of their time
on the job in order to provide for their families (Hewlett, 1991; Schor, 1992).

As the individual's control over life in the public sphere diminishes—and
as the institutions of the public sphere become less and less concerned with
providing meaning for their participants—the individual seeks more fulfillment, security, control, and meaning in the private sphere. The family and
other institutions of the private sphere are expected to provide a haven from
the public sphere and to compensate for what their members are deprived of
in the public sphere (Berger, 1967; Luckmann, 1967:106–114). Privatization
thrusts the search for personal identity out of the public sphere. The nature of
work and division of labor in modern societies make it difficult for people to
derive their primary sense of identity from their work role. Yet the family
cannot sustain the burden of providing all personal identity for its members.

The local community (as a social unit, not just a political or geographical
one) has also been undermined. Religion has traditionally been inextricably
intertwined with community; the boundaries of one were frequently coextensive with the boundaries of the other. Social and geographical mobility, to-

gether with the general acculturation of waves of immigrants, have changed the patterns of community in U.S. society. Local communities are increasingly voluntaristic, segmented, and irrelevant to institutions of the public sphere. The voluntary nature of community has the advantage of freeing individuals to choose their community of identification, but it also reduces their power relative to institutions of the public sphere and their ability to provide stable sources of identity for members. The process of privatization is not inexorable, but its impact on family and community is already evident.

Alternative Modes of Family and Community. One historical alternative to the dominant mode of family and community has been the ethnoreligious group that withdraws from the society in order to preserve its distinctiveness. Examples of this kind of community include the Jewish Hassidim, the Amish, Doukhobors, and Hutterites. The Hassidim are particularly interesting because their withdrawal is not to a rural refuge; instead, they have organized their life within urban areas to sustain their religious identity.

The Hassidim are an ultraorthodox group of Jews, mostly first- and second-generation immigrants who fled persecution in central and eastern Europe (see Kephart and Zellner, 1991; Poll, 1969; Shaffir, 1974). There are relatively few U.S. and Canadian converts to Hassidism, although the largest branch (Lubavitcher) proselytes among less orthodox Jews. More than 100,000 Hassidim live in the United States, and large Hassidic communities also exist in Canada. The Hassidic community is carefully organized to enable members to follow the 613 rules regulating everyday life for orthodox Jews. Members must live close enough to the synagogue to be able to walk there for services on the Sabbath. Food that passes the highest levels of kosher regulations must be available. Outside the home, men and women are segregated at all times, partly because of religious rules designed to protect men from defilement by contact with a menstruating woman. Thus, the men of the community arrange their work so that they have no interaction with women (e.g., by working in the neighborhood or in an all-male trade). The men further arrange their work time to be free for frequent daytime religious services.

The Hassidic community has its own system of law and its own shops, trades, and schools. The rebbe (i.e., chief rabbi) is the final authority in both spiritual and secular matters. Indeed, there is no distinction between spiritual and secular issues; religion pervades all spheres of life. The socialization of children into the group's approved way of life is relatively unproblematic because of the group's success in limiting outside influences. Children attend religious schools, are heavily occupied with their studies and (in adolescence) with their own religious duties, and are seldom exposed to outside media such as television or movies. The community enclave makes the Hassidic way of life a taken-for-granted reality to the child. Personal identity and a sense of belonging are similarly unproblematic. Roles and norms for appropriate behavior are fixed, with roles available to males varying considerably from those for females. Boys clearly learn their roles as men, and girls learn clear-cut expectations of women; both are rewarded by the close-knit community for meeting these role expectations.

A related kind of local community, the **intentional community,** is one in which members choose to live together communally. This arrangement turns the voluntary nature of its commitment into a virtue. Religious communalism has a long tradition in Western society; monastic communalism dates back to the sixth century A.D. A wide variety of groups dissatisfied with the values and structures of contemporary society have adopted a communitarian form of organization. Intentional communities are potential alternatives to the modes of both family and community life of the larger society. The intentional community allows the group to alter all aspects of social interaction—work, leisure, marriage, family, decision making, education, and so on. Thus, a group can integrate the spheres of work and religion for its members because, in the separate community, these spheres need not be segregated as they are for most members of society. The intentional community also provides a potentially stronger base of personal identity and a stronger sense of belonging than the privatized family and community can. Communitarian groups often resemble traditional communities with their close-knit, affective interpersonal relationships; they differ, however, in being essentially voluntary. Members cannot take their belonging for granted, although certain practices of the group may make it difficult for the member to consider leaving. The group's continual point of reference is the dominant society from which its way of life departs.

Twentieth-century commune movements in the United States, Canada, and parts of Europe illustrate some of these themes. Communes such as Twin Oaks, The Farm, Ananda, Brotherhood of the Spirit, and the Lama Foundation organized themselves as alternatives to the dominant mode of personal, family, and community life. Other intentional communities were outgrowths of specifically religious movements; twentieth-century examples include the Jesus People, Hare Krishnas (Krsna Consciousness), Unification Church, Catholic Pentecostalism (later called Catholic Charismatic Renewal), and the Bruderhof (Society of Brothers). Most Catholic Charismatic prayer groups, for example, consist of persons who live in typical middle-class neighborhoods and maintain ordinary middle-class family styles. Their prayer group attempts to provide additional sources of community support, bolstering families and individual members in their distinctive beliefs and practices. This movement also spawned a number of "covenant communities"—intensive groups whose living arrangements, finances, childcare, and other family functions are communally shared. Such communities enable members to have a much stronger base of social support for their way of life than would be otherwise possible.

Contemporary religious communalism can be interpreted both as a celebration of and protest against privatization of religion, family, and community. On the one hand, communitarian groups attempt to restructure social organization to allow religion, family, and community to influence other spheres of their lives such as work, education, decision making, and so on. Their protest is against the fragmentation of social life in society. At the same time, however, they are a celebration of privatization because they have removed themselves from society, presenting little challenge to the institutions of the public sphere. Privatization robs communitarian groups of their impact. Establishing a commune, no matter how dissident, is a somewhat expanded way of "doing your own thing" in the private sphere of family, religion, and community.

Adolescence

Whereas childhood is a period of initial socialization into a group's way of life and meaning system, the transition to adulthood is a critical transformation in most cultures. Childhood is important in establishing identity and group belonging in a general way, but the transition to adulthood means passage to a new identity involving responsibility, knowledge, ritual and symbolic roles, and acceptance into adult circles. The dramatic quality of this transition is ritually expressed in many cultures by **rites of passage** for adolescents.

Rites of passage are rituals that accompany a change of place, state, social position, and age. One interpretation suggests that rites of passage enable societies to effect an orderly, meaningful transition for individuals and groups who move from one socially recognized stable state (e.g., childhood) into another (e.g., adulthood). The stable statuses have clearly defined rights, obligations, and roles. The transition between them, however, is dangerous because rights, obligations, and roles are temporarily ambiguous and disordered. Both the social group and the individual member are thus served by these rituals that express, yet circumscribe, the ambiguity and disorder of this period.

These rites consist of three phases: separation, marginality, and aggregation. In the separation phase, ritual actions symbolize detachment of the individual from the previous stable state; the individual metaphorically dies to the old self. The phase of marginality symbolizes the ambiguity of the transition. It represents a structureless realm in which previously taken-for-granted roles and relationships are brought into question. The phase of aggregation unites the individual with others in the new status group. It includes the transmission of knowledge needed by the individual in the new role together with the symbolic features of the new role (e.g., clothing representing the new self). Rites of initiation in simple societies illustrate these processes (Van Gennep, 1960; see also Eliade, 1958; Turner, 1974a, 1974b, 1979).

Contemporary religious groups have rituals that are remnants of earlier rites of passage; baptism, for example, has many features of initiation rites. Confirmation among Christians and bar (bat) mitzvahs among Jews also represent the transition from childhood to adulthood in the religious group. These rituals are, however, relatively weak in effecting such transition in modern society. One study noted that, ironically, decline in adolescent religious observance and adherence in their groups' traditional beliefs generally began soon after confirmation, bar or bat mitzvah, or other rituals for adolescents—for most, the last obligatory religious training (Ozorak, 1989). The ineffectiveness of contemporary rituals of transition is partly because religious groups allow ritual adulthood to members whom they do not consider fully socially adult. A 13-year-old boy who had recently become a full member of the

Presbyterian church (having attended communicants' classes, made a profession of faith, and received his first communion) wrote to the elders (governing board) of the church suggesting changes in church activities. His letter was received with mirth: "Isn't that cute that he thinks he's old enough now to be telling us what to do." A stronger example of the ambiguity of adult status is the use of forcible "deprogramming," in which parents of legally adult "children" refuse to accept their offsprings' choice of religious identities.

Another more serious basis of this weakness is the quality of adolescence in this society. There is no clear-cut event that confirms adult status. Puberty marks the end of biological childhood, but social adulthood is many years away. The economic structure requires years of preparation for adulthood. A 20-year-old student who is not fully self-sufficient and independent of parental and school control is no longer a child but is not socially recognized as fully adult. Demarcation events such as leaving home, a first self-supporting job, and marriage are recognized as the beginning of adulthood for some individuals but not for others. Some observers suggest that the forms taken by adolescent rebellion against adult authority are themselves attempts to symbolize a transition (Berger, 1969).

Rather than the *medium* by which transition to adulthood is structured, religion in this society often ends up as the *content* over which adolescent rebellion is staged. Adolescence is often a time of identity crisis because the ambiguity is not culturally resolved. The assertion of one's identity (separate from the child's identity, which is defined by family and community values) sometimes takes the form of religious rebellion. The young person tries to assert an adult self by denying those aspects of life perceived as symbolic of the childhood self and tie to parents or family. An obvious example of this rebellion is the youth who joins a religious group that appears diametrically opposed to his or her childhood religion. A very different pattern—the young person who becomes superreligious in the childhood religion—is another way of expressing rejection of the family's "low level" of religiosity and their "hypocrisy." A holier-than-thou stance is comparable to changing religions as a way of asserting an independent identity and rejecting parental values (Greeley, 1979a). These seemingly dissimilar adolescent strategies may account for a number of contemporary religious expressions: the attraction of youth to countercultural religious movements, their seemingly fanatical adherence to both old and emerging religious groups, and the agonizing, often bitter struggles between parents and youth over a newly asserted religious identity.

During the period of transition from childhood to adulthood, the direct influence of parents on their adolescents' beliefs and practices often diminishes, whereas peer influences may increase (Potvin and Lee, 1982; see also studies cited in Hyde, 1990). Nevertheless, the parents' beliefs and practices are reasonably good predictors of certain aspects of their adolescents' religiosity. One study of a conservative religious group found that parents with relatively traditional religious beliefs and practices were likely to produce adolescents with relatively traditional religiosity, whereas less traditional parents were likely to raise less traditional adolescents. Both sets of adolescent children, however, were collectively less traditional than their parents (Dudley and Dudley, 1986).

Another study surveyed second-year high school students and later restudied them when they were approximately 27 years old. During high school, students exhibited a high degree of conformity to their parents' religious orientation, beliefs about God, and levels of church attendance. A decade later, the influence of parents' religious practices and beliefs was, at most, indirect. The religious involvement of one's spouse became far more salient for these young adults (Willits and Crider, 1989). Similarly, a study among Mormons found that one of the main enduring effects of youth religious socialization was channeling members into a social network of friends—a personal community of social support—which also served to sustain their religious worldview (Cornwall, 1987; see Benson et al., 1989, for a review of literature on adolescence and religion). Thus, both the original family and, later, personal community and new family through marriage served as significant plausibility structures (as described in Chapter 2), but none of these sources had fixed or deterministic influence.

Marriage, Sexuality, and Procreation

Marriage marks a status passage and the beginning of a new family of procreation. Especially for women in most cultures, marriage is the entry to full adulthood. A wealth of ritual and symbolism accompanies and defines this status passage. Religion often directly promotes the ideal of marriage by setting the norms for marriage and establishing the appropriate behavior of members before, during, and after the marriage ceremony. In societies where several religious or ethnoreligious groups coexist, religious groups also delimit the pool of acceptable spouses; members are not supposed to marry "outside the fold."

Images evoked in the marriage rituals of historic religions symbolize this status passage. The traditional transition of the woman from her father's possession to her husband's is symbolized by the father "giving away" the bride. Some denominations have attempted to change this symbolism to fit the realities of contemporary family life—for example, by having both parents escort the bride to the groom or by allowing the bride to present herself. Nevertheless, the symbolism represents a fundamental transition of status for both bride and groom. Other symbols (e.g., wedding bands, ceremonial binding of hands, exchange of ceremonial crowns) represent the unity of the new social group—husband and wife.

Although historically religion has been closely linked with family life and the regulation of marriage and sexual behavior, its impact in contemporary family life appears to be more indirect and perhaps weakened. Unfortunately, the data available are narrowly limited to the impact of religion on marriage in four particular areas of social control: endogamy (i.e., marrying within one's ethnoreligious group), sexual norms (e.g., restrictions of premarital, extramarital, homosexual, or "deviant" sexual activities), divorce, and reproduction.

Furthermore, these data describe only relatively recent attitudes and behavior. We have no comparable information about the impact of religion on family life in, say, 1830 or 1600. These data are also generally based on a narrow consideration of the term *religion;* correlations usually refer only to "religious affiliation."

Religious endogamy is decreasingly common in the United States and Canada, and it is of less importance in mate selection than even a few decades ago (Glenn, 1982; Larson and Munro, 1985). In this century, U.S. Catholics, Jews, and Protestants (except the most conservative denominations) have all experienced steady increases in the rate of exogamous marriage (McCutcheon, 1988). Religious exogamy may be correlated with denomination switching because often one spouse changes to the religion of the other (Bahr and Albrecht, 1989; Hoge, 1981). Religious homogamy (in which both partners are of the same religion) is not a significant predictor of marital commitment (Larson and Goltz, 1989).

Although all Jewish and Christian groups emphasize the ideal of permanent marriage, divorce rates and attitudes toward divorce have been changing. In the second half of the twentieth century, the rate of marital disruption of U.S. Protestants, Catholics, and Jews has risen steadily, as has the proportion of members who approve of divorce (and remarriage of divorced persons). Although the official teachings of the Roman Catholic church are strongly against divorce and remarriage, fewer than 20 percent of Roman Catholics disapprove of divorce (*N.Y. Times*/CBS Poll, reported in Berger, 1985; see also D'Antonio et al., 1989).

One study observed that religion's influence on marital commitment is often indirect. For example, it shapes the degree of emphasis on caring for one another and the couple's definition of gender roles, which subsequently have an impact on the marriage relationship and partners' commitment to that relationship (Scanzoni and Arnett, 1987). Another study found, however, that the linkage between religious involvement and marital commitment held mainly for male, religiously involved conservatives; by contrast, women's marital commitment was influenced far more by their degree of satisfaction with marriage and family life than by their religious involvement (Larson and Goltz, 1989).

Marital satisfaction may be positively associated with church membership and attendance (Bahr and Chadwick, 1985). A study of black families found that religious involvement and commitment to religious upbringing of children was strongly linked with positive subjective assessments of family life (Ellison, forthcoming). Subjective assessments of "satisfaction," however, are ambiguous variables; does religion produce an objectively happier marriage and family life or does it teach the person to define and interpret whatever exists as "happy?" (Ellison, 1991).

Interestingly, religious groups appear to be far more concerned about regulating families' procreative sexuality than with domestic violence: more upset about out-of-wedlock childbirth than about child beating, more outspoken against divorce than against spousal rape, and more cognizant of the national abortion rate than of the existence of incest in their own families. We must remember that religion is not always a source of marital happiness and well-being. It is sometimes linked with causing or at least condoning serious family and marriage problems, such as spousal or child abuse (Nason-Clark, 1995). Some studies show that father-daughter incest often occurs in morally rigid religious families (Imbens and Jorker, 1992), where daughters are expected to be totally submissive to parental authority. Similarly, some religious groups encourage parents and other adults to physically abuse children, if necessary, to break their will and force submission to "God's will" (Capps, 1992; Greven, 1991). Overall, it appears that religions emphasizing obedience in hierarchical relationships (e.g., father/husband exercising dominance over his children and wife) are more likely to condone, if not encourage, domestic abuse and violence.

Social science's emphasis on a limited range of religion's social control functions in marriage is imbalanced. It would be useful to know how religion (defined more broadly than religious affiliation) influences the extent to which people perceive themselves as having choices in life (e.g., the choice not to stay married). How does religion influence the individual's general orientation toward life (e.g., fatalism or hopefulness) and toward others (e.g., are others perceived as basically good or bad)? What impact, if any, does religion have on the individual's sense of self-worth and independence; does the individual's religion contribute to the sense of self outside the marriage or family? These broader influences of religion are probably more critical for our understanding of marriage and divorce than are formal moral norms of specific religious organizations (see D'Antonio and Aldous, 1983).

Similarly, it would be fruitful to explore what kinds of marriages result from different kinds of religious orientations. For example, are the religious orientations and perspectives of the spouses important in determining the power relationships between them? Their ideals of married love? The degree of interpersonal communication in their marriage? The allocation of tasks, resources, decision making, and rewards in the marriage and family? Probably religion is one important factor in shaping the quality of marriage relationships, but the precise nature of its influence has yet to be documented.

A related correlation is the influence of religion on gender roles. Various religious perspectives lead to different images of maleness and femaleness. Gender roles are extremely important in marriage, as in other institutional settings. What a man expects of himself and what others expect of him as a man directly influence the kind of activities and relationships in which he engages. If it is considered unmanly to care for babies, a man is likely to avoid situations in which he would do (or be seen doing) this. Religious beliefs, myths, images, and symbols have been important forces in shaping gender roles, as illustrated in the Extended Application section of Chapter 4.

Similarly, a broader approach to the influence of religion on childbearing, childbirth, and family life is needed. Numerous studies have correlated religious affiliation and fertility, but they have focused too much on the differences between religious organizations in relating their norms of assumed control over the reproductive behavior of members. Far more interesting would be information about how religion shapes the attitudes of men and women toward children. The meaning of having children is different for men than for women, for older couples than for younger ones, for Italians than for Germans, for Jews than for Hindus, and so on. What exactly is the impact of the individual's religion on his or her attitude toward children—especially one's own children? Also, what is the impact of religion on attitudes toward reproductive events per se? How does religion shape the individual's ideas of and feelings about menstruation, intercourse, pregnancy, and childbirth? How does religion influence the person's experience of these events?

One important step in this direction is a study of women's activism on the abortion debate. Female "pro-life" activists and female "pro-choice" activists were found to operate from totally different worldviews, which entailed radically different meanings of motherhood, family, women's roles, children, and the importance of the individual. Although religious affiliations informed these opposing worldviews, another explanatory factor was the difference in the two groups' socioeconomic situations. Pro-choice activists tended to have experience and educational credentials to pursue careers and higher-paid jobs; ability to control their reproductive functions was crucial in being able to achieve their career goals. By contrast, pro-life activists had chosen life patterns generally out of the paid workforce (and most were ill-prepared to succeed in that sphere); for them, childbearing and childrearing were central parts of their sense of purpose and self-esteem. Indeed, because of their values about children and motherhood, these women were typically opposed to the use of "artificial" birth control, even though its use would obviate the need for many abortions. From their respective worldviews, both groups of women had made life commitments that limited their ability to change their minds and life directions (Luker, 1984:158–215). Some of the fervor of the abortion debate is fueled by the stakes that persons on both sides have invested in their worldviews.

Finally, the narrow focus of research on attitudes toward sexual norms has failed to tap the broader issue of sexuality itself. All religions have attempted to interpret sexual themes or experiences. Religious symbolism frequently deals directly with themes of sexuality. Important parallels exist between spiritual and sexual ecstasy; sexual images are linked with religious images in the writings of famous Christian mystics (Bynum, 1991; Milhaven, 1993).

Religion's link with human sexuality is understandable because both are direct, personal experiences of *power,* which evokes a sense of chaos and need for control, pollution and need for purification, danger and need for protection (Douglas, 1966). Religion, with its capacity to give meaning and order, offers control, purification, and protection from the chaotic power of sexuality. The establishment of moral norms to regulate sexual behavior is only one aspect of this control. Other important aspects include religious interpretation of sexual experiences and events (e.g., interpreting marital intercourse as imitative of the marriage of the gods). Religion also provides rituals and symbols to deal with some of the more potent and awesome aspects of sexuality. Many religions have a ritual of purification for women after childbirth (e.g., the "churching" of women in some Christian groups). Indeed, the human body and its functions are potent symbols. Corporality (the quality of being a body) is a common metaphor for social and religious cohesion (see Turner, 1980, 1991). Religion also offers ways of channeling sexual energies into spiritual energies. Many religious groups believe that sexual activity reduces one's spiritual energies; thus, norms of abstinence from intercourse are based not on notions of right and wrong but on the idea of heightening one's spiritual powers.

These reflections about the interrelationship of religion and sexuality suggest some of the directions that further study could take. Existing research on religion and marriage, sexuality, and procreation is much too narrow in both its use of the concept "religion" and its definition of what is relevant in the study of these important human experiences.

Middle Age

Significantly, there are no rites of passage to middle age. In earlier times, fewer people lived in relatively good health much beyond the time their children became adults. Thus, there was no major gap between the years of active child-rearing and support and the years of relative retirement from work and old age. Now, however, with increased longevity and smaller families, a gap of twenty or more years between the time children leave home and parents reach retirement age is not uncommon.

Middle age is often a period of "identity crisis" for both men and women. Many women discover that they have invested their entire sense of purpose in serving their families, but their status as "mother" means little in middle age. The correspondence of this social loss with the time of biological change (i.e., menopause) can be doubly traumatic because some women consider their reproductive capacity a basic part of their sense of self and personal worth. Although the existence of a male biological equivalent to menopause is debated, the social transitions of men in middle age are often as pronounced as those of women. A man may have invested much energy and time in his career and, in middle age, begins to realize that his career goals are unlikely to be achieved. Or he may come to doubt the goals and values that had motivated him in earlier years of his career. Precisely at the stage when his children are becoming more independent of their parents, he often begins to want a closer relationship with them. The structure of family and work in this society makes middle age a problematic time: One's chances of achieving socially desirable goals in family and job are considered past or severely reduced, but one is not yet free to quit striving for those goals.

In Western societies, chronological age is a key criterion by which social roles are linked. It becomes an evaluative criterion when the individual or social group measures the age appropriateness of the person's achievements and activities. A 25-year-old student who returns to college after several years of apparent disinterest in educational goals is called a "late bloomer"; and statements such as "He is too young to be married" or "She is too old to be working full time" exemplify the social basis of these definitions of age appropriateness. The person may be physically capable of full-time work or marriage and childbearing, but the society has established its criteria of the appropriate ages for these roles (Kearl, 1980; see also Berger et al., 1973:73).

Age is also a key criterion in comparing oneself with others. Many social institutions encourage and even formally organize these comparisons. Schools compare the individual's achievements with others of the same age or grade; businesses compare the achievements of age cohorts of workers. Self-evaluation is also often based on the individual's comparison of self with others of the same age. As age progresses, however, opportunities for comparing favorably with others of one's cohort decrease. By middle age, many people sense that their number of remaining years and resources for competition severely limit their likelihood of "getting ahead." Middle age brings the peak of socioeconomic mobility; for many individuals, it may mark the beginning of the downward mobility characteristic of old age. Institutions, especially in the sphere of work, further solidify this sense; often workers' job statuses remain relatively static in later years of employment. Retirement marks the worker's entry into "old age"—again, a socially defined threshold of age-appropriate behavior. Generally, middle age is not a valued life period in our culture, the dominant values of which emphasize youthfulness: youthful standards of beauty, energy, bodily functioning, and carefree lifestyle. Contemporary religious groups do little to smooth this status passage with ritual or symbolism.

Old Age

Popular imagery holds that religion is more important in the lives of old people. Studies of religion and aging have found that, as people age, generally they do not dramatically change their religious involvement. While health problems may cause them to reduce public participation in religious services, there appears to be no reduction in private forms of religious involvement (for a review of the literature, see Idler, 1994). Indeed, it is possible that the homebound person might pray more frequently, remember religious experiences or events more intensely, or base more everyday activity on religious values (cf. Mindel and Vaughan, 1978).

Problems of meaning and belonging may be particularly acute for older persons. Difficulties of financial and physical limitations may compound the broader problem of the society's general devaluation of old age. While coping with their own diminished vitality, elderly persons must often cope with the difficulties of spouse's and friends' chronic illness, disability, and death. Some 35 percent of the persons giving care to the elderly are themselves older than 65 (cited in *The New York Times,* May 13, 1989). The extra burdens of caregiving and the very real losses, not only of death, but also of everyday abilities such as driving a car or playing a beloved musical instrument, pose serious psychological, social, and spiritual problems for many elderly persons.

One longitudinal study (Idler and Kasl, 1992) found that public religious involvement of the elderly has significant protective effects against disability, as well as helping believers cope with or avoid depression. Furthermore, persons who belonged to Christian or Jewish groups were less likely to die in the month before their respective religious holidays than the nonaffiliated. The authors concluded that religious involvement has a strong positive effect on health and, while some of this effect can be explained by social factors (such as social support or norms for healthy behaviors), other aspects of religion (such as ritual experience and meaning) also play a role.

Popular notions also suggest that the greater religiosity of old age is something of a last-chance "cramming for the final exam." Although there may be some validity to this conception, the religiosity of the elderly is probably more of a reflection of their social situation. Old people may emphasize spiritual and human relationship values in their lives because the society has relegated them to the private sphere—out of the world of work. Our society places great value on one's work role (especially for men, but increasingly also for women) throughout the individual's adult life, and old age brings the often abrupt end of that role. Retirement effectively means, for many people, leaving the public sphere. Thus, the elderly person must find all bases of identity and self-worth in the private sphere—family, leisure-time activities, religion, neighborhood. Perhaps because they have been characteristically confined to the private sphere, women often adjust better than men to old age. They have already developed more social roles in the private sphere, and society has not expected them to invest themselves in their employment roles (if any) as heavily as men (Myerhoff, 1978).

By contrast, some societies provide recognizable spiritual roles for elderly persons to assume. Hindu men who have raised and supported their children to adulthood are allowed to retire in honor to a life of contemplation and spiritual exercises. Upper-class, married Hindu women are allowed a similar freedom after menopause to perform *habisha*—rituals to protect their husbands and develop their own spirituality. Their freedom is temporary, however, ending when they become widows and must observe the social and ritual prohibitions for that status (Freeman, 1980). Elderly women in other cultures often have spiritual roles as healers and midwives—positions of culturally recognized spiritual power. Urban Western cultures, characteristically, do not provide such roles. Individuals may be recognized and honored for their "holiness" or "goodness" within their own immediate religious group, but such roles are privatized. The honor an elderly woman may receive at her Wednesday-night prayer meeting does not carry over to her treatment by the Social Security bureaucracy, hospital clinic, or other tenants in her apartment building.

One of the critical problems of meaning in middle and old age is modern society's sense of time. Primitive religions integrate all human action into cosmic time; the events of one's life can be interpreted as part of a larger cosmic drama. The individual's passage through the life cycle repeats and imitates the deities' birth, adolescence, marriage, childbirth, parenting, work, play, fighting, aging, and death. In this religious perspective, time has sacred significance. It collapses past and future into an eternal present.

Historic religions such as Christianity and Islam also give sacred significance to time. The past is full of the deities' self-revelation to humans; the present is important in the working out of the deities' will for humans; the future will bring the full realization of that will and celebration of a glorious reward. In this perspective, time promises immortality. Old age has sacred meaning, both as a fulfillment of divine will and as a threshold to higher levels of spiritual rewards. Modern society, by contrast, encourages a profane image of time. Time "passes," and its passage signifies decay or entropy. Time is a resource to be used but contains no special meaning; when the resource is depleted, life ends. In this context, old age has no special significance with reference to past accomplishments (e.g., social rewards for living a good life) or to future rewards (e.g., heaven, nirvana). Old age means merely the end of full life opportunities (Kearl, 1980).

U.S. society values economic roles especially highly. The individual's occupational role shapes the society's evaluation of that person and his or her own identity. Consumer roles are equally important. Social status is based partly on evaluations of the individual's ability to maintain a certain standard of consumption (e.g., quality of house, car, neighborhood, clothing). Elderly persons are often deprived of valued statuses in both kinds of economic roles; they are retired from their work roles, and, simultaneously, their fixed incomes leave them unable to maintain valued standards of consumption.

The legitimation offered for this loss of valued statuses is the idea of "retirement." Retirement is supposed to be an economic and moral vindication for growing old. It is described as a time for individualism (e.g., the freedom to move away to a retirement resort, play golf, lounge around, putter in the garden, and escape social obligations). The concept of retirement implies that the individual has earned this escape by having fulfilled life's social obligations. This individualism (if, indeed, the retired person could afford such luxuries in retirement) does epitomize what younger members consider to be freedom and a desirable reward. It cannot, however, provide meaning to life and death, a sense of belonging or self-worth, for the retired person (Kearl, 1980).

This life is made even more problematic in its ending. Often biological death follows the individual's social death by many years; the individual may become physically, financially, or mentally unable to sustain interactions that the society considers "alive." Yet the person is kept biologically alive, often by extreme measures of medical intervention. It becomes difficult to die "on time" (Kearl, 1993). The medical supervisors of death not only treat it as meaningless but also often segregate dying persons from family, neighbors, or friends who could support their personal meaning system. Religion has traditionally given meaning and dignity to old age and dying. Attempts to retain these values are not supported, however, by the structure of modern society. Old age and dying are generally perceived as times to fear.

CONVERSION

The capabilities of religion for providing the individual with a sense of both meaning and belonging are especially evident in the process of conversion. **Conversion** means a transformation of one's *self* concurrent with a transformation of one's basic *meaning system*.[1] It changes the sense of who one is and how one belongs in the social situation. Conversion transforms the way the individual perceives the rest of society and his or her personal place in it, altering one's view of the world.

This definition of conversion distinguishes simple changes in institutional affiliation from more fundamental alterations in the individual's meaning system. An Episcopalian who marries a Roman Catholic may join the Catholic church to accommodate the spouse's wishes. Such a change of affiliation is not necessarily a conversion. Similarly, a Presbyterian who moves to a new town and, finding no local church of that denomination, joins a Congregational church has probably not—strictly speaking—converted. Such denomination switching is relatively common in the United States and is strongly correlated with socioeconomic mobility (Stark and Glock, 1968:183–203). Several studies indicate that some 40 percent of the U.S. populace have switched denominations at least once (Roof and McKinney, 1987:165), and about one-third of those have switched more than once, mainly due to factors such as marriage, friendship, and family (Roof, 1989). Sometimes switching denominational affiliation symbolizes a more dramatic transformation of self and meaning system; typically, however, it is not a conversion but simply a change of affiliation from one organization to another.

If people's new affiliation brings them into a whole new cultural context, promoting values and behaviors different from the former affiliation, then coming to identify with the new group and participating in its distinctive practices eventually may have a real transformative effect. One study of colonial-era converts to Christianity among the Karo (of highland Sumatra) found that, although many sought baptism for purely pragmatic reasons (e.g., to gain access to the mission schools) and were at best nominal Christians, participation in the routine practices of the mission Christian life transformed their worldview and resulted in genuine conviction (Kipp, 1995).

Kinds of Conversion

There is considerable diversity among conversions. One distinction is the degree of personal transformation that takes place. How different are the new meaning system and self from the former ones?

---

1. Throughout this discussion we will emphasize the broad concept of *meaning system* more than the specific term *religion*. This usage is helpful because the processes described here apply to other comprehensive meaning systems as well as to specifically religious ones. The processes of conversion and commitment can apply not only to religious changes but also to psychotherapeutic and political transformation.

Radical Transformation. The extreme case is a radical transformation of self and meaning system such as when a highly committed Conservative Jew converts to a fundamentalist Christian worldview. Not only are such extreme conversions relatively uncommon, but they rarely occur as dramatically as popular imagery implies. The processes by which such radical transformations occur are actually similar in kind, though usually not in degree, to less extreme conversions (see Berger and Luckmann, 1966:157–163). These processes are described in greater detail in the next section.

Consolidation. Less extreme cases include conversions in which the new meaning system and self represent a consolidation of previous identities. Some young men who became *ba'ale teshuvah*—members of strict Orthodox Jewish yeshivot (i.e., commune-schools)—had come from non-Orthodox Jewish homes but rejected their Jewish way of life and had then tried various alternative worldviews (especially countercultural "trips" such as meditation, communes, and drugs). Becoming Orthodox was not a return to their former Jewishness because their new meaning system and selves were dramatically different from the old ones. Yet the way of life of the *ba'ale teshuvah* enabled these members to consolidate elements of both their former identities into a new, "superior" self (Glanz and Harrison, 1977; see also Danzger, 1989).

Reaffirmation. Another less extreme type of conversion is an alteration of self and meaning system representing a reaffirmation of elements of one's previous identity. Many adolescent "conversions" fit this model. It is difficult to specify how much change such conversions really entail. Often they involve no change in one's religious affiliation, yet exhibit real changes in the individual's personal meaning system and sense of identity. Roman Catholics who become Pentecostal (i.e., in the Catholic Charismatic Renewal) typically undergo a conversion experience, yet remain Catholics and often become even more active in parish life. They change their personal meaning systems and selves, but they interpret these changes as consistent with their former meaning system. They view their former religiosity as a vague groping for the truth, which they now have found. This type of conversion does not necessarily entail a total rejection of the previous meaning system.

Some identity-consolidating conversion experiences involve little or no change in meaning system and sense of self. Many religious groups expect young members to make a personal faith decision and to undergo a conversion experience as they approach adulthood. Such groups provide opportunities such as retreats or revivals in which the necessary conversion experience is more likely to occur. These experiences, although very real and meaningful to participants, are often not conversions (as defined) but rather rituals of reaffirmation of the person's existing identity and meaning system (Wimberley et al., 1975). Even these quasi-conversions, however, occur by some of the same social processes as more radical transformations of self; both are forms of religious socialization and are thus comparable in many respects. Some experiences are part of a process of commitment in which the individual's self-concept as a religious person becomes more central (see, for example, Staples and Mauss, 1987).

Conversion in the Context of Modern Societies. Modern societies are characterized by both the existence of more real options for choice and the expectation that more areas of life are matters of individual decisions. In modern societies, individuals believe they are free to make such significant choices as marriage partner, occupation, place of residence, and religion. At the same time, however, many people are genuinely ambivalent about such individual freedom and try to constrain these choices by other means. While parents may no longer arrange marriages, they still try to limit their children's range of choice of partners by, for example, sending them to small colleges attracting the desired ethnic, religious, and social class groups of students. Similarly, although in principle modern societies hold that religion is a matter of individual choice, few people would agree that all religions are valid "acceptable" options.

Ironically, unlike in traditional societies, religions in modern contexts must actively work at generating members' individual choices and commitments. In modern societies the religious training of youth, for instance, is aimed largely at keeping them in "the faith" as adults when they face many other options. By contrast, in highly traditional societies, young people's belonging to the group's religion in the future is taken so for granted that youth need only to learn how to perform their own roles in that group. In the modern context, some religious groups consciously orchestrate social occasions at which the "correct" individual choices will be made, yet be experienced as freely and fully chosen.

Social scientists are particularly fascinated with conversion and commitment, because these processes highlight important features of the relationship between the individual and society in the changing contexts of modern life. **Individuation** is the process by which cultural and social structural arrangements come to consider each individual as a separate entity—in relation to group entities such as the family, tribe, religious group, or political and judicial institutions. Modern societies are characterized by a much higher degree of individuation than traditional societies, but modern societies differ culturally in how they think about the individual and about individual choices and rights. For example, Protestant Christians have traditionally emphasized individual repentance and salvation, whereas Eastern Orthodox Christians have emphasized being saved in a community united by ritual practices. A social scientist would ask how these two groups adapt their understandings of the individual-to-society relationship in the context of largely urbanized modern social and economic conditions.

Because of this cultural diversity, however, we must be cautious not to confuse our own cultural "rhetorics" for the essential or definitive features of conversion. For example, most U.S. religious groups believe that conversion entails accepting a new set of religious ideas or beliefs; other cultures downplay religious beliefs and emphasize changed ritual practices. One anthropologist (Rosaldo, 1989) recounts his failure to comprehend the conversion of a tribesman of the Ilongot (of the highland Philippines) to evangelical Christianity. Christian beliefs—including beliefs about death or heaven—were unimportant to this man who was suffering greatly the death of seven children in a short span; what mattered was that the new religion offered religious *practices* that enabled him to cope with his grief and rage. In the traditional Ilongot way of life, the grief and rage experienced in bereavement had been ceremonially dealt with by head-hunting. After Marcos's declaration of martial law in 1972, the government forbade and severely punished head-hunting, so tribesmen had no way of dealing with these intense emotional experiences. In this example, the man's abandonment of the Ilongot way of life probably preceded his embracement of a Christian way of life, but Christian *beliefs* per se appear not to have been a significant feature of his conversion.

Conversion Accounts and Rhetorics. The main difficulty in distinguishing the degree of change that occurs in any given conversion is that the individual who converts reinterprets past experiences in relationship to the new meaning system. Therefore, it becomes difficult to determine what amount of the convert's description of the changes experienced represents the objective process of conversion and how much expresses the convert's subjective reinterpretation of those events. The convert constructs the story of conversion, drawing on a socially available set of plausible explanations, or *rhetoric.*

Several rhetorics are available to converts to use to "explain" their conversion (Burke, 1953). *Rhetorics of choice* emphasize how much the change resulted from a personal, often agonizing decision. Our society places much value on individual decision, so these rhetorics are prominent in explanations of conversion. In cultures where personal decision is less valued or even discouraged, rhetorics of choice are not emphasized; indeed, often converts do not even experience "making a personal decision" (Tippett, 1973). *Rhetorics of change* emphasize the dramatic nature of personal change in the conversion. Converts may compare the evil or unhappiness of their previous way of life with how wonderful their new way is. *Rhetorics of continuity* focus on the extent to which one's new meaning system and self are the logical extension of earlier beliefs and experiences. The convert might remember important past experiences as tentative steps toward the newfound truth.

Religious groups themselves often encourage the application of one type of rhetoric over another. Thus, new members of the Catholic Charismatic Renewal are encouraged to interpret their conversion as continuous with their former way of life, whereas many "born-again" Christian groups encourage new members to interpret their experiences as a dramatic change. Because the main source of information about conversion is the converts themselves—and because their explanations of events surrounding their conversions are reinterpretations consistent with their new meaning systems—evaluating evidence about conversions is difficult. Some sociological theories of conversion have mistaken these interpretations and the rhetorics that express them for the objective events of the conversion (Beckford, 1978a; Machalek and Snow, 1993).

Other problems in interpreting conversion accounts are the point in the convert's life history at which the account is given and the social context in which it is given. A psychotherapist hired by "anticult" parents to help their son or daughter to overcome a disvalued conversion experience would almost certainly hear a different story about that conversion than would a sociologist participant-observer talking to a satisfied group member. Even though both "stories" might be truthful, they would probably differ dramatically in what events were remembered and how those events were interpreted (Richardson, 1985a; Robbins, 1984). Just as there is some pressure from fellow members on believers to witness appropriately to their conversion experience (McGuire, 1982), there is also a set of expectations, especially from the family, for ex-members about how they should account for their experiences (Beckford, 1985). Thus, believers' accounts of conversion, although methodologically important and useful, cannot be taken at face value.

Explaining Conversion

A theoretical understanding of how conversion occurs is nevertheless worthwhile because it reveals much about the connection between the individual's meaning system, social relationships, and very identity. Because conversion consists in a change of the individual's meaning system and self, it has social, psychological, and ideational components. The social component consists of the interaction between the recruit and other circles of associates (e.g., parents, friends, co-workers). The psychological component refers to emotional and affective aspects of conversion as well as to changes in values and attitudes. The ideational component includes the actual ideas the convert embraces or rejects during the process. These ideas are rarely very philosophical or theological; they are simply a set of beliefs that both justify the new meaning system and negate the former one.

Factors in Conversion. An adequate theory of conversion must take all the aspects mentioned into account without overemphasizing any single component. Some theories give too much weight to social factors by creating the image of a passive person being pushed and pulled by various social forces. Although very real social pressures are exerted on the potential convert, the person who converts is not a passive object of these pressures. Conversion entails an interaction during which the recruit constructs or negotiates a new personal identity (Beckford, 1978a; Kilbourne and Richardson, 1989; Straus, 1979). Furthermore, only some of those exposed to such social pressures do decide to convert (Barker, 1983).

Some theories of conversion overemphasize ideational components of the process. These theories are consistent with the ideological claims of the religious groups themselves. Religious groups like to believe that the truth value of their beliefs alone is sufficient to compel a person to convert. The content of the belief system *is* a factor in conversion. Some beliefs are more appealing than others to people in certain circumstances; indeed, the potential convert will likely be recruited to a group whose perspective is consistent with that person's previous outlook, even though the specific content of the group's beliefs may be unfamiliar (Greil, 1977; Harrison, 1974). Also, we must acknowledge people's religious reasons for their religious behavior and not try to reduce every motive to some psychological function.

Nevertheless, ideas alone do not persuade a person to convert. Even in the scientific community where objective facts and the truth value of interpretations are supposed to be paramount, there is considerable resistance to change from an established interpretive paradigm to a new one—even when the old paradigm is inconsistent with the "facts" (Festinger, 1957; Kuhn, 1970). How much more are religious believers, with their emphasis on supraempirical reality, likely to resist changing their ideas? Thus, although the ideational component is important in the appeal of a new belief system, it is not sufficient to bring about conversion.

Other theories place too great an emphasis on psychological factors in conversion, explaining the change entirely in terms of the individual's personality, biography, and personal problems. Psychologistic explanations are attractive because they mesh with many of our individualistic cultural values. Nevertheless, they are too one sided, leaving out social situational factors and other important components. Also, some of these theories tend to assume that conversion to unusual religious groups entails "sick" behavior. Yet adherents' behavior is quite understandable and rational within their alternate meaning systems. If one believes that astrological forces influence human events, it is perfectly rational to act in accordance with those forces. Likewise, if one believes that the world is coming to an end in the very near future, it is not irrational or "sick" to give up one's possessions or career plans.

"Brainwashing."   One particularly misleading psychologistic model of conversion is the "brainwashing" metaphor. This model is based on studies during the 1950s of the processes by which certain U.S. military personnel in the Orient were pressed to convert to Chinese communism. The popular image applied to this process was "brainwashing," conveying the idea that the convert's mind was cleansed of prior beliefs, values, and commitment, then filled with a new belief system. Psychological studies of this process identified several factors contributing to conversion without cooperation of the converts (Lifton, 1963; Sargant, 1957). Various social scientists have subsequently generalized the interpretations of this drastic type of political conversion to other forms of conversion. Some accurate parallels do exist between forcible "brainwashing" and conversion, but these characteristics apply to all forms of resocialization. Thus, the training of soldiers for combat and rehabilitating juvenile delinquents also involve these processes. To say that conversion is a form of resocialization does not mean that it is therefore an extreme, involuntary form of resocialization.

The key problem with the "brainwashing" metaphor is its ideological use and potential application for abuses of civil liberties. Nonconverts often feel threatened by the conversion of someone close to them. The convert has rejected their own dearly held views and norms and has indirectly threatened the nonconverts' own meaning system. When people cannot understand why an individual would *want* to convert to an unfamiliar religious perspective, they find "brainwashing" an attractive explanation. This metaphor implies that the converting individual did not change voluntarily. The metaphor also allows people to negate the ideational component of the convert's new meaning system. A convert's parents can feel, "He doesn't believe those ideas because they are meaningful to him but because his mind has been manipulated." In its extreme form, the "brainwashing" metaphor has been recently used to justify the denial of converts' religious liberty on the ground that they do not know their own minds (Anthony, 1990; Anthony and Robbins, 1992a, 1995; Richardson, 1991, 1993c; Shupe and Bromley, 1980).

An interesting parallel with the current anticult charge of "brainwashing" is the nineteenth-century anti-Mason movement. Freemasonry is now a legitimate, middle-class form of fraternal organization, but it was severely attacked in the nineteenth-century United States as subversive to democracy. Other now respectable groups that were attacked (often violently) were Roman Catholics and Mormons. The key themes of the movements against Masonry, Roman Catholicism, and Mormonism emphasized that, unlike conventional denominations that claimed only partial loyalty of their members, these groups allegedly dominated their members' lives, demanded unlimited allegiance, and conducted some activities in secrecy (Holt, 1973; Vaughn, 1983).

This parallel suggests that the "brainwashing" controversy is an ideological issue at another level (see Richardson, 1981; Robbins and Anthony, 1979). The society defines as "deviant" one who is *too* committed to religion, especially authoritarian religion. The resocialization processes themselves are less of an issue than the legitimacy of the group's religion itself. To illustrate this discrepancy, two researchers compared conversion and commitment processes of the Unification Church of Sun Myung Moon and similar contemporary sects with the nineteenth- and early twentieth-century practices of the now socially acceptable Tnevnoc "cult." The Tnevnoc practices were essentially comparable and seem bizarre until we discover that the authors were actually referring to life in the convent—which, spelled backward, is Tnevnoc (Bromley and Shupe, 1979). The chief difference between many modern "cults" and groups such as Roman Catholicism, Freemasonry, and Mormonism is that the latter groups have now achieved social legitimacy.[2]

Keeping these cautions in mind, we can examine some of the factors in conversion. By emphasizing conversion as a *process* rather than an event, we take into account the fact that the convert has both a history and a future. Although an individual may experience conversion as a discrete event, numerous other experiences lead up to and follow that event that are also parts of the conversion. The following description examines the sequence of events in the process of conversion; but let us remember that no single step in the sequence is itself sufficient to "cause" conversion (Beckford, 1978a; Heirich, 1977; Machalek and Snow, 1993; Richardson, 1985).

---

2. To reduce the impression that conversion and commitment processes characterize only "weird" religions, I have drawn examples from both traditional and "new" religions.

Predisposition to Conversion

Several personal and situational factors can predispose people to conversion by making them aware of the extent to which their prior meaning system seems inadequate to explain or give meaning to experiences and events. By contrast, if individuals can satisfactorily "handle" experiences and events within the framework of their meaning system, they have no desire to seek alternative meanings for their lives. Sometimes the individual who acutely feels the need for a new set of meanings becomes a *seeker*—that is, a person who actively looks for a satisfactory alternative belief system. A seeker often tries many different alternative beliefs and practices (Balch and Taylor, 1977). One American (a business analyst) convert to Soka Gakkai, a recent Japanese form of Buddhism, described her previous path:

> I practiced Christianity [in] the Reorganized Church of Jesus Christ of Latter-day Saints [Mormons], which taught [that there was] no salvation outside the Church, yet there were some good people outside. My church was very rigid (drinking rules and such). In 1971, Transcendental Meditation—I was looking for balance, for a daily rhythm. It helped me feel good—computer work is very stressful. It was more powerful than praying to God. I did Transcendental Meditation for a year. I also met Gurdjieff teaching in 1974, and was part of the Gurdjieff Society from 1974 to 1977. I was against organized religion—protocol, ceremony. A lady I met in the Gurdjieff group was starting chanting. She said "Try it!" When she chanted I felt sick—I was against trying it. However, I felt the strength. I was searching, but blocked (quoted in Wilson and Dobbelaere, 1994:82).

Often, however, the individual experiences the desire for a more satisfactory meaning system as a vague tension, a malaise.

Many converts describe a crisis that they felt was a turning point in their lives. It is very difficult to evaluate the extent to which such crises actually precipitate the individual's conversion. Some crises may disrupt a person's life so completely that the individual has difficulty integrating them into the previously held meaning system. Natural disasters, war, and personal tragedy are particularly acute challenges. Serious illness or unemployment may be experienced as turning points. Social events such as an economic depression or anxiety over social conditions (e.g., crime or erosion of morals) may predispose people to conversion. Nevertheless, such crises do not *cause* religious conversion. Religious conversion is one among several possible resolutions of tensions and problems created by the crisis. Thus, serious illness might predispose one person to convert to a new meaning system; another person with a similar illness might find great meaning and comfort in his or her existing meaning system and have that belief confirmed by the crisis experience. Diverse other responses are possible, including alcoholism, political conversion, psychotherapy, suicide, and so on. Individuals converted to religious meaning systems are typically people whom previous socialization has predisposed to a religious perspective (see Greil, 1977; Lofland, 1966, 1977).

Determining to what extent the convert's description of a crisis experience is the result of after-the-fact interpretation of events is also difficult. Some religious groups encourage their members to witness about their conversion experience by telling the group how they came to "see the light." Whether the events thus described were really critical when they occurred is therefore difficult to reconstruct. Often the new group itself promotes the recruit's experience of a crisis. Much of the initial interaction between group members and the potential convert is devoted to making the recruit dissatisfied with the prior meaning system. Similarly, many groups raise the recruit's anxiety over social and personal problems. Jehovah's Witnesses, for example, often approach strangers with a message about common worries—war, inflation, crime. By "mediating" anomie, the group encourages the individual to convert. The group encourages an experiential crisis of meaning by emphasizing order-threatening conditions and magnifying the potential convert's feelings of dissatisfaction, fear, and anxiety, which the person may have previously felt only vaguely (Beckford, 1975b:174).

Initial Interaction

Most recruits are drawn to the group by friends or relatives. Besides introducing the newcomer to the group and its beliefs, these preexisting networks of friendship account for the plausibility of the beliefs and the attractiveness of belonging. Thus, a person might be impressed by a roommate's happiness in a religious group and be curious enough to "check it out." The fact that a person whom one knows and likes belongs to the group attests to the normalcy or desirability of the group's way of life (Gerlach and Hine, 1970; Lofland, 1966). Even groups such as the Nichiren Shoshu (a Japanese movement brought to the United States in 1960), which conspicuously proselyte in public and anonymous settings, typically recruit most of their members through preexisting networks of friends and relatives (Snow et al., 1980).

Through interaction with members of the group, the recruit is gradually resocialized into that group's way of life. This resocialization consists of the individual's reshaping of identity and worldview to become consistent with those considered appropriate by the group. Several social processes enable the individual to make this transformation. Group support is particularly important. The recruit enjoys warm, affective relationships with the new group. Members of the Unification Church, for example, shower the potential member with attention and affection (sometimes called "love bombing"). These bonds affirm the new self and meaning system. As the recruit gradually withdraws from competing social relationships, the new group's opinions become increasingly important (Berger, 1967:50–51). Intensive interaction and close affective bonds with group members are central to the conversion process because they link the individual's new identity with the organization's perspective and goals (Greil and Rudy, 1984).

At the same time, the recruit also weakens or severs those relationships that support the old self. Former attachments compete with new commitments, symbolize a worldview that the recruit wishes to reject, and are based on an identity that the recruit wishes to change. Imagine, for example, that through two good friends you are introduced to a group of U.S. Sufis (Sufism is a mystical sect within Islam). Suppose that you learned enough about the group that you decided you would like to join them. How would your non-Sufi friends and relatives react? Your roommate might say, "Oh, come off that mysticism kick. I liked you better when you were a drinking buddy." Imagine your parents' reaction when you announce at supper that weekend, "Guess what, Mom and Dad, I've decided to join this fantastic bunch of Sufis, and I'm going out to the West Coast to live in a commune with some of them!" Most converts find their former set of friends less than supportive of their newfound truth and new self. Relationships with the new group therefore become even more important to counteract opposition from others.

During this resocialization, recruits learn to redefine their social world. Relationships once valued become devalued, and patterns of behavior once undesirable become desirable. New believers may redefine their families to exclude the biological family and to include the new family of fellow believers. A Shaker hymn thus celebrates the severance of old family ties:

Of all the relations that ever I see
My old fleshly kindred are furthest from me
So bad and so ugly, so hateful they feel
To see them and hate them increases my zeal . . .

My gospel relations are dearer to me
Than all the flesh kindred that ever I see . . .

(quoted in Kanter, 1972:90).

This process results in a whole new way of experiencing the world and oneself. The individual comes to "see" the world with an entirely different perspective; indeed, the new believer may say, "I once was blind, but now I see." This phrase is not merely metaphorical because the new perspective actually causes the individual to *perceive* the world differently. That which was marginal to consciousness becomes central, and that which once was focal becomes peripheral. Every worldview entails the selective perception and interpretation of events and objects according to its meaning system. Conversion means adopting new criteria for selecting (Jones, 1978; Snow and Machalek, 1983).

Recruits also redefine their own biographies. They remember episodes that appear consistent with the newfound perspective and interpret them as "part of what led me to the truth." Events are reinterpreted in terms of new beliefs and values. Remembering how proud she was to have achieved scholastic honors in school, one young woman said, "What a fool I was back then to have put so much store on worldly achievements."

The actual interaction between the recruit and the group is especially important in bringing about this transformation of worldviews. The most obvious action of religious groups is **proselyting** potential converts. *Proselyting* means that an individual or group actively tries to persuade nonbelievers to become believers. Some religious groups (e.g., Jehovah's Witnesses) do much proselyting among nonbelievers; others (e.g., Conservative Jews) do virtually none. Although proselyting activities are relatively conspicuous to outsiders, they are important primarily as a commitment mechanism for already converted members (Beckford, 1975b; Festinger et al., 1956; Shaffir, 1978). One study of a growing Pentecostal denomination found that members with extensive religious experience (e.g., speaking in tongues, divine healing) were especially vigorous evangelizers of potential recruits (Poloma and Pendleton, 1989). Thus, a mutually reinforcing relationship developed between members' involvement in the group and their attempts to share their religious faith and experience.

Mutual witnessing within existing friendships appears to be especially effective in bringing about the recruit's conversion to a new meaning system. Thus, the newcomer may mention an apparent coincidence that had recently happened, and a group member may respond, "That was no coincidence. That was God trying to show concern for you so that you will change your life." Or a member might say, "I used to be just like you. I had my doubts and didn't know what to believe, but now it all fits together. Now I can see what I was missing." Through these informal interactions, the recruit may gradually "try on" the interpretations suggested by members and apply their meaning systems to personal experiences. Thus, the new believer comes to share their distinctive worldview.

Symbolizing the Conversion

The part of the conversion process typically identified as "the conversion" is essentially some form of *symbolizing* the transformation that has already been occurring. The convert affirms the new identity by some symbolic means considered appropriate in that group. In many Christian groups, baptism is meaningful as a symbol of conversion. Other ritual expressions of the new self include speaking in tongues (i.e., *glossolalia*) and witnessing. Some groups have very formal means for symbolizing transformation; others have more informal symbols that are not obvious to the nonbeliever (Gerlach and Hine, 1970; Lebra, 1972).

Although conversion is a gradual process, many recruits who have decided to convert adopt some of the symbols of conversion rather dramatically. Part of the resocialization itself is learning to act, look, and talk like other members of the group. Indeed, groups such as the Unification Church may focus on teaching recruits to display signs of commitment to the neglect of socializing them to actual commitment (Long and Hadden, 1983:9). The imitation of signs of commitment—which might be called "doing being converted"—may be one part of the resocialization process. It is, however, often mistaken by observers as evidence that a dramatic conversion has been accomplished.

Some groups may encourage the new member to seek a conversion experience. This special emotional and spiritual event thus symbolizes the person's full conversion. Such experiences, however, are only part of the larger conversion process but are valued ways of symbolizing the transformation in some groups. Often these conversion experiences are brought about in carefully orchestrated settings. In a revival meeting, the timing of the altar call is synchronized with music and spoken message to proclaim, in essence, "Now is the appropriate time to have that special experience you came for" (W. Johnson, 1971; Walker and Atherton, 1971).

Another symbolic expression of the new self is changing one's name. Nuns traditionally changed their names upon taking their vows. The new name both symbolized the person's new identity and helped confirm that identity every time she was addressed. Some groups also encourage converts to confess the wrongness of their previous way of life. Among the Society of Brothers (Bruderhof), such a confession event symbolizes the conversion and demonstrates how new members' views of themselves have been transformed (see Zablocki, 1971:239–285). In the confession, converts affirm their new selves by derogating the behavior of their old selves. These symbols of conversion illustrate the complexity of the larger process of resocialization (see Goffman, 1961, especially his comparison of the "mortification of self" in several different institutional settings).

COMMITMENT

The process of conversion does not end when the recruit formally joins the group and symbolically affirms the conversion. Rather, the conversion process is continued in the **commitment** process, by which the individual increasingly identifies with the group, its meaning system, and its goals. Groups that have highly effective recruitment strategies may not be able to keep and effectively deploy their members if their commitment processes are not also effective (Long and Hadden, 1983). Popular imagery somehow envisions the converted recruit as permanently changed. The turnover of members in contemporary religious movements belies this notion. A British study found that, in the tightly organized Unification Church, the majority of converts leave the group of their own free will within two years (Barker, 1983). Similarly, a Canadian study of numerous new religious and parareligious movements found that, while participation rates were relatively high, the proportion of adherents staying in these movements was extremely low (Bird and Reimer, 1982).

Commitment is a problem not only for contemporary movements. George Whitefield, an eighteenth-century Calvinist preacher, observed that his efforts were not as successful as those of John Wesley, the principal founder of Methodism. The key difference was that, although both preachers achieved many converts, Wesley insisted that the localities where he preached should establish "classes" to ensure the commitment of new converts (Snow and Machalek, 1983). Relative commitment of members is no indicator of the truth or "deviance" of a religious group. Maintaining commitment to *any* group is always problematic, and it is especially difficult in a modern, pluralistic, mobile, individualistic society.

Commitment means the willingness of members to contribute in maintaining the group because the group provides what they want and need. *Commitment* therefore implies a reciprocal relationship. The group achieves its goals by fulfilling the needs of its members, and the members satisfy their desires by helping to maintain the group. Persons who are totally committed to a group have fully invested themselves in it and fully identify with it. Commitment is the link between the individual and the larger social group. A person cannot be coerced into commitment but decides to identify with the interests of the group because of personal values, material interests, or affective ties (Kanter, 1972:65–70).

Conversion is a resolution of the individual's problems with former meaning systems and former self, but conversion alone is not sufficient to resolve new problems. The group's commitment processes help prevent the individual's doubts and new problems from undermining the conversion. The final result of the entire conversion process is not merely creating new members but creating members who will invest themselves in what the group is believing and doing. The same process also ensures the commitment of all members, new and old, to the group's values and objectives. *Commitment processes build plausibility structures for the group's worldview and way of life.*

The level of commitment that a group expects varies. Most major denominations in the United States do not expect intense commitment from their members or for that commitment to influence all aspects of their lives. Other religious groups (e.g., Pentecostal and Holiness groups, as well as many communal groups ranging from Hassidic Jews to the Bruderhof, Trappist monks, and the Unification Church) expect members to demonstrate intense religious commitment in all spheres of daily life. Commitment mechanisms in dissenting or deviant groups are especially important because of their difficulty in maintaining their worldview in face of opposition both from established religious groups and the larger society. All social groups, however, need some commitment from their members in order to maintain the group and achieve their goals (see Gerlach and Hine, 1970). And all social groups (including nonreligious groups such as the army) utilize commitment measures similar to those used by religious groups.

The processes by which the group fosters commitment are similar to processes of conversion. Both processes urge members to *withdraw* from competing allegiances and alternate ways of life, and both processes encourage members to *involve* themselves more deeply in the life of the group, its values and goals. These commitment mechanisms are used to some degree by all social groups. Groups desiring more intense or total commitment of members, however, are likely to use more extreme commitment processes. Groups gain greater commitment of members by asking them to sacrifice something for the group, but the degree of required sacrifice varies widely. Most religious bodies ask their members to give up some of their money and time for the group's goals and projects; and some groups expect their members to tithe a specific, substantial percentage of their income. Still other groups ask members to give up all belongings to the group and to live communally.

Withdrawal from Competing Allegiances

Any degree of sacrifice enhances the individual's commitment because giving up something makes the goal seem more valuable. Sacrifice gives observable evidence to the group that the member is committed, and it "weeds out" members who are not sufficiently committed. Religious groups further encourage sacrifice by signifying it as a consecration, so the act of sacrifice gains sacred status. Some groups ask members to sacrifice time and energy (e.g., devoting a certain number of hours each week to proselyting new members). Some groups expect members to abstain regularly or periodically from certain foods or from alcohol, tobacco, drugs, or sexual relationships. Several Christian groups encourage or require their members to fast during Lent. Either for all members or for an elite core group, many religious groups place special value on celibacy. Other groups may expect members to do without "worldly" pursuits (e.g., dancing, going to movies, wearing makeup or stylish clothing). Such sacrifices are demanded by most contemporary religious groups, especially marginal ones. Celibacy, for example, figures in the commitment process of such diverse groups as Roman Catholic clergy, some neo-Pentecostals, the Divine Light Mission, and the Unification Church. Vegetarianism and abstinence from smoking, drinking, and drugs are commonly required sacrifices (see Kanter, 1972, for examples from nineteenth-century communal groups; for contemporary examples, see Gardner, 1978).

Some sacrifices may also be interpreted as forms of **mortification,** the process of stripping the individual of vestiges of the "old self." Groups seeking to resocialize their members into a new identity consistent with the group's beliefs and values often encourage mortification. Members are asked to let go of those areas of life that compete with the new, desired self. They may have to wear prescribed dress and hairstyle, do without makeup or jewelry, and give up certain prized possessions. They are asked to sacrifice not because these things are wrong in themselves but because using them supports the "old self." Other forms of mortification include public confession, giving up control over one's time and personal space, and relinquishing personal choice in a wide range of matters (Goffman, 1961; see also Chidester, 1988; Zablocki, 1971). Some groups use rituals in which participating members must violate taboos pertaining to their former way of life. Thus, ritually consuming some formerly forbidden food or engaging in formerly taboo sexual practices, for example, serve as group commitment mechanisms (Palmer and Bird, 1992).

The group sometimes promotes further withdrawal of members from their former way of life by asking them to renounce competing relationships. Many sects and cults discourage members from interacting with the "outside world" and may adopt special social arrangements to insulate members from outside influences. Groups such as some Mennonites, Jesus communes, and Christian monks geographically separate themselves from the rest of society. Other groups insulate their members by operating their own schools, places of work, and social clubs. For instance, numerous conservative evangelical groups have instituted home schooling or separate sectarian schools (see Peshkin, 1986;

Rose, 1988; Wagner, 1990). The group may also exercise control over the communication media to which members are exposed or may limit interaction with outsiders.

More important than physical withdrawal from "the world" is the creation of psychic boundaries between the group and the outside. By use of these boundaries, members come to think of the group as "we" and the rest of society as "they." Furthermore, members perceive their in-group as good or superior and the outside as evil or degraded. Thus, the individual member's withdrawal from competing activities is motivated not only by controls that the group exercises but especially by the wish to identify with the in-group and to avoid the negative influences of the outside. Even groups that have not withdrawn fully from the larger society often create these kinds of psychical boundaries, for example, by urging members to join parallel religious organizations rather than secular organizations. Thus, groups like Christian Veterinarian's Fellowship or Fellowship of Christian Athletes provide some of the prestige and business networks of professional associations, while protecting members from secular society (Elzey, 1988).

Withdrawal from competing relationships often entails changes in the member's relationships with parents, spouse, and close friends. Typically, the individual identifies with outside relationships less and less, while simultaneously drawing closer to fellow group members. As we have seen, this disengagement from relationships that do not support the member's new identity and beliefs is partly necessary to maintain the individual's new worldview. Many religious groups try to exert some control over the member's choice of a marriage partner. Marriage to someone who does not support (or who even opposes) one's worldview can undermine the believer's meaning system. Also critical is competition between religious groups over the socialization of children from a "mixed" marriage.

The group frequently tries to guide or control even its members' relationships within the group. Close relationships among a small part of the group may detract from commitment to the group as a whole. The attachment of a married couple to each other or of parents to their children can compete with their involvement in the larger group. Groups that seek intense commitment from their members often have special structural arrangements to reduce this competition. The Oneida commune of the nineteenth century had a form of "open marriage" that diminished the pairing off of couples. Other groups (e.g., the Israeli kibbutzim) consider children to belong to the whole community rather than solely to the biological parents.

These processes all promote commitment to the group by encouraging individual members to withdraw themselves from those aspects of their former life that prevent them from being fully a "new self." The degree to which any given religious group asks its members to withdraw from nongroup loyalties depends largely on the type of group. The more marginal, sectlike groups typically expect high levels of attachment. Their commitment processes are therefore more intense and extreme than those of ordinary denominations.

Attachment

At the same time that groups encourage members to withdraw from other allegiances, they also urge members to become more and more involved in the group itself, drawing them into greater oneness with the group. This sense of unity is clearly related to the concept of belonging, discussed in Chapter 2. Some studies of fundamentalist and charismatic congregations suggest that the provision of a sense of belonging is as important, if not more important, than the specific belief content of the group in maintaining high levels of member commitment (Ammerman, 1987; McGaw, 1980).

Activities that draw the member into the fellowship and consciousness of the larger group promote both the cohesion of the group and the commitment of individual members. These commitment mechanisms make belonging to the group an emotionally satisfying experience. Commitment mechanisms for attachment are also likely to differ according to the intensity of the commitment desired by the group. Groups that expect intense commitment of members use stronger measures to promote attachment to the group.

The "we feeling" of group consciousness is promoted by homogeneity of membership. The more alike members feel, the easier it is for the group to gain a sense of unity. Established religious bodies typically achieve member homogeneity by self-selection. Individuals choose to join a church or synagogue with membership characteristics comparable to their own social status, racial, ethnic, or language group, and educational and religious background. Sectlike groups, by contrast, have more selective memberships, screening out or discouraging unacceptable members. Sectlike groups put greater emphasis on resocializing new members, thereby creating more homogeneity (Kanter, 1972:93, 94).

Group unity is also enhanced by the sharing of work and possessions. The extreme form of such sharing is full communal living, in which all possessions are held in common and all work is performed together. At the opposite pole are nominal forms of sharing, such as gathering Thanksgiving baskets for the needy or a painting party to decorate Sunday School rooms. Even these minimally demanding kinds of sharing promote a sense of unity in the group. Much sharing in religious groups consists of people taking care of each other. Group members may aid the family of a hospitalized member by caring for the children, preparing meals, and comforting the worried spouse. This kind of sharing promotes the commitment not only of the family receiving care but also especially of those giving the care.

Regular group gatherings also bring about greater commitment of members, and such gatherings need not be for overtly religious purposes. A church supper helps increase members' feeling of belonging to the group. For example, ethnographies of African-American churches note the centrality of preparing church suppers and of eating food together as an expression of group solidarity and sharing (Goldsmith, 1989; Williams, 1974). Communal groups meet very frequently, sometimes each day. Other sectlike groups also urge their members to meet often. Some groups hold prayer meetings three nights a week in addition to Sunday services and church socials. Not all religious groups, however, identify commitment to the congregation as critical. Some groups emphasize both family-level religiosity and supracongregational commitment as very important. Religious gatherings of the family (e.g., the family saying the prayers of the rosary together or praying special Sabbath blessings, *Shabbat b'rachot*) serve similar functions of commitment but not necessarily to the congregation.

The content of group gatherings can also promote the commitment of individual members. **Ritual** is one particularly important aspect of a group gathering. By ritual, the group symbolizes meanings significant to itself. Ritual gives symbolic form to group unity, and participating individuals symbolically affirm their commitment. Ritual both reflects and acts on the group's meaning system. Too often we tend to think of ritual as being empty and a matter of "going through the motions." Even going through the motions can promote a sense of unity, but in many groups the content of ritual is highly meaningful and especially successful in creating a sense of oneness. Rituals important in many religious groups include communion and other ritual meals, healing services, symbols of deference, embraces, special prayer postures, hymns, and rituals of purification.

Mutual **witnessing** continues to be as important in the commitment process as in the initial socialization or conversion of the believer. Through witnessing, members show themselves and others how their daily lives can be interpreted in terms of the group's meaning system. This kind of witnessing is prominent in Pentecostal and Jesus movement prayer meetings and in the *satsang* of the Divine Light Mission. Witnessing is a transformative process. All events, thoughts, and experiences are transformed into significant events, meaningful thoughts, and special religious experiences. Everyday and nonbelievers' interpretations of events are devalued during witnessing and replaced with religious interpretations. Witnessing can be relatively public or can occur in the setting of a small fellowship group or family. The public proselyting of Jehovah's Witnesses, Mormons, and Hare Krishnas is different from the relatively private witnessing of the evangelical Women Aglow movement or the pentecostal Full Gospel Businessmen's Association. Groups that consider themselves in opposition to the rest of "the world" are more likely to emphasize witnessing as a commitment mechanism (cf. similar functions in Alcoholics Anonymous, Weight Watchers, and psychotherapeutic groups; see Allon, 1973; Jones, 1975; Rudy and Greil, 1987, 1989; on witnessing in religious groups, see Ammerman, 1987; Kroll-Smith, 1980; McGuire, 1977; Shaffir, 1978; Tyson, 1988).

Commitment to a group can be strengthened if the group convinces the member that the group itself is extraordinary. If, for example, members become convinced that the group is in itself the exclusive path to salvation, they are more likely to remain with the group. Groups that expect the imminent end of the world typically portray themselves as the elect who will be saved. Other groups represent their rituals and practices as necessary for salvation in the next life. Many of these groups teach that fallen-away members will be even worse off than people who never knew the "right" way.

Group practices that promote a sense of awe further emphasize the significance of the group itself. These practices make the actions of the group appear more than mundane; mystery, magic, and miracles surround the group actions. When the leaders "receive directions" from God, as among Mormons and Pentecostals, the directions seem far more awesome than if members had voted on them. Social and symbolic distance also promote a sense of awe. Thus, medieval churches used physical barriers (e.g., rood screens) and space to separate the body of the congregation from the central ritual performance. Even today many religious groups have certain sacred spaces in their places of worship where ordinary members cannot routinely go. These practices may not have been deliberately created to generate commitment, but the production of a sense of awe does result in enhancing members' commitment.

DISENGAGEMENT

Although many members continue or increase their commitment to their religious group, others become less committed or drop out of the group. In many respects, the process of **disengagement** is the reverse of conversion. Like conversion, disengagement typically entails a transformation of self and basic meaning system. When a religious meaning system and identity have been especially central in a believer's life, the process of altering self and meaning system may be a wrenching transformation.

Both conversion and disengagement are forms of "status passage" in which the individual leaves one role and enters another (Glazer and Strauss, 1971). Persons in modern, mobile societies may go through such role-exit experiences, such as moving out of one's parents' home, divorcing a spouse, being discharged from the armed services, moving away from a community, leaving a hospital or prison, changing jobs or occupations or political parties, retiring, or quitting a social club. Some of these roles are more central to the individual's identity, and thus role exit may be more complex and emotion-laden than for other secondary roles. If the religious group has been the primary source of all other parts of the member's identity and social life, leaving can be particularly socially and emotionally wrenching. One Israeli who left the ultra-Orthodox *haredi* (Jewish) community said:

> You feel an emptiness, a very deep emptiness, and there's also confusion. You have nobody to talk to, nobody who really understands what you're going through. The loneliness can be overpowering. You're cut off. The close friends you've had since childhood, you never see again (quoted in Shaffir, 1995).

Stages of Disaffiliation

Like the commitment process, the disengagement process involves the pushes and pulls of various social influences. Individuals must weigh what they are accomplishing by being members against opportunities to reach those (or other) goals elsewhere. Typically, members reach the decision to leave the group only gradually, but just as members' retrospective accounts of their conversions are transformed to fit their new beliefs and image of self, so too are ex-members' accounts of the events and decisions that led up to leaving the group transformed. Four stages characteristic of role exit include (1) first doubts, (2) seeking and weighing role alternatives, (3) a turning point, and (4) establishing an ex-role identity (Ebaugh, 1988).

Disengagement is often the result of a breakdown or diminished effectiveness of the plausibility structure (described in Chapter 2) that supports the religious group's beliefs and practices. One study found that this breakdown of plausibility resulted from such factors as reduced isolation from the outside world, competing commitments (such as intimate dyadic relationships within the group or family links outside the group), lack of movement success, and apparent discrepancies between leaders' words and actions (Wright, 1987). Particularly if the social-emotional climate of a religious group was important in a recruit's initial conversion or commitment, those interpersonal experiences that fail to fulfill the believer's affective needs can be disconfirming (cf. Jacobs, 1989a).

Just as social networks initially brought members into the group and sustained their commitment to its worldview, so too competing social networks make disengagement plausible and attractive. One study (Aho, 1994) of persons who voluntarily left hate groups (such as the KKK and the Aryan Nations Church) found that voluntary exit was promoted by social pushes (e.g., harassment by colleagues, isolation from support group) and social pulls (e.g., falling in love with a nonbeliever, close ties with a pastor who undermined worldview of the hate group).

Just as conversion involves coming to take a new member role for granted as part of the recruits' identity, so too the "first doubts" stage of role exit typically involves experiences that call that taken-for-granted reality into question. Often the doubts are not about the group's beliefs; for example, many nuns who eventually left the convent were not doubting their Christian faith but rather their role commitments as nuns (cf. Ebaugh, 1988). Like the conversion experience, disengagement often entails a turning point, which is remembered more vividly in retrospect than many other parts of the gradual change. Just as the researcher must be careful when interpreting retrospective conversion stories, so too must tales of disaffiliation be treated with caution (Richardson et al., 1986).

Ex-members' recollections of their experiences with various contemporary religious movements were reconstructed differently (Solomon, 1981). Persons who had left voluntarily were more likely to remember their affiliation with the group as a growth experience, even though they had later changed directions. By contrast, people who were deprogrammed were more likely to recall their experience as one of deception and brainwashing (Lewis, 1989; Solomon, 1981).

The believer might respond to first doubts in a number of ways, such as seeking advice from other believers, trying to relieve doubts, reaffirming belief and allegiance to the group, or reconsidering commitment. Furthermore, persons who are disengaging from a group typically try to identify and weigh their alternatives to the member role. Whatever the outcome, neither conversion nor disaffiliation is a passive, mindless process.

Coerced and Voluntary Exit

The inadequacies of the "brainwashing" explanation of conversion become particularly acute in interpreting disengagement. If, as the "brainwashing" thesis holds, converts are coercively persuaded to belong to a religious group, then once converted and held in the group they would be unable to exercise the will to change beliefs and leave the group. This interpretation became the justification for the practice of forcible "deprogramming," by which believers were kidnapped, held against their wills, and subjected to a barrage of tactics designed to turn them against their former religious group and to convert them to an alternative perspective.

Sociological studies of disengagement from religious groups show, however, that the brainwashing/deprogramming conceptualization is a grossly inaccurate portrayal of these processes. Using this model, sensationalistic media convey the impression that converts are trapped, indeed "lost" indefinitely—or at least for a long time. In fact, much conversion (especially to demanding religious groups) is temporary (Wright and Ebaugh, 1993; see also Bird and Reimer, 1982), and most persons exiting religious groups—even intensely demanding ones such as Unification Church, Hare Krishna, and Children of God/Family of Love—do so voluntarily (Wright, 1991). For example, even socially "acceptable" conversions to well-established religious movements (like a Billy Graham crusade) are often very tenuous and short-lived (W. Johnson, 1971). A study of recruits to the Unification Church (one of the religious movements frequently accused of brainwashing) found, during the movement's growth period in the 1970s, that of some 1,000 recruits who were interested enough to attend a residential workshop, only 8 percent joined and remained as full-time members for more than one week. And of those few who did join, only about 5 percent remained full-time members for a year. Those recruits most likely to have joined with enthusiasm and later disengaged were those who had converted with idealistic expectations that the movement would make the world a better place, whereas those who continued in the movement were more likely to have joined for personal spiritual goals. The study concluded that it was "perfectly plausible" for the action-oriented idealists to decide to disaffiliate from the Unification Church when it appeared not to live up to its promise as an agent of social change (Barker, 1988a). Those converts who stayed in groups expecting high levels of commitment were hardly passive; they influenced and sometimes changed the group, even while the group influenced them (Richardson, 1993b). Mounting evidence from studies of the many "new religious movements" of the last three decades indicates that, far from being the malleable, passive, gullible dupes portrayed in the media, recruits to religious groups are both open to the spiritual alternatives and yet relatively resistant to conversions that entail high levels of commitment and drastic changes in identity.

Role exit from intensely committed religious groups in which much of each member's identity is invested can be extremely difficult. Similarly, disaffiliation in the context of a small, close-knit village is likely to be more difficult than in a large urban area. Most religious groups in modern societies get far less commitment and investment from their members. Religious disaffiliation from a casual or marginal religious role is simple and needs no social support.

Collective Forms of Disengagement

Doubts and unsatisfied needs may propel a member to leave a religious group, but individual disaffiliation is not the only response. One possible response to doubts and disillusionment is internal reform of the religious group. For example, during one of my own research projects, members of one large meditation group gradually realized that their guru was using his esteem and influence in sexual advances to several women members. Because the disappointed members still very much believed in their group's spiritual practices and ideals, instead of leaving the group they began discussing how to change the group. They decided to send the guru back to India and continue without him.

A related response is when a group of dissatisfied members collectively defect over movement politics, often staging their departure to make a strong statement to the leadership of the movement organization. For example, a group of former ISKCON (Hare Krishna) members formed the "Conch Club" to collectively defect from the national organization, which they rejected as unworthy of their commitment due to its direction after the death of its founding guru. They continued to retain their beliefs and identity as Krishna devotees, locating themselves within the larger movement but rejecting the organization of that movement (Rochford, 1989). Similarly, many former residents of Rajneeshpuram, the Oregon commune of disciples of Bhagwan Shree Rajneesh, had generally positive retrospective accounts of membership in the movement and living at the commune. Because the commune and movement organization were disbanded by the guru, they had not left as defectors, and many retained friends in the movement, so role exit was less drastic (Latkin et al., 1994).

Another historically important form of disengagement is for dissatisfied members to break off from the religious group and form a separate group. Rarely do schismatic groups consider themselves to be *leaving* their faith; rather, they view their exit from the group as keeping the true faith. They view their new ideology as more pure and their new practices as more faithful than those of the parent group. Thus, the earliest members of many of today's mainstream religious organizations had to cope with the decision and social agonies of disaffiliation. Such collective forms of disaffiliation are also occasions of some role transition and perhaps personal anguish, but they are rarely so wrenching as individual role exit because members have the mutual support and idealism of a group making the exit together.

SUMMARY

This analysis of the processes of religious socialization, conversion, and commitment illustrates the interrelationship of religious meaning and religious belonging. The individual's meaning system is socially acquired and supported through early socialization and interaction with other believers throughout life. If the individual changes meaning systems, it is through social interaction. And the processes that promote commitment to the meaning system and the group supporting it are fundamentally social processes.

Social factors are important in shaping the individual's religion, and examination of critical periods in the individual's life cycle suggests some of these factors. Early socialization in the context of the family, neighborhood, and ethnoreligious community is particularly important in establishing not only the basic beliefs and values but also the connection between the individual's belief system and very identity. Rites of passage to new statuses are often filled with religious significance. Passage to adulthood and marriage illustrates some of the ways in which religion shapes critical moments as the individual takes on a new social identity. Middle age and old age are, however, problematic in this society, and there are no satisfactory transitions into these devalued statuses. The society's secularized conception of time may be an important cause of some of these problems. Evidence about the nature of the interrelationship between the individual's religion and social factors is, however, generally limited either to nonmodern examples or to studies of narrowly defined, church-oriented religion. Thus, this chapter has only suggested some of the directions that further research into the individual's religion might take.

Conversion is essentially a form of resocialization similar to nonreligious resocialization. Through interaction with believers, the recruit comes to share their worldview and takes on a new self consistent with that meaning system. Conversion includes a range of changes: radical transformation of self and worldview, identity consolidation, and reaffirmation. Some changes are mere transfers of organizational affiliation and not real conversions of worldview and identity. The process of conversion funnels recruits from a general predisposition to conversion, through interaction with group members, to growing identification with the group and its belief system. The conversion process is generally gradual, although it may appear sudden and dramatic because of the way it is symbolized by some individuals or groups.

Commitment mechanisms promote the loyalty and attachment of all members, new converts and old members alike. Groups such as most denominations, which expect only partial commitment of members, typically use less extreme commitment mechanisms than do sectarian groups, which expect members' total commitment and immersion in the life of the group. The process of commitment involves simultaneously the individual's withdrawal from competing allegiances (e.g., by sacrifice) and greater attachment to the group (e.g., by frequent interaction with fellow members). Through these commitment processes, the group builds a firm plausibility structure for its meaning system. Like conversion, disengagement from religious commitment entails a transformation of self and worldview supported by a changed plausibility structure.

RECOMMENDED READINGS

Nancy Tatom Ammerman. *Bible Believers: Fundamentalists in the Modern World.* New Brunswick, N.J.: Rutgers University Press, 1987. This highly readable ethnography shows how members of a fundamentalist congregation maintain their worldview and commitment.

Eileen Barker. *The Making of a Moonie: Choice or Brainwashing.* Oxford: Basil Blackwell, 1984. Winner of the 1985 Distinguished Book Award of the Society for the Scientific Study of Religion, this book is a methodologically exemplary and highly readable description of the beliefs, recruitment, conversion, and commitment processes of the Unification Church of the Reverend Sun Myung Moon.

Lynn Davidman. *Tradition in a Rootless World: Women Turn to Orthodox Judaism.* Berkeley: University of California Press, 1991. This is a rich ethnography of two groups through which young, secular Jewish women make a transition to Orthodoxy, with its traditional patterns of family life and restrictive roles for women.

Jody Shapiro Davie. *Women in the Presence: Constructing Community and Seeking Spirituality in Mainline Protestantism.* Philadelphia: University of Pennsylvania Press, 1995. Based on ethnographic and interview evidence, this well-written book gives a real flavor of the personal spirituality of active Presbyterian laywomen.

Rosabeth Moss Kanter. *Commitment and Community.* Cambridge: Harvard University Press, 1972. Using evidence from nineteenth-century American communitarian ventures, this book presents a well-organized theoretical schema for analyzing commitment.

Mary Jo Neitz. *Charisma and Community: A Study of Religious Commitment Within the Charismatic Renewal.* New Brunswick, N.J.: Transaction, 1986. An ethnography of Catholic charismatic social reality, conversion, and community experience.

4

⊶◐⊷

Official and Nonofficial Religion

I n the United States, the mention of "religion" or "being religious" typically evokes the image of church-oriented religion. When we think of religion, we generally think of Protestantism, Catholicism, and Judaism. When
we think of being religious, we tend to locate that religiosity in the social
framework of Old First Church, Saint Mary's Church, or Temple Sholom.
Church-oriented religion is a prominent and important social form of religion in Western societies, but some other modes do not conform to the "official" model of religion.[1]

This chapter contrasts official religion and its characteristic expressions of
religiosity with "nonofficial" religion and religiosity. Both of these models illustrate the capacity of religion to provide meaning and belonging. A related
topic in this chapter is the difficult problem of how to measure or scientifically
explore the quality of an individual respondent's religion and religiosity. The
Extended Application at the end of this chapter describes the social definitions
of women's roles by both official and nonofficial religions and the concurrent
shaping of women's religion and religiousness.

THE OFFICIAL MODEL OF RELIGION

The development of a specialized official religion distinguishable from popular
or folk religion and other nonofficial religious elements is the result of the historical process of institutional differentiation. Earlier religion was relatively
diffused throughout all aspects of social life; the practice of religion was relatively unspecialized. Modern religion, by contrast, is characterized by **institutional specialization:** standardization of the worldview in a well-defined
doctrine, religious roles performed by specialists, and an organization to control doctrinal and ritual conformity, promulgate group teachings, and promote
organizational programs (see Luckmann, 1967:66; also Bellah, 1964). Interestingly, this same process of differentiation created the possibility of organized
irreligion—the institutional expression and consolidation of beliefs and practices of hostility or indifference toward religion. Organized irreligion (e.g.,
the Secularist, Positivist, and Ethical movements) closely parallels official models of religion (see Campbell, 1971).

Specialized institutions for religion thus become the focus for societal attention to religion. Each institutionally specialized religion typically consoli-
dates its beliefs, values, and practices into a coherent model. These official
models typically include a prescribed doctrine, set of ethical standards, cultic
expression, and institutional organization.

---

1. For sources on official religions in the United States, see Castelli and Gremillion, 1987;
   D'Antonio, Davidson, Hoge, and Wallace, 1989; Finke and Stark, 1992; Gallup and
   Castelli, 1989; Greeley, 1977, 1990; Haddad and Lummis, 1987; Heilman, 1990; Heil-
   man and Cohen, 1989; Hoge, Johnson, and Luidens, 1994; Hunter, 1983, 1987b;
   Liebman, 1988; McKinney and Roof, 1990; McNamara, 1992; Roof and McKinney,
   1987; Rosenberg, 1989; Seidler and Meyer, 1989; Wuthnow, 1988.

Doctrine

The official model of a religion is a coherent and consolidated meaning system, articulated by doctrinal experts (e.g., theologians) and typically promulgated only after official approval. Although church doctrine changes over time, a church can state its official model of religion fairly specifically at any one time: "This is what we stand for." Religions vary as to the extent of doctrinal sophistication and the importance of formal, prescribed doctrines. The Roman Catholic church has a vast, complicated doctrinal system, with some teachings having the weight of dogma (i.e., doctrines to which members are obliged to assent). By contrast, Judaism has no dogma but emphasizes the continual unfolding of religious truths through study of sacred scripture and texts; nevertheless, Judaism has some fairly explicit definitions of its shared meaning system.

Ethics

Official religion also characteristically prescribes a set of norms and regulations consistent with the group's doctrine. These obligations for behavior specify what actions are necessary to be a member in good standing. They include ideals toward which members should strive. Simultaneously, official religion usually uses specialized religious measures for the regulation and control of conduct so that misconduct can be noted and punished. In religions containing normative prescriptions that are complex, the organization sometimes supports specialized religious courts and ethical specialists to interpret the body of regulations. Other religions, whose ethical norms may be just as strong, rely on informal social control measures.

Cultic Expression

Each religion has a set of observances and devotions by which its meaning system is ritually expressed. Official religion tends to encourage standardized cultic expressions such as a liturgy or formal order of worship. Even less central expressions (e.g., devotions) are often standardized. Such standardized forms of expressing religious beliefs and sentiments include the Roman Catholic "Way of the Cross" and the Eastern Orthodox ritual of kissing icons. Some religions emphasize spontaneity in cultic expression; even so, however, the religion encourages certain regularized ways of being spontaneously religious. A minister might pray in his or her own words but choose words appropriate to the King James version of the Bible (e.g., "Almighty God, Thou hast showered us with Thy abundance"). Official religion also typically uses ritual specialists—priests, ministers, cantors, music directors, organists, choirs, and liturgical experts. Thus, official models of religion consolidate and standardize the cultic expression of religion.

Institutional Organization

Official religion typically organizes itself as a specialized association. It develops organizational specialists such as a professional clergy, organizational hierarchies (both lay and clerical), and often a host of auxiliary functionaries (e.g., religious teachers, church librarians, secretaries, program directors, and building service personnel). There are often formal specifications and procedures for membership and jurisdictional boundaries (e.g., parish boundaries). There are programs to be organized—Sunday Schools, "outreach" and "mission" programs, fund raising, youth clubs, study groups, and so on.

Especially important in the institutional organization of religious groups is the form of **polity** adopted. *Polity* refers to the arrangements for exercise of legitimate authority in the organization. The three characteristic forms found in most Western religious organizations are episcopal, presbyterian, and congregational polity. *Episcopal* polity refers to an organizational arrangement in which authority is centralized in a hierarchy (i.e., a pyramid of channels of authority). Thus, in the Roman Catholic church, local parishes are served by priests (sometimes hierarchically arranged even in the parish itself) who are responsible to bishops. The bishops place priests in the local parish, and the congregation has no official authority. Similarly, the bishops are responsible to the pope, using his delegated authority in their dioceses. Other examples of the episcopal form of polity include Eastern Orthodox churches and Episcopalianism; Methodism has a modified form of episcopal polity. *Presbyterian* polity is a form that uses a representative government by both clergy and laity. The local church is run by its clergy and lay-selected representatives. They, in turn, are responsible to a higher authority of elected representatives such as a presbytery, synod, or general assembly (as in Presbyterianism).

The *congregational* form of polity is characterized by high degrees of autonomy of the local congregation, including the power to call or dismiss clergy. This form is characteristic of Jewish congregations, Baptists, and the United Church. Denominations with congregational polity are less likely to conform with a uniform doctrine and ethical or ritual standards among congregations than are denominations with other kinds of polity.[2] Some observers (see Warner, 1994) suggest that most religious groups in the United States are converging toward a more congregational mode of polity, regardless of their historical patterns, because structurally religious organizations have become "voluntary gathered communities."[3] Similarly, as developed further in Chapter

---

2. It is therefore not surprising that sociologists trying to research members of these religious organizations have greater problems studying, as aggregates, the adherents of congregational polities such as Southern Baptists, in contrast to hierarchical polities like Mormons (cf. Hadaway, 1989; Cornwall and Cunningham, 1989). Because of the characteristic local autonomy of Southern Baptist polity, it is very difficult for a researcher studying members of many different congregations to identify who is "properly" a Southern Baptist.

3. Interestingly, this change appears to apply not only to Protestant groups but also to Catholic congregations, Pakistani Muslims, and Thai Buddhists in the U.S., suggesting that it results from a combination of modes of U.S. civic participation and religious groups' location in the voluntary sector of institutional life (Warner, 1994).

7, individual congregations in denominations with the congregational form of polity can be highly resistant to external pressures for change, ignoring the urgings of national denomination councils and firing change-oriented clergy. For the same reason, however, congregational and presbyterian forms of polity are more receptive than episcopal polity to the initiatives or needs of local congregations.

STRUCTURAL SHIFTS IN OFFICIAL RELIGION

There appear to be some structural changes in how and where official religions operate in modern societies. These developments suggest that, rather than focusing on denominations and other umbrella religious organizations as the unit of analysis, we should try to understand the religious units to which believers are actually committed. If the locus of religious identification has shifted, it is also likely that there are changes in the amount of authority vested in official religious groups by their adherents. Are there, simultaneously, shifts in how much autonomy in belief and practice is exercised by individuals?

Considerable evidence suggests that, at least in the last few decades, people's religious affiliation is not to a denomination or religious organization per se (Hoge et al., 1994; Hadaway, 1993). Warner (1994) argues that a growing pattern among Protestants and to some extent Catholics is for people to identify with a particular local religious organization rather than a denomination with which it is affiliated. When people move to another community, they often switch denominations in order to affiliate with a compatible local congregation. Often, other features of a particular congregation such as ethnic or class makeup are more important bases for affiliation than denomination-defined beliefs and practices. This pattern is corroborated by a Canadian study that found most (60 percent) active laypersons (i.e., regular church participants) believed switching congregations was warranted to be with "people more like yourself" in age, family composition, education, and income. Another feature of compatibility that warranted switching for 46 percent was to be in a congregation with "more people like yourself in terms of ethnic background" (Posterski and Barker, 1993:254–255). If, as some studies suggest (cf. Bibby, 1993), many congregations serve as homogeneous "clubs" or homogeneous enclaves—albeit with a religious component—their denominational affiliation may be a secondary feature.

Furthermore, except to religious specialists, the historical distinctions among many denominations have become virtually irrelevant. For example, one active Presbyterian said:

> When I was growing up in the fifties and sixties, I don't think there was a dime's worth of difference among Presbyterians, Methodists, or Episcopalians in terms of the way theology impacted on the people. If you get the ministers and priests together, they could have World War III arguing

about issues that nobody cares about. But as far as most people are concerned, they're really interchangeable. And I don't think there is any
particular value in having a Presbyterian heritage or ethos or way of
doing things. The only distinguishing characteristics in the church today
deal with our form of government. Anything specifically Calvinistic is
gone, and most Presbyterians aren't interested (quoted in Hoge et al.,
1994:120–121).

Thus, asking a respondent's "religious affiliation" may tell a researcher little or
nothing about members' actual religious beliefs, practices, commitments, or
experiences.

A related development among Protestants is the expansion of "nondenominational" churches. Organized by a dynamic preacher and not constrained by
a central denominational organization or belief system, these churches are
usually recognizably Protestant, yet often idiosyncratic. Even though they are
organized as churches (with buildings, Sunday Schools, budgets, choirs, clergy,
and committees), they are not fully "official religions"—as historically defined. Thus, researchers cannot make any general assumptions of what affiliation with such a church means; each congregation is independent of external
control, so learning that a respondent belongs to a nondenominational church
tells the researcher only that the respondent has a Protestant affiliation and that
it is *not* to any identifiable denomination. For example, one member of one
nondenominational "megachurch" (drawing more than 15,000 people each
week) said that was precisely what attracted him: "a lack of tradition" (quoted
in Niebuhr, 1995).

Another factor diminishing the explanatory power of denominational
identification is the increasing prominence of special interest groups, voluntary associations organized across denominational lines and focused on specific
issues such as abortion legislation, civil rights, or the role of religion in the
public schools (Wuthnow, 1988). If deeper loyalty goes to the special interest
group, members of a congregation or entire denomination may become profoundly split. Special interest groups have considerable potential for mobilizing religious activism and some sense of group identity, but they are not likely
to be a stable locus for religious ritual or experience. Nevertheless, overworked
or overcommitted members may feel the need to choose between investing
energies and social networks in the religious movement organization or in the
local religious congregation. If some of these hypotheses of structural shifts are
accurate, then it becomes far more difficult for sociologists to research and interpret what believing or practicing an official religion means in the actual
lives of individual members.

INDIVIDUAL RELIGIOSITY IN THE
OFFICIAL MODEL

Official religion, then, is a set of beliefs and practices prescribed, regulated, and socialized by organized, specifically religious groups. These groups set norms of belief and action for their members, and they establish an official model of what it means to be "one of us." Nevertheless, the actual religion of the individual member may not correspond very closely to the official model. What is the *operative faith* of the individual? What norms, ritual actions, and beliefs hold priority in the individual's life?

The specialized religious institution mediates the official model of religion to the individual member. Thus, for example, being a Presbyterian might be experienced in terms of attending Presbyterian Sunday worship services and Sunday Schools in a particular building of one's congregation; receiving communion in the Presbyterian church and being a member of the committee that washes the communion cups; interacting with Presbyterian clergy; singing in the church choir; having a grandfather who is a member of the Session (i.e., elected Board of Elders); having once memorized but now forgotten the Westminster Confession of faith and answers to the catechism; and learning other Presbyterian beliefs and norms in Sunday School and adult discussion groups or from sermons. In other words, the individual member is socialized into the model of his or her church.

For this reason, individual religiosity in official religion can be described in terms of its conformity to the official model. The individual's beliefs are somehow related to church doctrine, and the individual's ethical standards are measured by conformity to the ethical teachings of the official religion. Cultic expressions of the religion are translated in terms of individual observances and devotions. The individual's relationship to the institutional organization is expressed as membership and participation in that organization (see Luckmann, 1967:74).

Thus, it is possible for a sociologist studying an official religion to operationalize—to create a working definition of—individual religiosity in terms of the officially defined model. The Roman Catholic church (especially before Vatican II) used a fairly explicit official model of expected individual religiosity. Thus, in his studies of Roman Catholic parishes, Fichter (1951, 1954) was able to operationalize individual religiosity for research purposes, using criteria such as Mass attendance, reception of communion, sending children to parochial schools, participation in parish activities such as Holy Name Society, and interaction with clergy. This approach to operationalizing the concept of religiosity works, to some extent, *only* because the official model of Roman Catholic religion was precise and mediated to individual members.

Nevertheless, we cannot infer that these aspects of official religion explain *why* the individual member participates in any aspect of it. We cannot infer that those who attend religious services each week do so in precise conformity with what the official model of that religion puts forth as the purpose and meaning of those services. Some attenders may do so. Others may be interested more in sociability, entertainment, or a sense of tradition; some may not even know the "official" version, yet may get a strong subjective sense of meaning and belonging all the same. Thus, although the official model of religion has implications for individual religiosity, there is little indication that all aspects of official religion hold equal importance for each individual believer.

A number of factors bring about this discrepancy between the individual's religion and the official model. One factor is how the individual has been taught and socialized into the official model. Some persons have not learned the official model of their group's religion. The official belief systems of Christian and Jewish groups, for example, generally include a set of moral norms called "the Ten Commandments." According to Gallup Poll data, however, a majority of U.S. adults could not name more than half of those commandments (Gallup Poll, 1984:14). Many members may not know items of official belief such as catechisms, creeds, and confessions of faith. Furthermore, members vary in how adequately they have internalized the official model. It is one thing to know that one's group holds certain beliefs; it is another to believe them oneself.

The official model is often overlaid with nonofficial religious themes, perhaps drawn from folk religion, mythology, popular culture, or the teacher's own "misinformed" view of the official model. For example, one woman said: "I was in my thirties before I realized that a lot of the stuff I learned about religion in grade school wasn't the official church teaching. I'm sure the church has some theology about sin and grace, but to this day whenever I hear those words I think of what my religion teacher in fifth grade taught us about good angels whispering in our right ears and bad angels urging us in our left ears." Although official religion is characteristically a coherent, consistent body of beliefs and practices, the individual's version of it is not necessarily coherent or consistent. Individuals vary in how much consistency they expect in their own lives, but many people appear to be comfortable with highly incoherent assortments of beliefs and practices (see Bibby, 1983; Hertel, 1980).

Representatives of religious organizations may decry this misinformation or lack of knowledge, but it is an understandable product of socialization. In socialization, the official model is mediated (if taught at all) through other persons who are significant in the child's life: parents, friends, teachers, clergy. The child may be exposed to a relatively incoherent assortment of stories, admonitions, images, threats, and examples.

Individual members may also deviate from the official model of religion in their priorities, often emphasizing aspects of belief and practice that the official model holds as relatively unimportant. Thus, a religious group may emphasize norms against cheating and stealing, whereas an individual member may downplay those norms while emphasizing norms against drinking and gambling.

Sometimes individual members may disagree with the official beliefs and practices. Of U.S. Roman Catholics, a sizable proportion does not agree with the official church teaching on birth control, abortion, divorce, women clergy, and clerical celibacy. In a 1987 poll, 77 percent of U.S. Catholics said they would be more likely to follow their own conscience rather than the pope's teachings on these matters (Gallup and Castelli, 1989:99). Another study found that most Catholics surveyed felt that a person could be a good Catholic without going to church every Sunday (70 percent), obeying church teachings on birth control (66 percent), going to private confession at least once a year (58 percent), or obeying the church's teachings regarding divorce and remarriage (57 percent) (D'Antonio et al., 1989:57). In the 1950s, when official national organizations of major Protestant denominations made strong statements about Christian responsibility on the issue of racism, many individual members chose to disagree with these teachings (Hadden, 1969). Similar recent disagreement has occurred in both Protestantism and Roman Catholicism over teachings on homosexuality, women's roles, abortion, war, and divorce.

Adherence to *any* official religious group—church, denomination, or any of the alternative organizational forms described above—is complicated by what some are calling the "new voluntarism" (Roof and McKinney, 1987). This thesis suggests that, in the modern context, individual believers are and believe they should be free to choose all significant components of their religious belief and practice. Accordingly, denominational affiliation is utterly optional, and selection of a local congregation within that denomination is a matter of personal tastes and "preference" (Marler and Roozen, 1993). Church attendance and ritual participation are viewed as up to the individual.

High proportions (80 percent, according to one Gallup survey) of Americans believe that the individual "should arrive at his or her religious beliefs independent of any church or synagogue" (Gallup Poll and Princeton Religion Research Center, 1988). Many active church members feel free to choose components of their belief system, combining elements of their official religious tradition with other culturally available elements such as New Age spirituality, Asian and Native American religions, popular psychology, and popular religion. Many others acquired a greater sense of religious autonomy in response to a growing gap between what official religions' spokespersons taught and the experience of the laity. For example, one woman (a full-time volunteer worker in a Catholic peace-and-justice center) said, "I think it would be easier to change the color of my eyes or to get a new genetic code than it would be to stop being a Roman Catholic." Later in the interview she added:

> I really set aside exclusive Roman Catholicism. . . . I remained a religious woman and I have never stopped thinking of myself as a religious woman, ever, not for a moment. But I began to weave in understandings of a variety of different religious and . . . spiritual traditions, and . . . ways of behaving in the world—concrete actions. So the traditions that I looked most closely at were Native [American] traditions, the Jewish tradition (and I—because of intermarrying in my family—have a number of men and women who came out of a Jewish faith tradition) and Zen Buddhism and Tibetan Buddhism (quoted in Spickard and McGuire, 1994).

This sense of religious autonomy, documented among a sizable proportion of Americans (Hammond, 1993; McGuire, 1988; McNamara, 1992; Roof, 1993;

Roof and Gesch, 1995) suggests that many official religious organizations have lost much of their authority over their own members. If this voluntarism is characteristic of an emerging pattern of affiliation with official religious groups, then the main distinction among groups (as discussed further in Chapter 5) may be whether they allow voluntaristic membership or demand totalistic conformity with official group beliefs, norms, and ritual practices.

RESEARCHING OFFICIAL RELIGION AND RELIGIOSITY

Researchers of official religion have the methodological advantage of being able to clearly identify authoritative institutional spokespersons and decision-making bodies, approved teachings, and ritual practices.[4] Thus, it is possible to compare the Episcopalian position with the Methodist position on some issue (e.g., the ordination of women) or to analyze the liturgy of Lutheran churches or to study the development and effects of a specific institutional change within a religious organization (e.g., a denomination deciding to require candidates for the ministry to obtain a college or post-college degree preparation for ordination). Using historical data, interviews, archival records, participant observation, and other intensive research methods, many useful research projects have illuminated our understanding of official religions.

It is far more difficult, however, to methodologically tap the religion of individuals. Standard sociological sources such as census data, opinion surveys, and institutionally generated figures (e.g., membership rolls) can be highly misleading. Survey methods often describe an individual's "religion" by asking respondents their religious affiliation or "preference." A questionnaire may ask the respondent to indicate whether he or she is Protestant, Catholic, Jew, other, or "none." A further refinement of this is specification of the denominational affiliation (e.g., Presbyterian, Southern Baptist, Conservative Jew, Mormon, etc.). These categories then become the basis for broad comparisons of behavior (e.g., college attendance) or attitudes (e.g., approval of desegregation).

Because we need to know more about the values, attitudes, feelings, and experiences that motivate the individual believer, this emphasis on understanding individual religious opinions is worthwhile. The methods and measures used, however, have serious shortcomings that cannot be overcome by minor refinements. One problematic issue is that most of this research has not been interested in the individual's religion per se but in generalizations about the "sum" religiosity of larger groups (e.g., denominations or ethnic groups).

---

4. Groups vary, however, as to how much authority they vest in the hierarchy and official group agencies and how much autonomy is left to individual congregations and members.

Thus, the research does not really try to grasp the total picture of each individual's religion and commitment to it; instead, it isolates certain readily measurable aspects of each individual's religion, then lumps these into a composite picture of the larger group.

This methodological approach might be useful *if* members of the larger group (e.g., all Congregationalists, all Nazarenes) were highly homogeneous. That assumption does not hold true for many religious groups, however. Within-group variance among Catholics, Baptists, or Quakers, for example, is often *greater* than between-group variance. If so, what does it tell us to find that "on average" persons who identify themselves as Baptist are more likely to believe in "faith" item X than those who identify themselves as Catholics?

Similarly, institutional data tell us little about the actual religion of members. Some churches count the infants of members as full members at baptism; others require an act of commitment in the teen years; yet others admit only adults as members. Some religious organizations enthusiastically encourage attenders to "join"; others neither actively recruit nor readily accept new members. Some religious groups consider a member to be permanently a member unless that person actively quits or is excommunicated; other groups drop inactive members from their rolls; yet others have strict "tests" of membership and expel those who do not live up to their expectations. If religious organizations vary so widely in their definitions of membership, researchers must be very cautious in how they use those figures.

The same is true of official measures of individual religiosity. As official religion sees it, "religiosity" refers to the intensity of commitment to the official belief system as expressed in institutionally identifiable attitudes or behaviors (e.g., receiving communion or agreeing with the group's moral condemnation of certain actions). As with all methodological approaches, this "way of seeing" is also a "way of not seeing" (Burke, 1935:70). A focus on official religiosity highlights certain beliefs, commitments, and practices but simultaneously makes others invisible.

This has certain consequences for sociological research. Standard survey approaches are extremely limited in their ability to tap the depth and complexity of each individual's religion and religiosity, because they require respondents to fit their answers into codable (or even precoded) categories determined by official religious conceptions of religious belief and practice. It is entirely possible that such limited categories *never* adequately described individual religiosity; recent changes in the shape and location of religion in modern—especially religiously plural—societies make survey methodologies even less effective in understanding the quality of individual religiosity. More successful approaches include combinations of ethnographic methodologies, intensive interviews, narrative analysis, and related methods of letting respondents name their own meaningful categories. Whether we are studying religiosity in the official or nonofficial forms of religion, it remains important to leave the definitive categories for individual religiosity open—as part of the research question—rather than assume them to be usable, simply because they are defined as "religious" by historical official religious organizations.

Researchers studying religiosity in the official model recognized that it was a complex feature: No single quality could be used to describe the individual as "religious" or as relatively "more religious" than another individual. Glock (1965) distinguishes five dimensions of religiosity:

1. The **experiential dimension** includes feelings or sensations that are considered to involve communication with "divine essence." It refers, for example, to a feeling that one has been saved or healed, a feeling of deep intimacy with the holy, or a sensation of having received a divine revelation of some sort.

2. The **ritualistic dimension** includes religious practices such as worship, prayer, and participation in certain sacraments.

3. The **ideological dimension** refers to the content and scope of beliefs to which members of a religious group are expected to adhere.

4. The **intellectual dimension** encompasses the individual's knowledge-ability about the basic tenets of the group's beliefs and sacred scriptures.

5. The **consequential dimension** includes the effects of religious belief, practice, experience, and knowledge upon the individual's behavior in institutional settings that are not specifically religious (e.g., business, voting, family, and leisure-time activities).

These distinctions are useful to remind us that a person could be highly religious in one dimension (e.g., go to church regularly, pray often) and yet not know church teachings (intellectual dimension) or have had any religious experiences. Unfortunately, these dimensions are not all amenable to quantification and measurement through standardizable indices. We need to use different methods to understand, for example, religious experience than we would to study how strongly adherents of official religions agreed with their group's ideology.

No research has yet defined these dimensions in terms that are not biased toward the orthodox position of certain historic (i.e., traditional Judeo-Christian) expressions of religiosity. In their study of U.S. church members, Stark and Glock's operational definition of these five dimensions illustrates such a bias (Stark and Glock, 1968:22–80). Thus, a person who assented to a higher proportion of the following beliefs in an index of the ideological dimension was considered more religious than someone who assented to only some:

1. Unqualified certainty in the existence of God
2. Belief in a personal God
3. Belief in miracles as described in the Bible
4. Belief in life after death
5. Belief in the actual existence of the devil
6. Belief in the divinity of Jesus
7. Belief that a child is born into the world already guilty of sin (i.e., original sin)

By these measures, liberal denominations appear less religious than conservative ones, and liberal members of any denomination appear less religious than conservative members. Devout Quakers would, by these measures, probably score very low on religiosity. Subtle cultural and gender biases may also influence these measures. Apparent differences between men's and women's religiosity and between blacks' and whites' religiosity may be, in part, an artifact of how religiosity is operationalized and measured (Feltey and Poloma, 1990; Jacobson et al., 1990).

Recognition of the cultural biases of some ideological items tapped in these scales has led Glock and others to formulate parallel measures for emerging religious patterns in the United States. In an ambitious study of nontraditional religious affiliation in a California community, Glock and his co-workers included several non-Christian options such as Eastern mysticism, astrology, Transcendental Meditation, Satanism, and Yoga (Glock and Bellah, 1976; see also Glock and Wuthnow, 1979; Wuthnow, 1976b). This research asked about religious practices other than specifically Christian ones—whether respondents meditated, for example, and what kinds of techniques they used in meditation (e.g., a mantra, drugs, breathing techniques, prayer, etc.). Similarly, a Canadian study tapped several alternative meaning systems beyond the usual official or "visible" religions; researchers identified the "invisible" religious cultural themes as mystical, paranormal, familist, feminist, and positivist (Bibby, 1983). Other researchers have attempted to overcome possible gender biases by eliciting respondents' images of God and by including a measure of perceived closeness to God (see Feltey and Poloma, 1990; Nelsen et al., 1985; Roof and Roof, 1984).

Although such studies greatly expand the range of religious beliefs and practices considered, they are nonetheless limited to discrete, identifiable clusters of beliefs. It may not be possible to devise a research instrument to tap all possible religious beliefs, practices, knowledge, and experiences held by any individual in a culture as diverse as that in the United States. Survey research is predicated on measuring the distribution of opinions, attitudes, or attributes that can be specified in advance (i.e., in wording the questions of the survey). Such a specification is only possible by using the criteria of official religion. A number of features of religion and religiosity in contemporary societies make it problematic to assume that an individual's religion can be adequately described by measuring its correspondence to orthodox official religion (Luckmann, 1973).

Let us now consider some of the ways in which the religion of the individual may differ from official religion.

NONOFFICIAL RELIGION AND RELIGIOSITY

Alongside or overlaying official religion is another pattern of religious belief—
nonofficial religion. Nonofficial religion is a set of religious and quasi-
religious beliefs and practices that is not accepted, recognized, or controlled by
official religious groups. Whereas official religion is relatively organized and
coherent, nonofficial religion includes an assortment of unorganized, inconsistent, heterogeneous, and changeable sets of beliefs and customs (Towler and
Chamberlain, 1973). Nonofficial religion is sometimes called "common,"
"folk," or "popular" religion because it is the religion of ordinary people rather
than the product of religious specialists in a separate organizational framework.

HISTORICAL PERSPECTIVES: THE SOCIAL CONSTRUCTION OF "RELIGION"

Most sociology and historiography has assumed that, at least since the beginnings of Christendom, there has been a radical disjunction between official
and popular forms of religion. This disjunction was taken for granted largely
because, at the time of the early development of sociology, the official religious organizations had successfully achieved politically legitimated cultural
dominance throughout Europe and North America. Social scientists merely
accepted official (i.e., Protestant, Catholic, and later, Jewish) groups' definitions of religion's boundaries as "given" (see Trexler, 1984). Without questioning the *social construction* of those boundaries, the social sciences
inadvertently supported official religions' definition of their practices and beliefs as "pure" and linked with the "sacred"—that which was to be protected
from the "profane" world of everyday life. The residual category, "popular religion," came to be identified as a tainted or impure form of religion, or—
worse—as downright pagan.

Thus, the notion of religious syncretism—the blending of diverse cultural
elements into one religion—came to have a powerfully negative connotation.
"Syncretism" was used to imply inauthenticity or contamination, the insidious
infiltration of a putatively "pure" religious tradition by meanings, symbols, and
ritual practices borrowed from an alien, impure religious tradition. Accordingly, guardians of "pure" Christianity have decried most New World religious
expressions such as Umbanda, Santería, and Vodou as syncretism in this negative sense. By contrast, however, recent anthropological thought has debunked
the idea of "cultural purity" or "authenticity." Some critics have argued that in
contemporary anthropology, "syncretic processes are considered basic not only
to religion and ritual but to 'the predicament of culture' in general" (Shaw and
Stewart, 1994:1; for illustrative case studies, see Badone, 1990; Espín, 1994;
Shaw and Stewart, 1994, and Stevens-Arroyo and Pérez y Mena, 1995).

Through identifiable historical processes, then, official religion was linked with social and religious elites and nonofficial with the uneducated and superstitious "masses" (Brown, 1981, Obelkevich, 1979). The elites promoted "virtuoso" religiosity (conveniently defined in terms of participation in official religious practices) and decried the inferior religiosity of the hoi polloi. Religiosity in the official mode was assumed to have "sacred" objectives—holiness, spiritual blessings, and salvation. By contrast, popular religiosity was linked with the profane—for example, giving attention to people's pragmatic, quotidian needs (such as healing, fertility, protection from adverse fortune, obtaining desired material goods, etc.).

In the light of recent scholarship, however, we must reject this assumption for two reasons. First, the sacred/profane dichotomy, upon which the distinction of official, elite religion is based, may produce a misleading conception of religious life. One study of religious beliefs and practices in preindustrial Germany suggests, on the contrary, that the sacred is experienced from *within* the profane, within the very human context of the historical, cultural, and socially shared situation of believers (Scribner, 1984:17–18). This generalization finds more recent corroboration in the anthropological analysis of "Los Pastores," a popular religious drama that has been performed in the Mexican barrios of San Antonio, Texas, every Christmas season since 1913. As performed in the backyards and homes, this devotional ritual, with its fulfillment of *promesas* (i.e., pledges of piety in thanks for answered prayers or blessings), meals, and surrounding festivities creates communal sociability (Flores, 1994). It is precisely because "el Niño Dios" is experienced in the context of the profane—a neighbor's yard, where only last week the neighborhood teenagers were fixing their low-rider cars—that it can speak profoundly to people's lives.

Furthermore, the sacred is experienced in everyday life as it is embedded in human social practices such as calendric rituals, use of space, and meaningladen postures and gestures (Bourdieu, 1977; Hart, 1992; Scribner, 1984). The religious dimensions of people's everyday experience and the so-called profane dimensions are often not merely linked, but interpenetrate (Parker, 1993:245; 1994).

A second reason sociology of religion should reject the radical disjunction of official and nonofficial religion is that not only the content but indeed the very definition of religion is a *social construction*. The difference between official and nonofficial religion is not an *essential* quality of religion but is rather the result of a concrete historical process in which certain social groups used their power and authority to privilege certain forms of religious practice over others (cf. Droogers and Siebers, 1991).

The establishment of an "official" religion by definition excluded—sometimes violently purged—nonofficial "people's" religion. The rhetoric legitimating this hegemonic boundary maintenance spoke of delineating and protecting truth against error, pure practice against the impure or sacrilegious, proper authoritative leadership against nonapproved, and so on. Sociology of religion has already documented the regularity with which these ideas were linked with the class interests of the elites that articulated and enforced them.

The process of establishing "official" religion thus generally excluded religious expressions characteristic of the poor, the women, various minorities, indigenous peoples in colonized lands, and other powerless groups.

We seem to assume that once the elites effectively defined nonofficial religion out of existence—as profane—the rest of society gradually became fully incorporated into official religion, albeit with occasional temporary expressions of "heresy" and "superstition," widespread "ignorance" of official teachings, and "laxness" in expected religious practice. Interestingly, for example, some of the earliest ethnographies of New World indigenous peoples[5] were produced to provide the guardians of the official religion (e.g., the Inquisition in Mexico) with a way to distinguish the people's Catholic faith from vestiges of indigenous religiosity (Martin, 1994). Official religion thus often proceeded hand in glove with colonial domination.

Nonofficial religion did not disappear, however, although its expression may have been hidden in various times and places. Rather, it exists alongside official religion in a number of fascinating patterns. Latin American popular religion illustrates that, while belonging to the official religion, people can simultaneously engage in nonofficial beliefs and practices that are at least as meaningful or useful to them as are their official religious beliefs and practices. For example, elements of their popular religion address practical needs such as healing, successful harvests, divination, and fertility. This traditional, premodern pattern of religion, although maintained by a very different structural relationship to official religion and to the lay community, is not utterly unlike the pattern of "voluntarism" sociologists describe as characteristic of contemporary U.S. society.

Another reason for studying popular religion is that it often expresses (although sometimes counterproductively) dissent and resistance. For instance, in some popular religious rituals the official order is temporarily inverted or mocked; in some popular religious celebrations, wealth is somewhat redistributed; in some, nonpowerful persons can temporarily refuse to cooperate with the work or other social demands of their bosses, husbands, village rulers, and so on (see Lancaster, 1988:27–54; compare Benavides, 1994; Comaroff, 1985).

Even in relatively rigid traditional societies, popular religion can express and sometimes activate the concerns of subordinate people. Interestingly, recent historical studies of the interaction between representatives of "official" religions and whole regions experiencing popular religious beliefs and practices show that the *influence was reciprocal*. Often official religion deliberately accommodated popular religion, absorbing its beliefs and practices into orthodoxy and orthopraxis (see, for example, Luria, 1991). Nor were subordinate peoples such as colonized indigenous groups necessarily mere passive recipients; sometimes they were active agents in creating new "shared" meanings in

---

5. The most famous is Bernardino de Sahagún, *História general de las cosas de Nueva España;* others include Diego de Landa, *Relación de las cosas de Yucatán,* and Hernando Ruiz de Alarcón, *Tratado de las supersticiones.*

a mestizo (mixed) culture (Carrasco, 1995). Often too, "official" religion was successful at reshaping popular religion, as by inserting observance of approved Sacraments into the celebration of a popular pilgrimage.

At the same time as emphasizing the importance of more study and appreciation of popular religion, however, we must remain critical in our research approaches. One of the most difficult methodological tasks is to avoid romanticizing popular religion. Just because it is a religion "of the people" does not mean popular religion is necessarily any more benign or beneficial than official religion. Some popular religion is intolerant, oppressive, or even violent. For example, both in the Middle Ages in Europe and in the twentieth-century United States, much popular Christian religion promoted anti-Semitism (Lippy, 1994; Vauchez, 1993); popular religion is highly likely to reflect popular prejudices, making it blatantly racist, misogynist, homophobic, or politically chauvinistic. Some popular religion is an obstacle to adherents' educational, occupational, or economic successes; some is antidemocratic and suppressive of certain freedoms, including the free expression of religion itself. Just because popular religion may challenge the status and power arrangements of official religion does not mean that it does not *also* serve to reproduce hierarchical arrangements, such as men's power over women or one tribe's over another.

To the extent that sociology of religion has made paradigmatic the characteristics of such "official" religion, it marginalizes all other forms of religion, all other religious expressions, all other patterns of religiosity. Thus, using this approach, sociologists simply fail to *notice,* much less comprehend, the complex and diverse ways religion is constructed and practiced in people's everyday lives in the modern context.

This is not to imply that popular religion is limited to traditional societies or to marginalized groups. My 1988 study of nonmedical healing groups in a suburban middle-class New Jersey area found many people drawing on a wide range of nonofficial religious beliefs and practices. The Extended Application at the end of Chapter 5 discusses some of these groups and how they may be linked with the location of religion in contemporary societies.

POPULAR RELIGION IN MODERN SOCIETIES

Since the main distinction between "official" and "popular" religion is whether religious beliefs and practices are under the control of official religious organizations, the United States has a plethora of popular religious expressions. This abundance is partly due to its history of never having allowed an established church—a legally enshrined official church of the nation (although, as shown in Chapter 8, not all religions achieved status of being legitimate "official" religions). Other historical factors that multiplied unregulated popular religion were the sheer size of the country and the geographical dispersal of the people, especially during frontier expansion. Many Protestant religious groups have their roots in popular religious expression. The camp meetings and tent revivals of early U.S. Protestantism exemplify the process. Anyone on the frontier who could attract an audience could preach, regardless of training, knowledge, or quality of content (see Lippy, 1994). The official denominations tried to co-opt this religiosity by sending out traveling ministers or by enrolling the local ministers who had attracted followings. They never quite succeeded, though; on the frontier and later in urban revivals, there was almost continual emergence of new religious expressions outside of official control. While perhaps never matching the United States in sheer diversity and numbers of popular religious expressions, Europe has a long history of popular religion, from effervescent religious movements to vestiges of pre-Christian practices, from popular inspirational literature to pilgrimage sites and shrines.

"Pop" Versions of Christianity

In addition to nonofficial preachers, another source of popular versions of Christianity is religious publishing. While much religious publishing is done under the control of official religious bodies (e.g., Sunday School educational materials), the vast majority of religious publications are not edited or reviewed for their "orthodoxy." A large body of inspirational literature exists that is nontheological and unrelated to the official religion of its readers, and the religion promulgated in this literature is largely quasi-magical, emphasizing religious techniques. One author of many such books, Dr. Norman Vincent Peale, urges, "Learn to pray correctly, scientifically. Employ tested and proven methods. Avoid slipshod praying" (quoted in Schneider and Dornbusch, 1957:479; see also Meyer, 1965). Values and ideological perspectives embodied in popular literature, from *Reader's Digest* to *Gentlemen's Quarterly*, find their way into popular religion (Elzey, 1976, 1988; Moore, 1994).

A parallel and even more popular source of popular religion is inspirational programming on radio and television. Some mass-media preachers gain a sufficient following to build up a large-scale organization. Oral Roberts, for example, began as a popular tent revivalist and faith healer and subsequently developed a large, well-financed sectlike organization. Televangelists (i.e., television preachers) exemplify nonofficial religious independence from doctrinal, organizational, and fiscal controls.

There is a complex relationship between religion as promulgated by popular culture and that of official religious groups. On the one hand, studies have found that the "electronic church" (especially television religious shows with popular preachers) is watched almost exclusively by already converted, relatively committed conservative or fundamentalist Christians (Stacey and Shupe, 1982). On the other hand, many organized religious groups (notably including some highly conservative Christian denominations) have criticized the "electronic church" mainly for the freedom from outside evaluation and control enjoyed by the independent and often highly charismatic evangelists—especially control of their finances and the content of their preaching (Hadden, 1993).

Since televangelists are entrepreneurs, their success hinges on achieving an audience response that can be converted into contributions of money. This necessity is often connected to the beliefs they teach: For example, many entrepreneurial preachers tell their viewers that success, prosperity, and answered prayers come to those who contribute, especially to that particular evangelist's organization (Hadden and Swann, 1981). Thus, while popular evangelists present a version of religion, that version often differs in content and emphasis from orthodox teachings of official religions.

Because of its material and pragmatic emphasis, popular religion is particularly open to commercialization. Thus there have long been markets for religious objects, from the time of early Christian purchases of relics (e.g., bones of dead saints) to the contemporary sales of Christian video games, bumper stickers, amusement parks, and jewelry (McDannell, 1995; see also Moore, 1994). Religious publishing and mass-media evangelism both have clear commercial niches. Official religions also participate in such commercialization (e.g., sales of wedding services and cemetery plots), partly in an effort to control both the religious ideas and the economic resources.

The ideas and practices from popular culture religion are widespread. Artifacts of popular religion bear some resemblance to those of official religion: a wall plaque announcing the presence of angels, a plastic Jesus on the car dashboard, gold-plated crosses on neckchains, bumper stickers proclaiming "I Found It" or "My God Is Alive; Sorry About Yours," plaques, plates, and posters with messages such as "Smile, God Loves You." One shrine of popular religion is Holyland, U.S.A.—a large sculpture garden where busloads of tourists visit various scenes depicting biblical events such as Noah's Ark and the Crucifixion (cf. Elzey, 1975, 1988). Popular religion also includes religious or quasi-religious themes from patriotism or nationalism (explored in detail in Chapter 6). National symbols such as the flag or Statue of Liberty form parts of some individuals' belief systems and are often indistinguishable from elements of official religion.

Superstition and Magic

Another overlay from common religion onto the official model consists of a variety of superstitious and magical beliefs and practices. Some superstitions pertain directly to practices of official religion. Asked why she had all of her babies christened, one woman responded, "Don't know. People have it done. It keeps them safe like wearing a St. Christopher" (quoted in Towler and Chamberlain, 1973:23). Another somewhat magical use of official religion is *stichomancy*—the practice of divining or getting a "message" from random opening of the Bible. Similarly, quasi-magical attitudes often accompany religious actions. At the end of a healing service observed in a mainline denomination, the minister offered to bless salt and oil for home prayer for healing. With a rustling of brown paper wrappers the congregation opened the spouts of salt boxes and lids of cooking oil bottles so the blessing could "get in." This overlay suggests a difficult methodological problem: how to interpret a situation in which members participate in rituals of official religion with attitudes and motives inconsistent with the official model.

Much superstition is related to more mundane matters and represents an attempt to control or explain one's environment. One study in England found superstitious beliefs and practices to be fairly common. Thus, 22 percent of the sample believed in lucky numbers and 18 percent in lucky charms, 15 percent avoided walking under ladders, almost 50 percent threw spilled salt over their shoulders, and over 75 percent touched wood for protection. Some superstitious actions were not serious, but a core of 6 to 8 percent of the sample said they became uneasy if they did not perform the appropriate actions. This study found persons who characterized themselves as religious more likely to be superstitious than those characterizing themselves as nonreligious; churchgoing "religious" persons, however, were much less superstitious than nonchurchgoing "religious." This evidence suggests that for some religious persons, official and common religious beliefs and practices become indistinguishably merged but that churchgoing may filter out nonofficial elements of belief (Abercrombie et al., 1970:98, 113; see also Jahoda, 1969, regarding the affinity between belief patterns of superstition and official religion).

The Paranormal and Occult

Other beliefs and practices concern **paranormal occurrences** (i.e., events outside the usual range of experiences). Belief in paranormal occurrences is fairly widespread. Approximately one-fourth of U.S. adults believe they have seen a ghost or spirit of the dead (Greeley, 1975:36). A 1994 *Newsweek* poll reported 13 percent of Americans have sensed the presence of an angel; 33 percent have had a mystical or religious experience (Kantrowitz, 1994). One U.S. national survey (Gallup and Castelli, 1989) reported that 42 percent of respondents felt they had undergone an experience of being really in touch with someone who had died. In the same study, 67 percent reported experiences of déjà vu (i.e., the sense of having already seen a place where they had never been before), and 67 percent reported experiences of ESP (extrasensory perception). Similarly, surveys of European beliefs and values found about 25 percent of British, German, and French respondents believed they had communicated with the dead, and about 35 percent reported experiences of ESP (Gallup Poll/Social Surveys, 1984). Using a broader definition, several studies in Italy found 65–85 percent of respondents reporting extraordinary religious experiences (Acquaviva, 1993). Some persons attribute great significance to paranormal experiences; whole belief systems (e.g., Spiritualism) allocate special meaning to such experiences. Other persons who have had such out-of-the-ordinary experiences recognize them as unusual but place no special interpretation on them.

A study in an English industrial city found comparably high rates of extraordinary religious experiences in a random sample in which very few claimed active membership in an official religious institution. Despite the fact that many of these experiences were described as personally very meaningful, the majority of respondents kept the experiences private or altogether secret for fear of ridicule or being labeled mentally imbalanced (Hay and Morisy, 1985). Thus, some elements of nonofficial religion are "invisible," not so much because of their deviance from official religion, but more because the general culture treats them as inappropriate.

Extrascientific explanations and techniques are also part of common religion. These are ways of interpreting and manipulating the natural and social environment that are not accepted by this culture's official religious or scientific groups. Included are beliefs in astrology, hexing, palmistry, numerology, amulets and charms, water divining, UFOs, divination (by Tarot, pendulum, rod, I Ching, entrails), and witchcraft. Science fiction, for example, deals with a number of quasi-religious themes about humanity and its boundaries, which have been embodied in religious groups such as UFO cults and groups that believe they communicate with nonterrestrial beings (see Mörth, 1987). One study found disproportionate numbers of young, highly educated males among those attracted to deviant (or occult) sciences and occult quasi-religious practices (Ben-Yehuda, 1985). It asks, Why this prevalence of nonofficial religion (and nonofficial science) among precisely those social groups most favored in modern technological society?

Beliefs and practices pertaining to paranormal occurrences and extrascientific explanations and techniques are part of what is generally termed the *occult*. **Occultism** is a set of claims that contradict established (i.e., official) scientific or religious knowledge. This focus on the anomalous makes the occult seem strange and mysterious (Truzzi, 1974a:246). Some observers have suggested that the 1960s and 1970s were periods of occult revival in U.S. popular culture (Marty, 1970; Truzzi, 1972). The esoteric and occult have long been part of countercultural currents in Western societies. Many wide-ranging innovations have begun in the "seedbed" of the esoteric, and modern medicine, chemistry, and psychology owe much to the esoteric sciences of earlier times (Tiryakian, 1974:272, 273).

Although some of the recent interest in esoteric beliefs and practices may be merely countercultural experimentation or fads, many people take their belief in the occult seriously. Witchcraft exemplifies beliefs and practices followed by adherents with varying degrees of seriousness and involvement. At one end of the spectrum are people for whom witchcraft is a fad or form of entertainment. Others believe in witchcraft and therefore seek occasional spells for health, protection, and success; they might also ascribe special importance to astrological events such as solstices and eclipses. At the other end of the spectrum are people who are fully involved in the beliefs and practices of witchcraft, belong to a group (i.e., coven), and organize their lives around witchcraft as a counterofficial religion.

There is a distinction between white and black magic. White witches employ their magic for only beneficial purposes such as healing. They often emphasize the continuity of their tradition with the medieval European white witches—"wise women" and men who combined pre-Christian symbolism and lore. It is not uncommon for white witches also to belong to church groups of official religion. One interesting contemporary development of witchcraft and neopaganism has been the adoption of some of their symbols and organization for the expression of feminism. More so than official religions, these nonofficial religious beliefs envision feminine features of the divine, and their ritual practices emphasize the empowerment of women. Some women are proud of the deviant heritage that the white witch represents (Christ, 1987; Griffin, 1995; Neitz, 1990, 1994). By contrast, practitioners of black magic use their powers for evil as well as good ends. They believe they have allied themselves with the forces of evil (i.e., Satan), from which they obtain power and direction (Alfred, 1976; Bainbridge, 1978; Moody, 1971, 1974; Truzzi, 1974b).

Secret societies have long been a part of nonofficial religion. Mystery cults, kabbalistic groups, Freemasonry, and Rosicrucianism exemplify their variety. Mystery and awe surround the esoteric knowledge protected by the secret society, and initiation rituals symbolize admission of the member to each level of secret knowledge. The protection of this secrecy promotes the solidarity of the group (Carnes, 1989; Kertzer, 1988; Simmel, 1906; Tiryakian, 1974:266, 267). Seldom is there any clear border between secret societies that are overtly religious and those that are not specifically religious. For example, many groups of the "Women's Spirituality" movement and the "Men's Movement" use secrecy and rituals of initiation (McGuire, 1994). Some explicitly religious movements (e.g., early Mormonism) have similar features because they are organized around a *gnosis* (i.e., special knowledge).

Another strand of esoteric culture that is a much more familiar part of common religion is **astrology,** a set of beliefs and practices predicated upon the idea that impersonal forces in the universe influence social life on earth. Believers try to read and interpret zodiacal signs (e.g., the position of the stars at the time of one's birth) in order to predict outcomes of the influence of those forces, to choose the best moment for certain actions or the best probable course of action, and to discern when humans bear responsibility or blame for certain outcomes. Under the general category of astrology fall a wide variety of beliefs and practices ranging from highly sophisticated, complex speculations to pop versions such as newspaper horoscope columns (Fischler, 1974).

Astrological beliefs and practices are widespread in modern Western societies. According to a 1990 survey, 25 percent of the U.S. population believe in astrology (Gallup, 1990). Similarly, a Canadian survey found that 34 percent of respondents believe in astrology, and 88 percent consult their horoscopes as least once a month (Bibby, 1993). A recent study of highly committed adherents found them to be disproportionately well educated, occupationally and economically stable, white, and female compared to the U.S. population (Feher, 1992).

The nature of belief in astrology illustrates the methodological difficulty of studying nonofficial religion. Astrological beliefs are widespread, significant parts of many people's worldviews. Some people hold them in combination with official religious beliefs; for others, they are substitutes for official religion. There is no "church" of astrology. Connections among committed adherents are not commitments to a group but are more like being part of a milieu or "lifestyle enclave" (Feher, 1994). Although there are some individual astrological practitioners, most adherents learn their beliefs and practices through relatively anonymous media—books, newspapers, magazines. Some evidence suggests that, like superstition, commitment to astrology is strongest among traditionally "religious" persons who are not involved in official religion (Wuthnow, 1976a:166).

While popular, folk, or occult religious beliefs and practices may be defined as "deviant" by arbiters of official religous or scientific knowledge, in everyday life they are in the range of "normal" and plausible to many people (Campbell and McIver, 1987).

INDIVIDUAL RELIGIOSITY IN THE NONOFFICIAL MODE

One interpretation of nonofficial religion in contemporary societies suggests that modern society is increasingly characterized by a new mode of religiosity in which individuals select components of their meaning systems from a wide assortment of religious representations. But while traditional "orthodox" representations from official religion may be among the components selected, other elements are drawn from popular culture—newspaper advice columns, popular inspirational literature, lyrics of popular music, popular psychology, horoscopes, night-school courses on meditation techniques, and so on (Luckmann, 1967:103–106). This interpretation may somewhat overstate the contrast of this eclectic religiosity with that of earlier times; it is entirely probable that earlier generations held widely varying versions of official religion, melding together combinations of folk beliefs, superstition, pagan customs, and misinformed interpretations of the official model of religion to form individual meaning systems. The sharp contrast between official and nonofficial religion is probably itself a product of relatively recent social processes by which society came to define religion as the domain of official religious institutions and by which religious organizations articulated their distinctiveness from popular religion.

The complex threads woven together to comprise the individual's operative religion make it difficult to describe that religion by reference to any single model such as an official religion to which the person may belong. For example, in my own research (McGuire, 1988), one respondent had experienced multiple healings and he devoted two evenings per week to helping others with healing needs; he described himself as "very spiritual" but only "somewhat religious" (a distinction made by many persons whose religiosity was in the nonofficial mode). As part of his spirituality and healing efforts, this person used primarily meditation with visualization (derived not only from Christian sources), as well as yoga, crystals, and reflexology. *And* he was a practicing Methodist, indeed an active clergyman.

This amalgam of beliefs and practices is not at all unusual. Evidence on the distribution of nonofficial religious beliefs and practices shows that many adherents are also adherents of official religion. Only a small portion of nonofficial religion is practiced in lieu of official religion. As suggested in the first part of this chapter, individual religious autonomy and voluntarism may be profoundly affecting official religiosity; those same forces make nonofficial religiosity even more plausible in the modern context. Chapter 5 examines further the organizational effects of these patterns of religiosity.

Merely listing the elements of a person's religion does not adequately describe it. We also need to know which beliefs and practices are central and which are peripheral. How does this complex mixture change over time and how do some parts recede in relevance? Similarly, we need to know when the individual applies which beliefs and practices. The believer may turn to official religion to help socialize children but seek nonofficial religious help for healing a chronic back problem. We need to understand when and how parts of this eclectic religion generate commitment or community. We need more information about the effective religion of individuals in its complexity and richness.

RESEARCHING NONOFFICIAL RELIGION
AND RELIGIOSITY

What does this overview of the varieties of popular religion imply about how we should study religion today? It suggests that official forms of religion—while still of interest—are far less important for understanding the larger religious situation than we had been led to believe. Specifically, we need to give more research attention to those nascent movements (no matter how small) through which religious symbols are being proffered and mobilized, especially in tension with those of religious and other established institutions. For this we need a richer appreciation of the cultural factors in social and political change. Exactly *how* is religion employed as a cultural resource? In a structural situation where religion is flexible and relatively autonomous from authoritative control in any institutional sphere, we must focus upon *ongoing processes* by which believers create, maintain, and change their symbols for making sense of their worlds. These symbol systems are sometimes enduring, sometimes transformed into new meanings, and sometimes completely ephemeral. Nevertheless, given the modern situation of religion, it is the *process,* rather than whether it results in an institutionalized form with a quantifiable membership, that is important for research and analysis.

For study of such flexible and ever-changing processes, we should pay particular attention to rituals, symbols, and other cultural vehicles for world images (whether toward change or preservation) and for shared religious experiences and sense of community. We need greater attention especially to language—such as shared symbolic language, narratives of personal meanings and group experiences, and the structure of discourse and how it develops and changes. Concomitantly, we need less emphasis on quantitative methods such as surveys, opinion research, and formal organizational analysis, because these methodologies presume a relatively fixed, institutional form of religion. They depend upon shorthand identification of people with their institutional location—such as religious affiliation or political party.

We must also drop the assumption that traditional institutional religious symbols, language, rituals, and practices are used for the reasons an institution once intended them. We cannot assume continuity between the functions served by certain religious expressions in another time and place and the meanings of the same expressions today. For instance, we cannot assume that our interpretation of Mass attendance in rural Ireland in the eighteenth century applies, even vestigially, to 1990s Mass attendance of, say, Irish-Americans in Dallas, Texas. Rather than treating traditional meanings as given, we should make them research *questions.* Toward this end, we need to ask people about what is meaningful to them and about what they experience when they participate in official and nonofficial religious practices.

Methodological tools such as ethnography, intensive interviews, participant observation, and ritual analysis are a good start. Anthropological methods are also limited for our task, however, because anthropological research typically has been done among peoples whose beliefs and practices change far more slowly, for whom tradition is more authoritative, and who are relatively isolated from the pluralism that characterizes the modern religious situation. Nevertheless, good ethnography tries to represent the voices and the meanings of the participants themselves. It attempts to derive analytic categories from the culture itself rather than impose ideas, including the researcher's scientific meanings.

We need better methodological tools for studying the richness of religious ritual and experience—whether in the context of official or nonofficial religion. Although it may be one of the most compelling components of many people's religion, it eludes our methodologies based on words and ideas. For many, religious experience is—almost by definition—ineffable. How, then, can we methodologically tap this critical religious dimension and begin to take it more seriously in our interpretations of people's religion as practiced, as lived?

Our study of religiosity in the nonofficial mode is challenged by the distinction between religiosity and spirituality that many people themselves make. Perhaps we would learn far more about both official and nonofficial religiosity if we included both together in our researches, as complementary and often overlapping resources for individual religious expression. In order to tap both aspects of an individual's religiosity, we need to look with great depth at how people put together all the aspects of their personal religiosity and/or spirituality. Simple catalogs of beliefs and practices may be a start, but they do not adequately tell us how and when the believer draws on one element and not another. One way to focus questions about spirituality is to center on a theme, such as health and illness, childbearing and parenting, dying and loss of a loved one, or other critical experiences for which a respondent could reflect on the whole range of religious responses that was part of each experience. Chapters 3, 4, and 5 report several studies that are promising efforts toward such methodological sophistication.

Expanding our research methods to include nonofficial religion and religiosity makes the research far more complex and requires far greater researcher skill than the seemingly tidy quantitative methods for studying official religion. Nevertheless, both our understanding of official and nonofficial religion and our appreciation of individual religiosity and spirituality will be greatly deepened by the effort.

EXTENDED APPLICATION: WOMEN'S RELIGION AND GENDER ROLES

As we have seen, the individual's religion is a personally meaningful combination of beliefs, values, and practices that is usually related to the worldview of a larger group into which that individual has been socialized.[6] In socialization, the individual typically receives many of these beliefs, values, and practices from representatives of the larger group such as parents and teachers. Much of this received meaning system becomes internalized—that is, made a part of the individual's own way of thinking about self and others. The meaning system received by the individual includes a number of beliefs, images, and norms about that group's definitions of maleness and femaleness. All religions have addressed the theme of human sexuality and gender roles because sexuality is a potent force in human life and because gender is, in most societies, a major factor in social stratification.

This analysis will focus first on how women's religion influences their gender roles and identities, illustrating some of the concepts in Chapter 3 on the individual's religion. A similar application could be drawn for men as males, showing how their gender roles are linked with religion, although this application is not pursued here.

Second, this analysis will illustrate why focusing only on official religion and religiosity is sometimes misleading. As religion became a differentiated institutional sphere, many of women's religious roles and expressions were excluded from official religion and, if continued at all, were kept alive in nonofficial forms such as healing cults, popular religiosity, mediumship, witchcraft, and spiritual midwifery. We need to examine the structural bases for women's status; official religious institutions have historically epitomized the structural and ideological suppression of women. The religious legitimation of their gender role, then, raises some interesting questions about women's religiosity. Do they come to think of themselves as the official model that their religion portrays of them? Impressive historical evidence exists that nonofficial

---

6. This Extended Application exemplifies concepts in both Chapters 3 and 4.

religion is one vehicle for women's assertion of alternative religious roles. A related question is whether official religions are capable of changing to legitimate and express gender equality even if they were to adopt these goals.

Gender Role Definitions

The definitions of maleness and femaleness are culturally established. On the basis of these definitions, a group develops and encourages certain social differences between men and women. U.S. society expects and encourages little boys to be more active and aggressive than little girls; little girls are expected and encouraged to be more polite, passive, and nurturant. In socialization, males and females are taught their culturally assigned **gender roles**—the social group's expectations of behaviors, attitudes, and motivations "appropriate" to males or females. Historically, religion has been one of the most significant sources of these cultural definitions of gender roles; and religion has been a potent legitimation of these distinctions.

Through socialization, the group's definitions of appropriate women's roles become part of the individual woman's self-definition. She evaluates herself in terms of these definitions: "I am a good girl/daughter/woman/mother/wife." The group's definitions of gender roles are subtly interwoven into the individual's learned "knowledge" about the social world. They are embodied in the language and imagery of the group and thus indirectly shape the members' thought patterns. The culture's use of words such as "bitch," "slut," "sweetie," "doll," "housewife," and "mother" has strong evaluative connotations. These words imply qualities that some people attribute to women and for which there are no obvious male equivalents. Numerous words (e.g., "slut") exist for promiscuous women; there are no equivalent words for promiscuous men. Similarly, a man who is said to "father" a child performs a single biological act; a woman who "mothers" a child provides continuing nurturant care. The language embodies the different standards of the society for men's and women's roles (Lakoff, 1975). In learning the group's language, the individual internalizes these images and symbols. Specifically religious symbols and images also shape the individual's gender role concept.

Although both men and women belong to the same official religions, the individual's religion is not necessarily a small carbon copy of the group's entire official stance. Women's versions of a certain religion are probably very different from men's versions; a woman may focus on those aspects of the group's worldview that speak to her social situation. Thus, an Orthodox Jewish man's personal religion might focus on his public ritual roles (e.g., carrying the Torah and forming part of a minyan, i.e., the ten men necessary for a prayer gathering to be official). He might emphasize the intellectual development, discussion of sacred texts, and careful application of religious law expected of men in the Orthodox community. Women, by contrast, have few ritual duties (e.g., the ritual bath after menstruation, making the Sabbath loaf, and lighting Sabbath candles). Their religion is more likely to be focused on the home, in instructing the very young, and in enabling the men of the home to participate

in their public religious roles. The very meaning of being Jewish is thus probably very different for the men than for the women. Central religious activities
(e.g., attending Sabbath services at the synagogue and family worship around
the Sabbath candles) also probably mean something different to each sex.[7]

A woman's religious experience and what she holds religiously most important are qualitatively different from men's religious experience and focus.
Women's religion is nevertheless shaped heavily by the larger religious group
because it is not a separate religion. The larger group attempts to form the individual's role through its teachings, symbols, rituals, and traditions. Thus, in
the example of the Orthodox community, women's religion is strongly influenced by men's ideas of what a properly religious woman should do and be.
Men interpret the Law and run the congregation and the religious courts. In
most historic (i.e., official) religions, women have had considerably less power
than men in establishing social definitions of appropriate gender roles. In the
Roman Catholic church, at least since the time of the early church, men have
held all significant positions of authority to set and interpret religious norms,
practices, and beliefs. The beliefs, ritual expressions, norms, and organizational structure of official religion, in short, effectively subordinate women.

The religious legitimation of women's roles is very similar to the religious
legitimation of other caste systems. As shown in Chapters 2 and 7, religion is
often used to explain why certain social inequalities exist. These explanations
justify both the privileges of the upper classes or castes and the relative non-
privileges of the lower ones. A **caste system** is a social arrangement in which
access to power and socioeconomic benefits are fixed, typically from birth, according to certain ascribed characteristics of the individual. In the Hindu caste
system, an individual of the highest caste has certain permanent prerogatives,
benefits, and responsibilities merely by virtue of having been born into that
caste. Medieval feudalism was a similar system, in which individuals born into
a given socioeconomic stratum could move up or down relative to others of
that stratum but could not aspire to the power and privileges of higher social
strata.

Modern societies typically emphasize individual achievements as a basis of
socioeconomic status more than traditional societies; however, elements of
caste are still important. Race is one ascribed characteristic that still figures
into socioeconomic status in most modern societies. Gender is another major
basis of caste in modern societies. Women are generally excluded from access
to power and socioeconomic status available to men. The expectation that
housework is "women's work" exemplifies women's caste status. No matter
how great a woman's achievement in art, business, or scholarship, the society
still expects that she is the one responsible, by virtue of her gender, for the

---

7. There is little empirical evidence about exactly how men and women experience religious rituals and events, but some clues appear in Myerhoff, 1978:207–241. Some parallels may also be drawn from studies of how different racial castes, socioeconomic strata,
   and ethnic groups differently experience similar religious rituals and belief systems.

menial tasks of housecleaning. Such gender role expectations have socioeconomic implications, becoming the justification for denying women opportunities for economic achievement (Kemp, 1994). Women's status in most religious groups is also circumscribed by caste. Gender is far more important than theological or spiritual qualifications in determining whether an individual can perform certain rituals such as carrying the Torah or consecrating the communion elements.

Caste status confers necessary, but not sufficient, advantages to men's chances for recognition, power, and prestige. Not all men obtain these privileges in the social system, but maleness is virtually a prerequisite for them. Men's superior caste status does not necessarily make life comfortable and pleasurable for all individual men. Some men also suffer exploitation and discrimination—but not because of their gender. Religiously legitimated caste distinctions thus do not empower all men; they do, however, disempower all women by virtue of their gender identity (see Ruether, 1975, for an analysis of sexism as a prototype of domination in Western society).

Religious Legitimation of Caste

Symbolism and Myth. Religion contributes to, or legitimates, a caste system in a number of ways. Often religious symbolism embodies a caste-stratification system. It may depict the higher deities as male and the lower deities (or even negative spiritual forces) as female. The Hindu goddess Kali represents "female illusion," symbolizing in the story of Shiva's entanglement with her the eternal struggle that men must wage against the evils of orientation toward material existence (Daly, 1973; Hoch-Smith and Spring, 1978:4).

Similar distinctions are embodied in the language of religious ritual. The ritual words of the Catholic Mass had, for centuries, stated that Christ's blood was shed "for all men." After several years of internal pressures, in 1985 the Vatican approved changing the words of the Mass to the gender-neutral term "for all" (see also Wallace, 1988). The language of Orthodox Jewish prayer makes gender-caste distinctions even more explicit. Daily morning prayer includes "Praised are you, O Lord our God, King of the Universe, who has not created me a woman." Conservative Jews have changed this wording to "Praised are you, O Lord our God, King of the Universe, for making me an Israelite"; whereas Reformed Jews do not use this blessing at all (Priesand, 1975:57–60).

Some biblical language lends itself to gender-caste distinctions between men and women, although many such distinctions are derived from translations, which embody cultural assumptions of the language of translation overlaid on the distinctions made by the original writers of the many parts of the Bible. Nevertheless, besides the hierarchical, patriarchal language used to legitimate caste distinctions, both the Old and New Testaments include radically antihierarchical images of human relationships with the deity and each other (Ruether, 1989). Religious groups that have become relatively hierarchical or patriarchal tend to "remember" those biblical images that legitimate their status and power arrangements. For example, Christian worship services and hymns traditionally referred to God by the images of "Lord," "Father," and "King," terms clearly full of masculine and hierarchical—even feudal— imagery. Because the language of worship is linked with social power, potential changes are politically volatile, so much so that the scholars writing the *Inclusive Language Lectionary* (readings for Sunday services) received anonymous death threats (Niebuhr, 1992). Far from being a minor matter of political correctness, language changes aim at reshaping as well as reflecting emerging changes in believers' images of God. Some changes, for example, encourage participants to envision not only powerful males but also women, people of color, infirm or handicapped persons, and powerless people as being "in God's image." Such a reenvisioning has profound theological implications.

Creation stories in male-dominated cultures often assign to women the responsibility for the presence of evil or the troubles of the present world. In these myths, women's presumed characteristics of sexual allure, curiosity, gullibility, and insatiable desires are often blamed for both the problems of humankind and for women's inferior role. The Hebrew myth of Lilith describes that the Lord formed both the first man and the first woman from the ground. The first woman, Lilith, was equal to Adam in all ways, and she refused what he wanted her to do (including his sexual demands). In response to Adam's complaints, the Lord then created Eve from Adam's rib, thus making her inferior and dependent. In one version of the story, Lilith persuaded Eve to eat from the Tree of Knowledge; thus, both the "good" woman through her gullibility and the "bad" woman through her wiliness and willfulness brought about the expulsion of humankind from the Garden of Eden (see Daly, 1978:86; Goldenberg, 1974).

In general Christianity, Judaism, and (to a lesser extent) Islam and official Buddhism allowed that women were, in principle, equal to men before God, even though they were socially unequal. Thus, unlike other religions of their time, these religions held that women might aspire to salvation or spiritual perfection as much as men. This idea was not only revolutionary but also quite tenuous, however; eight centuries after the founding of Christianity, church authorities continued to express doubts about whether the souls of women were equal to those of men (Weber, 1963:105; see also McLaughlin, 1974).

Moral Norms. Religion typically creates or legitimates the **moral norms** that define what is appropriate to male and female gender roles. Thus, each culture has different ideas about appropriate clothing for males and females, and religious norms of modesty often give these concepts a moral connotation. Earlier Christian missionaries in Polynesia and Africa introduced different norms for clothing (especially for women) to the native peoples not merely because they thought Western customs of dress superior but especially because, according to their moral norms, the natives' state of dress or undress was sinful.

Historically, religious moral norms have been very important in defining the appropriate gender behavior of men and women. Many moral norms of religious groups are gender-specific (i.e., separate expectations for women and men). Religion shaped norms pertaining to sexual behavior (e.g., regarding premarital and extramarital sexual activity). It influenced norms of dress, physical activities, entertainments, and drink. Religion also legitimated gender distinctions in work roles, home responsibilities, childcare responsibilities, education, marriage responsibilities, political responsibilities, and legal status.

Organization. Religious organization is also a framework that supports gender-caste systems. Access to positions of authority and to central ritual and symbolic roles (e.g., the priesthood) is limited or denied to women in most historic religions. Control of ritual and symbolic roles is an important source of social as well as "spiritual" power. Women's lack of power is often symbolized and communicated through their inferior ritual status (see Wallace, 1975, 1988).

Ritual Expression. Ritual and symbolic roles (e.g., the role of celebrant of the Mass or reader of the Torah) have become the focus of some women's dissatisfaction with their religion's treatment of women because these roles are important symbols of power and status. Most official religions have allotted women a ritual position in the home. Women light Sabbath candles, lead family rosary, arrange family altars, participate in family Bible readings and mealtime grace, and lead young children in their prayers. Women are typically responsible for arranging the observance of religious holidays in the home, including keeping a kosher kitchen at Passover, preparing feast and fast meals, and bringing home blessed palms and holy water for home use. These ritual activities are seen as consistent with her gender roles as mother and homemaker.

In public rituals, women's traditional roles have been clearly subordinate. Men have traditionally held all important leadership and symbolic roles; women were allowed to be present (under certain circumstances) but silent. Women and children were historically segregated in the congregation, sitting in separate pews or separate parts of the room from men. Sometimes they were further hidden by screens or veils. One explanation of these arrangements is that they were intended to prevent women's presence from distracting men. Another explanation is that because women are not regarded as real *persons* in the central action, they are merely allowed to observe as unobtrusively as possible. Whatever the historical background, the segregation and veiling of women clearly communicated their inferior status (Borker, 1978). In many religious groups, women were not allowed in certain sacred places (e.g., on the altar platform or behind the screen where central ritual actions occurred). Most religious groups traditionally held that women required special purification after menstruation or childbirth before they could return to participate in group worship or key rituals (e.g., receiving communion).

Religious groups are inherently conservative because they base their beliefs and practices upon the tradition or scripture produced in an earlier era.

Thus, Christian groups often justify their exclusion of women from central ritual roles by reference to Saint Paul's views of the proper place of women:

> I desire then that in every place the men should pray, lifting holy hands without anger or quarreling; also that women should adorn themselves modestly and sensibly in seemly apparel, not with braided hair or gold or pearls or costly attire but by good deeds, as befits women who profess religion. Let a woman learn in silence with all submissiveness. I permit no woman to teach or to have authority over men; she is to keep silent. For Adam was formed first, then Eve; and Adam was not deceived, but the woman was deceived and became a transgressor. Yet woman will be saved through bearing children, if she continues in faith and love and holiness, with modesty (1 Tim. 2:8–15, RSV).

Similar texts exist in other historic religious traditions. Although the Buddha's approach to women's roles was revolutionary relative to Hindu tradition (e.g., he allowed women to become monks and considered them capable of aspiring to spiritual perfection), he prescribed a greatly inferior role for them in the *samgha* (religious order), where they were forbidden authority and leadership roles. When his favorite disciple Ānanda asked him why women should not be given the same rank and rights as men in public life, the Buddha replied, "Women, Ānanda, are hot-tempered; women, Ānanda, are jealous; women, Ānanda, are envious; women, Ānanda, are stupid" (cited in Schweitzer, 1936:95).[8]

Changing Religious Roles for Women. Acceptance of changes in religious roles for women frequently hinges on whether a group accepts religious traditions as literal truths and divinely ordained. Groups that emphasize orthodoxy to literally interpreted religious traditions are resistant to any kind of change. Thus, Orthodox Judaism, Eastern Orthodoxy, some strains of Roman Catholicism, orthodox Islam, and fundamentalist Protestantism generally oppose changing the ritual and symbolic roles of women. Other religious groups interpret their traditions or scriptures differently. Thus, some groups believe that Saint Paul's restrictions, as previously quoted, are merely the product of his culture's gender role distinctions rather than part of the central message of the Christian tradition. Generally, those religious groups that are least open to change in women's roles are also more conservative toward other social changes as well. A group's approach to defining women's roles merely reflects its general attitude about the relationship of its tradition to the conditions of modern society. (We will further describe these diverse stances in Chapter 5.) For example, the turn-of-the-century controversy in many U.S. synagogues over gender-mixed seating (i.e., a whole family sitting together on a pew, in con-

---

8. No full biographies of Gautama, the historical Buddha, were written until several generations after his lifetime. Early Buddhist scripture suggests that his message was radically egalitarian, but later legends and tales, adopted into the canons of Buddhist schools, portray misogynist attitudes (Barnes, 1987).

trast to traditional segregation of women and men) was an important symbolic struggle over accommodation to wider U.S. cultural patterns (Sarna, 1986).

In many religious groups, women are pressing for changes in rituals and symbols in order to reflect their experiences as women and to change the images of women held by the whole group. For example, some Jewish women substitute new, women-affirming blessings for the traditional ones. Others have developed an annual Seder ritual that expresses women's voices in the story of the Exodus (Brozan, 1990). Similarly, in 1990 the Mormon church adapted some of its central ritual, dropping the oath of obedience to their husbands that women previously had taken in place of the men's oath to obey God and the church (Steinfels, 1990).

Thus, the religious heritage of a group does not determine its arrangements. Many contemporary religious groups have integrated women more into the central ritual actions; most Protestant and Reformed Jewish congregations allow laywomen the same privileges as laymen. Roman Catholic and Conservative Jewish congregations are more divided in their treatment of laywomen. There is some resistance in Roman Catholic churches to allowing laywomen to serve as lectors, altar servers, and Eucharistic ministers (who distribute communion, especially to homebound and hospitalized parishioners). Similar controversy exists in Conservative Jewish congregations over whether to admit women to central lay ritual roles, such as being called to the Torah (Umansky, 1985).

The issue of the ordination of women is the most controversial because of its great symbolic importance and because the role of the clergy is more powerful than lay roles. The significance of the ordination of women is that it presents an alternative image of women and an alternative definition of gender roles. Thus, the ordination of women has an impact on laypeople as well as clergy.

Women began to press for ordination in the mainline Protestant denominations beginning in the 1880s (about the same time that the Women's Movement began, demanding women's right to vote, own property, etc.). Although some women were ordained in a few denominations in the early part of the twentieth century, only after the 1950s were a substantial number of women granted ordination. The Methodist and Presbyterian churches allowed women clergy beginning in the 1950s, and in the 1970s the Lutheran and Episcopalian churches voted (after considerable controversy and some defections) to ordain women (Hargrove et al., 1985; see also James, 1980; Ruether and Keller, 1981; Ruether and McLaughlin, 1979). Despite increasing numbers of women in seminaries, women are still greatly underrepresented in the profession; the proportion of women in the ministry (4.2 percent in 1980) was less than half that of women in medicine or law (cited in Carroll et al., 1983). The proportion of women clergy in England is even smaller (see Nason-Clark, 1987a).

At the same time, studies in the United States and England show lay church members in mainline Protestant and Catholic churches to be generally positive about women in the ministry (Hoge, 1987; Lehman, 1985, 1987; Nason-Clark, 1987a). Laypersons' opposition to the ordination of women in the

Roman Catholic church has dropped dramatically since Vatican II; indeed, 63 percent of U.S. Catholics surveyed agreed that ordination of women "would be a good thing" (Gallup Poll, 1993). The Catholic church retains its ban on women clergy, however. In 1976 the Vatican issued a "Declaration on the Question of the Admission of Women to the Ministerial Priesthood," basing continued refusal to ordain women on theological grounds that the sacramental expression of Christ's role in the Eucharist required the "natural resemblance" of Christ to the (male) minister (Sacred Congregation, 1977). Catholic theologians in the United States and elsewhere roundly ridiculed the theological basis of the document (Iadarola, 1985). In 1992, after nine years of work and four controversial drafts, the U.S. bishops failed to agree on a statement on women's roles in the church and remained deeply divided (Steinfels, 1992). In an effort to close further debate on the issue, the Vatican's Congregation for the Doctrine of the Faith announced in 1995 that the church should consider Pope John Paul II's pronouncement forbidding women's ordination to be "infallible" (Steinfels, 1995).

Nevertheless, the experience of women who have been ordained so far (e.g., in the Presbyterian and United Methodist churches and in Reform Judaism) is far from equality. Women are underemployed, paid lower salaries, and are less likely to be considered for the better positions. A church may employ a woman as "assistant minister" with special responsibility for youth programs. Although she may occasionally lead worship services, she is not seen as having an important leadership or decision-making role; being in charge of youth activities is consistent with traditional roles of women (Carroll et al., 1983; Umansky, 1985; Verdesi, 1976).

Women who have had the opportunity to serve in pastoral capacities have often redefined those roles; their style is generally more relational, less authoritarian, and more democratic, involving the laity in leadership (Nason-Clark, 1987a; see also Adriance, 1991, on women's religious and pastoral roles in Brazil). Although the Catholic church does not allow women to be ordained, in many regions nuns and laywomen have taken over many functions of the clergy (due to insufficient numbers of ordained clergy). Similar redefinition of leadership roles appears among these women pastors, who develop collaborative relationships with parishioners rather than hierarchical ones (Wallace, 1991, 1993; compare Lehman, 1993; Lummis, 1994).

Resistance to women's roles as clergy or ritual leaders is not due merely to the weight of traditional beliefs and practices. Some groups are afraid of harm to the local congregation or denomination that might be caused by internal controversy over the issue (Lehman, 1985). Other organizational factors figure into the resistance to hiring ordained women clergy as head or sole minister. For example, many churches traditionally assumed that by hiring a married minister they would also receive the (unpaid) labor of his wife. If they hired a married woman minister, in all likelihood her husband would have a career of his own to pursue. In some denominations, fellow clergy appear to be more of an organizational barrier to women's ordination and hiring than are laypersons. For example, in the Church of England (Episcopalian), the opposition has come from the House of Clergy, rather than the House of Bishops or House of Laity. One study of Church of England clergy found that these attitudes were more strongly correlated with people's notions about women's proper roles in family and society than with any theological or scriptural foundation (Nason-Clark, 1987b). In those denominations that have ordained women, however, women clergy have been increasingly accepted by their congregations and colleagues (Carroll et al., 1983).

Power and Sexuality

Evidence from relatively simple societies suggests that religious distinctions between males and females accompany social distinctions between them, especially in the division of labor. An interplay of power is involved in establishing social inequalities. Religion and human sexuality are both important sources of power, partly because of the mystery, awe, and intensity of the experiences they engender. Those who possess religious power in a social group frequently attempt to control the use of sexual power because they view it as a threat to their power base. Similarly, on the individual level, many religions view the use of sexual power as diminishing or distracting from spiritual or religious power. The regimen of the 3HO movement (Happy-Healthy-Holy— a Western version of Sikh/Hindu religion) severely restricts the frequency of sexual intercourse of members, not because members believe there is anything intrinsically wrong with intercourse but because sexual activity is believed to drain the individual's spiritual energy (Gardner, 1978:120–133; Tobey, 1976).

Female sexuality has historically been perceived as a source of dangerous power to be feared, purified, controlled, and occasionally destroyed by men. General ideas of carnal female sexuality have existed for centuries. Some Western cultural forms of these ideas include images of the enchantress, the evil seductress, the movie queen, and the prostitute. Other cultural images of dangerous female power include the witch, the nagging wife, and the overpowering mother (Hoch-Smith and Spring, 1978:3).

There are several plausible hypotheses for such emphasis on the powers of female sexuality. Douglas (1966) suggests that female sexuality is perceived by males as dangerous and impure because it symbolically represents females' power to counteract the dominance of men. Societies in which there is ambiguity about the legitimate sources of male dominance over females are more likely to develop beliefs in sexual pollution by females than are societies in which women's position is unambiguous; that is, if formal and informal social power is unambiguous (whether or not a society has established a subordinate position for women), men are less likely to fear sexual pollution. Fear of female sexuality is particularly characteristic of societies in which male dominance over females is a primary feature of social organization. The Pygmies, for example, have a division of labor and power structure in which gender is relatively unimportant and have little concern over sexual pollution (Friedl, 1975; Sacks, 1974).

Douglas (1966) cites the contrasting example of the Lele tribe of Zaire. The men of this tribe use women as something of a currency of power by which they claim and settle disputes. They fight over women and use their dominance over women as a symbol of their status relative to other men. Women of the tribe challenge total male dominance by playing off one man against another and by manipulating men. The men of the Lele have considerable fear of pollution concerning sex and menstrual blood. Women are considered dangerous to male activities, and contact or intercourse with a menstruating woman is believed to be contaminating. Fear of the power of female sexuality is related to the disruptive role that women sometimes play, upsetting the order of a male-dominated system of rewards and privileges in the society. Female sexuality is therefore symbolized as chaotic, disordering, and evil.

Another feared power related to female sexuality is regeneration—the power of life and death. Many cultures observe with awe and some fear the power of female sexuality to create and sustain new life and, by implication, the dangers of female renunciation of this role. For example, the celibate Hindu goddesses are those with feared malevolent powers, whereas goddesses with male consorts are not so dangerous because their power *(sákti)* can be regenerative and benevolent (Marglin, 1985). The association of female sexuality with fertility, regeneration, and the power of birth makes it both dangerous and potentially fruitful. Thus, religious rituals and social sanctions try to control this source of power.

Sexuality and Religious Images of Women

The theme of sexual pollution is prevalent in more developed historic religions as well. The dominant religious images of women in Western societies are built on a dualistic image of female sexuality: The evil side is the temptress/seducer/polluter, and the approved side is the virgin/chaste bride/mother. Religiously approved roles for women are rather narrowly limited to expressions of these images and are defined and legitimated by reference to Scripture, tradition, and role models (e.g., biblical women or saints). Approved roles emulate the motherhood of a biblical Sarah or Mary, the wifely faithfulness of a Rebekah or Saint Elizabeth of Hungary, or the virginal chastity of Mary or Saint Agatha (sources of female images in the Judeo-Christian heritage are explored in Bynum, 1986; Gold, 1985; Hyman, 1973; Ruether, 1974). In nineteenth-century Catholic piety, Mary was elevated to considerable importance—for example, as a source of miracles and prophetic apparitions. Her power, however, derived from attributes of motherhood and virginity, not fertility or sexuality (which figured into some earlier Mary cults). The promulgation of the controversial dogma of the Immaculate Conception (1854) promoted the new piety for a religious image of Mary as a pure and passive vessel (Pope, 1985).

The narrowly sexual base of these dominant symbols of femaleness contrasts dramatically with representations of men in ritual and imagery. Whether a man is a virgin matters little in his religious status or participation in ritual.

Women are represented as one-dimensional characters in ritual and imagery; their images are based almost exclusively on sexual function (Hoch-Smith and Spring, 1978:2–7). The Roman Catholic church honors a number of both female and male saints by dedicating the Masses of certain days to their memory and commemoration of their particular virtues. Men are remembered as martyrs, popes, bishops, doctors of the church (i.e., recognized scholars and teachers), kings, abbots, and confessors of the faith (i.e., typically missionaries). Women are remembered as virgins, virgin martyrs, martyrs, holy women, and queens. Saint Agatha is one woman so commemorated because she was tortured and killed for resisting the sexual advances of the Roman governor. The prayer said on a virgin martyr's feast day reads: "O God, one of the marvelous examples of your power was granting the victory of martyrdom even to delicate womanhood" (Maryknoll Fathers, 1957:807; recent changes have altered the significance of these commemorations but not the relative treatment of men and women). Even women renowned for their organizational skills, intellectual ability, literary accomplishments, and spirituality (e.g., Saint Theresa of Avila, a famous mystic who also brought about the reform of her order and administered its convents) are venerated as "virgins."

Women's Religious Identity

Sense of Self-Worth. Such imagery serves to reinforce the distinctions of the gender-caste system. On the one hand, it places some value on "good" women's roles. Such symbolism can enhance the individual's sense of self-worth by showing her existence as part or imitative of some larger reality. She may think, "Although my position is lowly, it is part of something bigger"—a larger cosmic picture. Religion often explicitly links spiritual rewards with the fulfilling of caste obligations. A Hindu woman learns that she may be rewarded in the next life with a higher status (perhaps as a man) if she obediently fulfills the role requirements of a woman in this life.

On the other hand, although the religious legitimation of women's roles gives meaning to their subordinate position, it does not necessarily make them happy or give them a positive identity. In the Sinhalese religion in Sri Lanka, for example, both the Buddhist doctrine and the folk religious imagery describe women as vehicles for impermanence and sorrow. The law of karma holds that all actions and thoughts have results in the future, including future rebirths of the individual. The desire for life and the illusions of the world perpetuate rebirths and thus suffering. Suffering refers not only to pain but also to the impermanence, emptiness, and insubstantiality of this existence. Buddha taught that escape from this suffering is possible only by detachment from desire, especially for life in this world.

The Buddhist imagery of women portrays them as particularly caught in this web of rebirth. Because women represent birth, they are themselves a symbol of the force of karma that maintains suffering in the world. They are also seen as the source of men's karma because they are among the things that men most desire. Yet women's desirability (i.e., sex appeal) is as impermanent and illusory as the rest of this world. Thus, women are especially entangled in the web of karma and have greater difficulty detaching themselves (Amara-Singham, 1978). In Buddhism, as in Western religions, women are used to symbolize to men and women alike the undesirable qualities of human existence. Women represent suffering and evil.

We know very little about the actual impact of these perspectives on women's identities. Such imagery and symbolism surely influence how a woman feels about herself. Does she, by participating in such a religion and culture, come to perceive herself as inferior and impure, more prone to sin, and a cause of males' sin and suffering? If participation in such rituals and symbols does not lower a woman's self-image, it must be because she is able to separate herself from them or experience this part of religion in some unofficial way. Do women selectively relate their identities with different aspects of rituals or with different rituals altogether? Do they use gender-linked symbols differently from men?

One study of the writings of religious people during the later Middle Ages—a period rife with misogynistic theological, philosophical, and scientific ideas—found that female and male spiritual leaders used gender images very differently. Male spiritual writers were more likely to view the genders as dichotomous, whereas women writers were more likely to use androgenous images, if they addressed gender at all. Men's spiritual ideals focused on conversion and renunciation of the world of wealth and power (for which they invoked gender inversions to symbolize giving up those statuses), whereas women's religiosity emphasized continuity and a relationship to Christ's humanity (and thus physicality). Gender distinctions created genuine restrictions on how women could express their opinions and experiences or achieve heights of spiritual "achievement." For example, both men and women sought renunciation of "the world," but women did not control enough power or wealth to give them up. Thus, women's renunciation was often limited to giving up food or other aspects of daily life over which they did have control. Nevertheless, the religiosity of these outstanding female medieval theologians and mystics was not merely the passive product of the misogynistic ideas of their time. Rather, they appropriated cultural symbols differently, creating a positive spiritual identity for themselves and their followers (Bynum, 1986).

At the same time, religious traditions have provided women with some positive roles and images (albeit severely limited ones) that bring reward and recognition in their fulfillment. For a woman to aspire to achievements in men's arenas would be deviant and punished by the social group. She may, however, aspire to achievements in women's spheres with no competition from men and be rewarded. These values, too, are internalized. The woman who is successful in fulfilling the mother role comes to perceive herself positively in those terms. She feels good about herself because she has achieved a desired status or has performed valued roles.

In Victorian society, the cultural and religious legitimations of motherhood and women's role in the home achieved a new synthesis. The Victorian ideal of womanhood, as reflected in women's magazines and public statements in the nineteenth century, was one of "piety, purity, submissiveness, and domesticity" (Welter, 1966).[9] The emerging capitalist–industrialist society of the nineteenth century made both the home and religion increasingly marginal to the arenas of power and production. Ideological responses of many religious organizations and revivalists to this privatization struck a responsive chord among middle-class women who were enmeshed in the privatization of the home. An "evangelical domesticity" resulted that emphasized the home and women as the primary vehicles of redemption. The rhetoric of preaching and the themes of gospel hymns of this period portrayed the home as the bastion of tranquillity in a turbulent world. Woman's image was that of the keeper of this refuge who, through her piety and purity, was the foremost vehicle of redemption, in opposition to the aggressiveness and competition of the public sphere (Sizer, 1979; see also McDonnell, 1986; Ryan, 1983; Taves, 1986; Welter, 1976). Thus, the religious ideology of evangelical domesticity celebrated the privatization of the home and women's roles, further reinforcing the caste distinctions between men and women. At the same time, it enhanced this narrow role in women's eyes by claiming redemptive significance for it.

Sanctions for Deviance. The religious definition of gender roles is supported by religious sanctions (i.e., punishments) for deviance from them. Men as well as women are subject to negative sanctions for failure to fulfill gender role expectations, though conformity to the broader, more powerful male social role is less onerous than complying with the restrictive, less powerful female role. Religion, as we shall see in Chapter 7, is a powerful force for social control. One of the most effective sources of such social control is the individual's own conscience. The person who has been socialized into a certain way of thinking about gender roles feels guilty if he or she fails to fulfill these expectations. One young woman who had learned that "women naturally and instinctively feel great affection for their children" felt guilty that her initial reaction to her newborn child was not a great outpouring of maternal love.

Those who control religious power (especially in official religion) frequently use specifically religious sanctions to enforce gender roles. Recent history provides the examples of a Mormon woman excommunicated for her active support of the Equal Rights Amendment to the Constitution; and Roman Catholic priests and nuns silenced by Vatican officials for challenges to the church's treatment of women (among other issues). In Iran, the political power of fundamentalist Islam has resulted in strong restrictions of women's public roles and appearances, as symbolized by the mandate of returning to the *chador* (i.e., veil).

Historically, the intertwined power of religious organizations with other social institutions gave religion more influence in the control of deviance than in modern society. The persecution of witches in the Middle Ages (and until as recently as the end of the seventeenth century) was focused primarily on

---

9. The instance of Victorian gender roles illustrates the fact that men's roles too are often limiting and burdensome. In this case, men were given the full burden of providing for the family by work away from the "comforts of home." The moral expectations of the husband/father as sole provider for a family are relatively recent in Western history.

women, who were presumed weaker and therefore more vulnerable to the influences of the devil. Authorities typically singled out women who deviated from religiously established norms for females—wise women, healers, midwives. Suspected women (e.g., widows and "spinsters") were disproportionately likely to be relatively independent of men. Witch-hunts such as the massive medieval persecutions were primarily concerned with purification. Women in particular were considered threats to social purity because of cultural assumptions about their sexuality and because anomalous (i.e., independent) women were perceived as dangerous and disruptive. A substantial number of women were put to death or tortured as witches; scholarly estimates suggest that several hundred thousand witches were killed during the Inquisition.[10] There has been no parallel persecution of men *as men,* although men were persecuted as members of other minorities (e.g., Jews).

Alternative Definitions of Women's Status and Roles

Women's Religious Groups. One of the most effective social controls of women has been the limitation and prevention of separate, autonomous women's religious groups. Religion has historically had the potential for promoting social change among colonial peoples and subordinated racial and ethnic groups because, in organizing around a religious focus, such groups have often mobilized to press their social and economic concerns. Unlike racial and ethnic minorities, however, women have not generally formed separate religious groups in Western societies. Male-dominated religious organizations have been understandably threatened by the possible power of autonomous women's groups.

Medieval sisterhoods (including religious orders, lay orders, and numerous unofficial bands of religious women) offered socially approved roles for women who could not or would not fit the primary gender roles of woman as wife and mother. Socially anomalous women (again, typically widows and unmarried women) found in these sisterhoods both security and opportunities for achievement not available to other women in society. Medieval abbesses, for example, were often relatively powerful leaders of large organizations. Church officials, however, kept these groups under relatively tight control (e.g., by regulations restricting sisters' freedom to come and go in "the world"). Like other dissenting, change-oriented groups (e.g., early Franciscans), women's groups that asserted alternative roles were often co-opted and controlled.

---

10. Figures vary greatly from 30,000 to 9 million (sources cited in Daly, 1978:183). See especially Daly's essay on European witch-hunts, as well as Szasz (1970); and Anderson and Gordon (1978:2). Daly notes the widespread religious legitimation of violence against women. The Indian *suttee*—cremation of the living wife along with the body of the dead husband—and African clitoral circumcisions are among the ritual "gynocides" examined in her provocative book. She notes that male-dominated scholarship has not shown as much concern over such historical events as the European witchcraze and institution of the *suttee* as over comparable genocides such as the Nazi Holocaust, observing that scholars have glossed over gynocide such as the *suttee* by calling it a "custom." Daly doubts that they would ever refer to comparable atrocities in the Holocaust as "Nazi customs."

Feminism had a specific role in the development of sisterhoods in the Anglican church during the nineteenth century (M. Hill, 1973a). The Victorian image of women, as we have seen, emphasized domesticity. Furthermore, the middle- and upper-class Victorian "lady" was educated to be an ornament in the home; her enforced leisure was evidence of her husband's social status. Women such as Florence Nightingale who asserted any other role for themselves were severely criticized. The Anglican church had dissolved monasteries and convents in its initial break with the Roman church in the sixteenth century, and in the nineteenth century strong voices were raised against reestablishing brotherhoods or sisterhoods in the church. Nevertheless nascent feminism, combined with a considerable demographic surplus of women, exerted pressures to create acceptable social roles for women who would work in "unladylike" jobs such as nursing, running settlement houses, teaching the poor, and running hospices for derelicts. The creation of acceptable deviant roles for women was perceived as an attack on the Victorian home and women's proper duties. One bishop stated:

> The rules which I have myself laid down as most necessary in my dealing with such communities [the sisterhood] have been the following:—To point out that the first of all duties are those which we owe to our family. Family ties are imposed direct by God. If family duties are overlooked, God's blessing can never be expected on any efforts which we make for His Church. Every community, therefore, of Sisters or Deaconesses ought to consist of persons who have fully satisfied all family obligations (cited in M. Hill, 1973a:277).

Even among those who approved of the idea of women's religious groups, there was disagreement about the degree of autonomy to allow them. Religious sisterhoods were much more problematic than brotherhoods to church authorities because sisterhoods allowed the possibility of women directing their own work and running their own organizations. The less radical alternative of the office of deaconess (in which women workers would be under the direct supervision of male clergy) was proposed because "women need more support than men" (cited in M. Hill, 1973a:276). Women were gradually allowed into "unladylike" areas of work, partly through the transitional institution of the sisterhood. The potential for feminist protest was, however, somewhat co-opted, especially as women's groups were brought under greater control by authorities and their work was gradually redefined as "women's work." Nursing, for example, was consistent with the serving and nurturant image of women as dependent on (male) doctors for decision making. As women's professions became more respectable, women's religious groups became less important as vehicles for feminist aspirations.

Specifically, women's religious groups still remain potentially expressive of women's social and religious interests. Within Roman Catholicism, much of the pressure for the ordination of women comes from the ranks of women's religious orders. Some women's groups have creatively used their greater education and structural independence. While dissenting from the patriarchal stratification of their churches, some women are "defecting in place" (Winter et al., 1994). They are forming their own small group supports for women's spirituality without leaving their churches. They are exploring alternative rituals and religious symbolism that embody their gender and their sexuality (Neitz, 1995).

The main arena for the expression of feminist concerns has become public-sphere organizations. The privatization of religion means that religious achievements are similar to domestic accomplishments: Neither really "counts" as a symbol of achievement in the larger society. Conflicts within religious collectivities over appropriate roles for women primarily result from religion's tendency to continue legitimating traditional roles for women in *both* public and private spheres.

New Religious Movements. New religious movements are often a vehicle for the assertion of alternative religious roles for women. Religious movements of nonprivileged classes have typically allotted equality to women, at least in their formative years (Weber, 1963:104–105). Many of the medieval millenarian movements initially allowed women virtual equality. The Protestant Reformation's principle of the "priesthood of all believers" was initially interpreted as opening important roles to both laywomen and laymen. Although their contemporaries were scandalized by the fact, numerous seventeenth- and eighteenth-century religious movements allowed women equal roles in public preaching and prophesying (Garrett, 1987; Mack, 1992; Stuard, 1989).

Many religious movements of the eighteenth century, and later the Great Awakening, granted equality to women. The Shakers believed that their founder, Mother Ann Lee, was the Second Coming of Christ, thus completing the manifestations of Christ (male and female). The Society of Friends (Quakers) was emphatically egalitarian, encouraging women to equal education, leadership, and public-speaking roles. Quaker women such as Susan B. Anthony, Alice Paul, and Lucretia Mott were leaders in the women's suffrage movement; they were also prominent in movements for abolition of slavery, prison reform, fair treatment of Native Americans, and other human rights causes (Bacon, 1986). The late Victorian period was one of nascent feminism, and alternative women's roles were an important part of the beliefs and practices of Pentecostal and Holiness religion, Salvation Army, Seventh Day Adventists, Spiritualism, Christian Science, Theosophy, and New Thought (Baer, 1993; Bednarowski, 1983, 1992; Braude, 1985, 1989; Gilkes, 1985; Robertson, 1970:188, 189; Wessinger, 1993).

New religious movements are more amenable to alternative gender roles because they are based on an alternative source of authority. Traditional ways of doing things are protected by the established religious collectivity, but new religious movements are often based on charismatic authority (see Chapter 7), which is not bound by tradition. The charismatic leader says, in effect, "You have heard it said that . . . , but I say to you. . . ." This new source of authority allows a break from tradition. Weber (1963:104) points out that as the emphasis on charisma fades and the movement becomes established, these movements tend to react against keeping women in roles of authority, typically redefining women's claims to charismatic or other authority as inappropriate or dishonorable behavior. This process is illustrated by changes in the Assemblies of God. In its early years, the Assemblies of God was relatively open to women assuming preaching and pastoral roles because of its emphasis on the "gifts of the Holy Ghost" as a basis for authority and leadership. As the movement became more formalized and bureaucratic and as it came more under the influence of other conservative religious groups' Biblical literalism and attitudes toward women, the Assemblies of God greatly restricted women's leadership and decision-making roles. The proportion of women pastors declined from a high in 1915 of 13.5 percent to 1.3 percent in 1983 (Poloma, 1989). Similarly, while always subordinate to male authority, women in the early years of the Mormon church (Latter-Day Saints) had numerous leadership opportunities, but as the movement became established there was greater emphasis on the (male) priesthood line and priesthood authority, thereby greatly increasing the value of men (Cornwall, 1994).

The role of women in developing religious movements is thus a result of alternative sources of authority and power. Many of these movements are understandably appealing to women who are excluded from status and privilege in the dominant religions. Weber (1963:104) accounts for some of the appeal of Christianity and Buddhism by noting that relative to the religious competitors of their times and regions, their subordination of women was less extreme. Similarly, the religious movements of the late Victorian era appealed to women because of their alternate definitions of women's roles.

Several emerging religious movements also focus on gender roles but generally reassert traditional rather than new ones. Thus, the Jesus People, neo-pentecostal movements, evangelicalism, Hare Krishna (Krsna Consciousness), and the Unification Church (of Reverend Moon) define women's roles very conservatively. One study found that many women who "converted" to Orthodox Judaism wanted a traditional religion that offered unambiguous and unchanging gender role expectations and legitimated their desires for a family (Davidman, 1990b). Thus, some newer religious movements may attract women precisely because they reassert traditional, nonmodern gender roles, whereas other religious movements attract other women because they offer totally new ways of thinking about gender and spirituality.

Another emerging religious movement, the goddess movement, explicitly redefines women's spiritual, social, emotional, and physical roles. The goddess is an image, or symbol, of the divine that is contrary to the image used by patriarchal religions (such as much of the Judeo-Christian heritage). Although much of the goddess movement is self-consciously outside traditional religious groups, some adherents consider themselves Christians or Jews who are seeking new bases for spirituality in their traditions. Some theologians and groups use the conception of the goddess to balance the gender of the deity as both male and female, often combining the idea with a call to a nonhierarchical religion. Others emphasize the goddess as a basis for women's power. One theologian writes:

> The basic notion behind ritual magic and spellcasting is energy as power. Here the Goddess is a center or focus of power and energy: she is the personification of the energy that flows between beings in the natural and human worlds (Christ, 1987:127).

Some groups take the image of goddess literally, trying to enact older pagan practices dedicated to goddess worship. Others use earlier pagan goddess-imagery more loosely, eclectically weaving old symbols and practices together with new ones to create a new, nonpatriarchal women's spirituality (Neitz, 1990; see also Jacobs, 1989b, 1990). Interestingly, certain forms of goddess movements also promote political activism as well as spiritual development (Finley, 1991).

Can official religion create or encompass new or changing roles for women? Sociological analysis suggests the answer is a qualified yes. On the one hand, religious images and symbols are elastic; they can be changed and reinterpreted. On the other hand, traditional images and symbols of women are generally negative and resistant to change. Because much of the power of symbols and ritual is in their reference to tradition, they are essentially conservative. Although religious movements are able to express women's dissent and create new social roles, religious movements have historically returned to traditional, hierarchical, or bureaucratic forms of authority as they become settled—and in so doing have reverted to less innovative and more submissive roles for women. Women's religious groups have the potential of focusing and asserting their social concerns; yet they have been historically controlled and co-opted by (male) authorities of the dominant religion.

Nonofficial Religion. While the dominant religious groups have allocated women few if any important ritual roles, women have often found important symbolic roles outside the approved religious structure. Women have been prominent as healers, mediums, and midwives. These roles often paralleled approved religious or medical roles and functions. Just as men have been prominent in official religion as it became a differentiated institution, women have been important in "common" or nonofficial religion. Nonofficial religion has provided women with a *chance to express their own specific concerns for meaning and belonging*. Midwifery, for example, has traditionally been a highly spiritual role (see Dougherty, 1978).

In nonofficial religion, women have opportunities for leadership and power, as exemplified by the status of women mediums, faith healers, and astrologers (Bednarowski, 1983; Braude, 1985, 1989; Brown, 1991; Haywood, 1983). Nonofficial religion has likewise offered black and Latina women opportunities for ritual leadership and power. For example, Latina women served in nonofficial religious roles as *partera* (midwife), *rezadora* (prayer leader), and *curandera* (healer) in addition to their religious roles in the home (Díaz-Stevens, 1994). Some nonofficial religion is organized as specifically feminist dissent (e.g., some contemporary witchcraft covens). Much of the nonofficial religion discussed in this chapter may be understood as a way in which individuals assert alternative definitions of reality. Women throughout the ages have asserted alternative roles for themselves through religious expression.

Conclusion

Religious groups' treatment of women's roles and women's sexuality is essentially an issue of *power*. If women obtained greater power than they have traditionally had, they could redefine themselves religiously and socially. On the other hand, religious legitimations and organizations presently work against their obtainment of that power.

This Extended Application, while focusing on women's roles, illustrates more broadly how religion comes to embody and further legitimate caste stratification and power. Our analysis suggests that nonofficial religion sometimes represents one form of counterassertion (albeit often incoherent and fragmented) of power and self worth by those excluded from power in official religion. If this is true, we need to pay more serious attention to nonofficial religion and its complex relationship with official religion.

SUMMARY

This chapter has contrasted official religion and its characteristic forms of individual religiosity with nonofficial religion and religiosity. Official religion is a set of beliefs and practices that are prescribed, regulated, and socialized by organized, specifically religious groups. Nonofficial religion, by contrast, is a set of religious and quasi-religious beliefs and practices that are not accepted, recognized, or controlled by official religious groups. Official religion is relatively coherent and organized, but nonofficial religion includes an assortment of unorganized, inconsistent, and heterogeneous beliefs and practices.

The official model of religion is clearly identifiable. Its elements—doctrine, cultic expression, ethics, and organization—are mediated by religious institutions. These institutions consolidate and control the religion's definitions of what it stands for. The relative specificity of these official models of religion enables researchers to develop measurable criteria of religiosity. Such measures are, however, bound to the orthodox official models of religion, making them of limited use in understanding religious change and nonofficial religion. Nonofficial religion is more widespread than generally recognized, and it overlaps official religion in both content and adherents.

The Extended Application demonstrates how religion influences women's gender roles and identities. Religious organizational structure, myths, and symbols embody and legitimate subordinate caste status for females—a point of conflict in many religious groups today. As the Extended Application suggests, some aspects of nonofficial religion involve ways by which persons who are unrecognized or even disempowered by official religion have asserted themselves. Thus, nonofficial religion and religiosity are both serious and relevant to our understanding of religion as a whole.

RECOMMENDED READINGS

Articles

Meredith B. McGuire. "Gendered Spirituality and Quasi-religious Ritual," *Religion and the Social Order* 4 (1994): 273–287.

Mary Jo Neitz. "In Goddess We Trust." In *In Gods We Trust: New Patterns of Religious Pluralism in America,* T. Robbins and D. Anthony, eds. New Brunswick, N.J.: Transaction, 1990, pp. 353–372.

R. Steven Warner. "The Place of the Congregation in the Contemporary American Religious Configuration." In *American Congregations,* J. P. Wind and J. W. Lewis, eds. Vol. 2. Chicago: University of Chicago Press, 1994, pp. 54–99.

Books—Nonfiction

Hans A. Baer. *The Black Spiritual Movement: A Religious Response to Racism.* Knoxville: University of Tennessee Press, 1984. An ethnography of a widespread syncretic religion in the context of U.S. urbanization and racism.

Dean R. Hoge, Benton Johnson, and Donald A. Luidens. *Vanishing Boundaries: The Religion of Mainline Protestant Baby Boomers.* Louisville, Ky.: Westminster/John Knox Press, 1994.

Wade Clark Roof (with the assistance of Bruce Greer, Mary Johnson, Andrea Leibson, Karen Loeb, and Elizabeth Souza), *A Generation of Seekers: The Spiritual Journeys of the Baby Boom Generation.* San Francisco: HarperSanFrancisco, 1993. Liberally illustrated with long quotes and vignettes of respondents, this volume describes some patterns of religiosity of the "Baby Boom" generation.

Wade Clark Roof and William McKinney. *American Mainline Religion: Its Changing Shape and Future.* New Brunswick, N.J.: Rutgers University Press, 1987. Using survey and historical data, this book describes the socioeconomic location, demography, contemporary moral issues, and social changes of mainline Protestant denominations in the United States.

Books—Fiction

Margaret Atwood. *Handmaid's Tale.* New York: Fawcett Crest, 1985. A dystopian vision of a society governed by a fundamentalist theocracy that has drastically reshaped gender and family roles.

5

The Dynamics of Religious Collectivities

R eligious groups vary dramatically in organizational form and the kind of commitment they expect of their members. Some have loose bound-aries and relaxed authority structures; others have clear-cut boundaries and rigid authority structures. Some groups are large and relatively anony-mous; others are small and face-to-face. While some religious groups are in conflict with the rest of society, others are socially comfortable and thor-oughly integrated into the larger society. Some groups begin as dissenters from some feature of the established religion(s) of their society but later change structurally and ideologically to become more like the very groups they had criticized.

All religious groups change over time: Some become more stable, while others become more volatile; some grow or stabilize or decline; and some be-come more strident in their dissent, while others become more quiescent. Many collectivities expect their members to organize their day-to-day life around their religious commitment, whereas many others expect only partial and segmental commitment. The task of classifying organizational patterns is not an end in itself, but understanding the varieties of religious collectivities as they change in time helps us analyze the internal dynamics of a religious group and its relationship with the larger society.

RELIGIOUS GROUPS IN RELATION TO THEIR SOCIAL ENVIRONMENT

One of the earliest systems of classification arose out of the realization that re-ligious groups orient themselves differently to their social environments. Max Weber distinguished between two basic stances. Whereas the **church** organi-zation accepts the masses, according to Weber, the **sect** is an association that accepts only religious qualified persons. By this distinction, the church-type group embraces the society in which it lives, while the sect-type group sets it-self apart from the larger society; it exists in tension (indeed, often conflict) with its social environment. The religious community of Jesus and his follow-ers, for example, was qualitatively different from the religious community within the established Jewish religion of his time. The former was sectarian; the latter churchly in stance. Weber observed that religious sects were poten-tially dynamic sources of social change, but their effectiveness depended upon the organization of a community of followers (Weber, 1963:60–65; 1968:1204).

Weber's distinction between churchly and sectarian orientations was ampli-fied by his contemporary Ernst Troeltsch, a theologian and church historian. Troeltsch was interested in tracing the social impact of certain ideas or "teach-ings" of Christian churches in Europe. He thought that three types of religious orientation were implicit in Christian teachings, and that a given group devel-oped along one of these three lines depending upon its relationship to its social environment (Steeman, 1975:202–203). Troeltsch's three types were church, sect, and mysticism. Each corresponds to a different spiritual and ethical approach and each results in a different kind of religious organization.

Troeltsch characterized the **church** as an orientation and organization that is essentially conservative of the social order and is accommodated to the secular world. Its membership is not exclusive but incorporates the masses. The **sect,** by contrast, is exclusive, aspiring to personal perfection and direct fellowship among members. The church is characteristically an integral part of the social order, whereas the sect stands apart from society in indifference or hostility. Thus, the church tends to be associated with the interests of dominant classes, whereas the sect is typically connected with subordinate classes. The church has an objective institutional character. One is born into it, and it mediates the divine to its members. By contrast, the sect is a voluntary community of fellowship and service; relationships with fellow members and the divine are more direct. According to Troeltsch, both types are logical results of Christian beliefs.

The church type, exemplified by the Roman Catholic church of the late Middle Ages, is of diminishing significance in contemporary societies. The medieval synthesis of church and society was undermined, partly because of the challenges of the sects. Later church-type organizations like Calvinism in Geneva or the free churches (e.g., Congregationalism) in England and the United States achieved considerable synthesis with their societies, but the synthesis was based on a general Protestant consensus rather than upon a monopoly of legitimacy (Steeman, 1975:194).

Troeltsch (1960:381) forecast that modern society would be increasingly characterized by his third type of religious association: "idealistic mysticism" or "spiritual religion" (see Garrett, 1975). This is essentially a radical religious individualism. It puts little emphasis on the relations between believers; any association among them is based on a "parallelism of spontaneous religious personalities." Such groups are indifferent to sacraments, dogmas, ethical norms, and organization. Mysticism tends to resist control by authorities and is thus a threat to organized religious forms. Nevertheless, because it tends to be uninterested in changing the church or the world, it does not have the social impact of sect-type groups. While Troeltsch developed the category of "idealistic mysticism" to describe a particular historical phenomenon—eighteenth- and nineteenth-century German Lutheran mysticism—he suggested that it also described the religion of contemporary educated classes. He viewed this trend with pessimism because he considered some organizational and historical continuity necessary to the continued existence of religion (Campbell, 1978; Troeltsch, 1960:798, 799).

Troeltsch's categorization was intended primarily for comparing and contrasting concrete historical developments rather than for creating a classification scheme. Later formulations, built uncritically on Troeltsch's, inherited the historical limitations of his focus. Thus, in Troeltsch's scheme, the church-type and sect-type are polarized extremes, reflecting the historical situation of late medieval Christendom, in which class structure was generally polarized and religious institutions were relatively undifferentiated from economic and political ones. This lack of differentiation made it possible for religious dissent to be simultaneously politico-economic dissent. Any attempt to apply this dichotomy to modern societies stumbles because the modern situation is vastly more complex (Robertson, 1970:117).

The Development of Church–Sect Theories

This difficulty of applying the church–sect dichotomy to the U.S. situation prompted Richard Niebuhr (1929:25) to add the concept **denomination,** which was characterized by accommodation to society comparable to that of the church but lacking the ability or intention to dominate society. Niebuhr's main interest was the process by which sects were transformed into other organizational forms. To him, the initial enthusiasm of the sect appeared limited to the first generation (i.e., to those who consciously chose to belong). After the initial fervor, the sect must either compromise with the "world" or be organizationally weakened. A sect that compromises moves toward a denominational stance—organized religion accommodated to society. Niebuhr's typology was less neutral than Troeltsch's. As a theologian, Niebuhr used the schema to criticize some religious groups for their accommodation to the stratification system—a symptom of secularization and divisiveness in his view (Martin's 1962 clarification of the concept of denomination eliminates much of the value-laden quality of this schema).

Further qualifications were added by Howard Becker (1932), who developed the category **cult.** The cult is characteristically a loose association of persons with a private, eclectic religiosity. The concept of *cult* is very useful sociologically because it describes the form of social organization of much popular (nonofficial) religion. Historically, many cults arose from the popular beliefs about particular local holy persons and places. For example, St. Nectarios of Greece, a bishop on the island of Aighina, was widely recognized as a holy man even before his death in 1920. After his death, his veneration gradually became nationwide and numerous miracles were attributed to him. By the end of the twentieth century, his icons could be found in houses, hospitals, cars, and buses throughout the country, and his shrine on Aighina had become the pilgrimage site for hundreds of thousands of devotees. Widespread public and private devotion to this saint *preceded* his canonization by the Orthodox Patriarchate in 1961 (Kokosalakis, 1987).

Although many cult devotions historically took place in a group (e.g., a whole village making an annual pilgrimage), they tended to be highly individualistic. Participants sought personal intentions, and the desired religious experience was not expected to bind members into any long-lasting sense of community. Structurally, cults and cultic devotions such as pilgrimages and healing rituals could be threatening to church authorities' control. Analyzing several historical pilgrimages, Turner and Turner (1978:32) observe:

> There is something inveterately populist, anarchical, even anticlerical, about pilgrimages in their very essence. They have at times been linked with popular nationalism, with peasant and anticolonial revolt, and with popular millenarianism. They tend to arise spontaneously, on the report

that some miracle or apparition has occurred at a particular place, not always a place previously consecrated. Pilgrimages are an expression of the communitas dimension of any society, the spontaneity of interrelatedness, the spirit which bloweth where it listeth. From the point of view of those who control and maintain the social structure, all manifestations of communitas, sacred or profane, are potentially subversive.

The response of the authorities was, if not to suppress the cult, then to routinize its practices; for example, by putting church officials in charge of shrines and by inserting approved rituals of penance and Eucharist as the culmination of a pilgrimage.

Historical cults also had loosely defined boundaries: Doctrine was absent or minimal and membership was loose, undefined, and voluntary. These groups tended to be very pluralistic, so one might belong to several different cults depending on one's needs at the time. For example, a person might adhere to one saint's cult for healing arthritis, use another set of devotions for the blessing of crops, make a pilgrimage to a holy well to seal a vow to stop drinking, participate in a different saint's cult for fertility, and use icons and amulets of yet another cult to protect loved ones from harm.

Because of this pluralistic tolerance among historical cults, the groups coexisted—indeed flourished—alongside the churches. Part of the historical strength of Roman Catholicism and Eastern Orthodoxy may be due to their flexibility in absorbing and, often, co-opting these cults. By absorbing local popular cults as optional devotions, these churches were able to build organizations that transcended familistic, localistic, or tribal barriers. For example, early Christian bishops encouraged the cults of the saints but restructured the practices, then prevalent among wealthy Christians, of privately controlling the remains (relics) of martyrs for sanctifying family burial plots and familistic rituals with them, such as feasts celebrating the dead. By placing the relics in church shrines and altars and transforming these saints into mediators for the whole congregation, the bishops shifted the balance of social power in the community, transcending the narrow kin-based patronage system of Christian communities in late antiquity (Brown, 1981).

Furthermore, the cults of the saints and shrines at holy sites, such as wells, caves, mountains, and stones, enabled Christianity to absorb and transform pre-Christian, pagan elements of indigenous cultures. It was only with the gradual consolidation of *official religion* (see Chapter 4), with its characteristic boundaries around approved, "pure," orthodox religious belief and practice, that cults and cultic adherence acquired a more negative status in Western societies. Sociological analysis needs to pay particular attention to periods of high levels of cult-type religious expression as evidence of certain social changes. Might there not be parallels between today's cultic activity and that of earlier times, for example, in how these religious groups make possible a particular mode of individuation, changes in gender and family roles, and a breakdown in older social barriers embedded in the orthodoxy from which cultic groups deviate?

Since the 1960s, the use of the concept *cult* has been confused by the media's pejorative misapplication of the term to all unfamiliar religious movements. In sociological terms, most of these movements are more accurately characterized as sects, but the concept of *cult* has become a negative label for religious associations considered deviant by the larger society (and by the highly sectarian anticult movement).[1] The sociological concept of *cult* is, by contrast, neutral, merely describing characteristics of organization and individual orientation of members.

Usefulness of the Typology

It is debatable whether church–sect categories are useful outside the Christian context. Some sociologists argue that the basic theme of the category (i.e., religious group versus the "world") is specific to the development of Christian churches. The theology of the Incarnation (i.e., God-as-human) illustrates this tension. Thus, Christianity is more likely to develop a wide range of religious collectivities (e.g., church, sect, cult, denomination) than other religions such as Hinduism, which does not have the same tension with the "world" (Berger, 1967:123; Martin, 1965:9, 10, 22).

By contrast, other sociologists point out that it is worth studying situations in world history when non-Christian groups adopted similar stances to their social environments. For example, although the Buddha preached nonengagement in sociopolitical activities, which would seem to discourage a churchly orientation, Theravada Buddhism (one major strain of Buddhism) became closely linked with most Southeast Asian monarchies (i.e., Siam, Burma, Laos, Ceylon, Cambodia) from about 1000 A.D. to the mid-1800s. Its churchly hegemony was interrupted not by sectarian strife but by European colonialism. The impact of colonialism was to put Buddhism—still the majority religion—into a position of marginality relative to the dominant colonial society. Buddhism then became more sectlike relative to society and thus a vehicle for nationalist dissent (Houtart, 1977; Tambiah, 1992).

We can, with care, refine our concepts so as to avoid the historical, cultural, and ideological problems that have limited the usefulness of church–sect typologies (see Johnson, 1963; Robertson, 1970). While doing so, we need to resist the tendency to use these categories rigidly. Classification for its own sake is meaningless; however, it is worthwhile to conceptualize different patterns of religious collectivities if those concepts are useful in helping us to understand how those collectivities come into being, develop, have a social impact, and change—both internally and in relationship to the larger society.

First, we can distinguish a religious *collectivity's stance* from the *individual orientations* of members. The concepts *church, sect, denomination,* and *cult* have been

---

1. Because of this pejorative distortion of the concept, I seriously considered substituting another term in this edition of the book (see Richardson, 1993a). The term *cult* is retained, however, because it captures the historical continuity with earlier Christian and non-Christian cults and highlights the social processes leading to an era such as our own, in which highly individualistic popular and other nonofficial religion is both prevalent and suspect.

used variously to describe both of these aspects. This usage leads to confusion because within the same religious group (e.g., a denomination or a congregation) may be individual members with very different orientations; some may be sectarian, others denominational or cultic/mystical in orientation. For this reason, we will separate discussion of collectivity stances and member orientations.

Second, conceptual distinctions should be based on *sociologically* important characteristics. Two criteria used by sociologists appear especially useful: the relationship between the religious group and the larger society (Johnson, 1963); and the extent to which the religious group considers itself to be uniquely legitimate (Robertson, 1970:120–128). The relationship between religious collectivity and its social environment has been a key theme since Troeltsch and Weber. Furthermore, it reflects a basic tension between values derived from religious sources of authority and other values in the society. No organization could neutralize this tension, though each type responds differently to it. The tension derives in part from the peculiar nature of religious sources of authority, but the general patterns also apply to nonreligious organizations (Figure 5.1). The dominant medical system has churchlike characteristics in contrast to sectlike organizations of alternative healing systems (Freund and McGuire, 1995). Similarly, political splinter groups often display sect-type characteristics relative to churchlike or denomination-like dominant political organizations (Jones, 1975; O'Toole, 1976)

Types of Collective Stance

Basically the church–sect typology is a general developmental model for any organization defining itself as uniquely legitimate and existing in a state of positive (i.e., churchly) or negative (i.e., sectarian) tension with society (Mayrl, 1976:28). When the collectivity does not claim unique legitimacy or the society prevents it, its stance may be characterized as denominational (i.e., positive tension) or cultic (i.e., negative tension). Because there is considerable variation among groups on both these criteria, we should think of them not as fixed and dichotomous but rather as fluid, changing, and on a continuum. This model is depicted in Figure 5.1.[2]

Churchly Stance. Churchly collectivities consider themselves uniquely legitimate and exist in a relatively positive relationship with society. They do not recognize the legitimacy of claims of any other religious group. Churchly organizations proclaim *Extra ecclesiam nulla salus* ("Outside the church there is no salvation"). The response to competing claims may be to ignore, suppress, or co-opt the competing group. The churchly stance is basically accepting of society but—to the extent that religion is differentiated from other institutions—

---

2. This schema is similar to that of Swatos (1975, 1979). His model refers to monopolistic or pluralistic *societies,* whereas this model refers to pluralistic or monopolistic *claims to legitimacy*. Part of the tension with the dominant society depends on the extent to which the collectivity attempts and succeeds in getting its claims to legitimacy recognized.

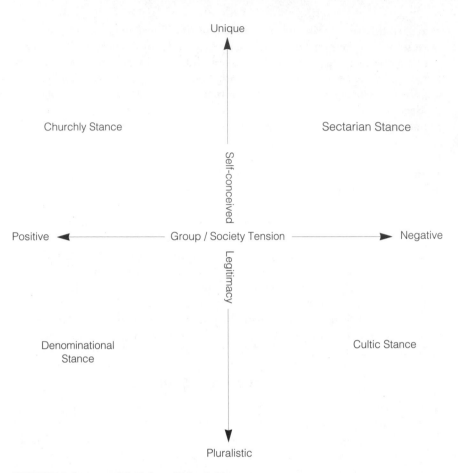

FIGURE 5.1 Stances of Religious Collectivities

it exists in tension with other institutions. Overall, however, a churchly stance tends to support the societal status quo.

At the same time, the model emphasizes the extent to which the society also accepts the churchly collectivity. Membership in this kind of group is not considered deviant or unusual. Examples of churchly stances include the Roman Catholic organization of fifteenth-century Europe and the Theravada Buddhism of feudal Southeast Asia. In both instances, the churchly organization claimed an effective monopoly of legitimacy (although Buddhism was less aggressive toward minority religions), and there was positive tension between the religious collectivity and the society. The churchly collectivity legitimated political power and the economic system, accepted the goals and values of society, and, in turn, was approved and protected by society.

Sectarian Stance. Sectarian religious collectivities consider themselves to be uniquely legitimate; they are in a relatively negative relationship with the dominant society. The sectarian stance does not accept the legitimacy claims of other religious groups; like the churchly stance, the sectarian stance proclaims itself the only way. The sectarian stance may result in any number of strategies to deal with competing religious groups: withdrawing from their influence, aggressively attacking, or trying to convert and absorb them. Because the sectarian stance places religious collectivities at odds not only with other religious groups but with the larger society, it tends to produce organizational patterns that support, confirm, and protect believers as a **cognitive minority**—a group of people whose worldview differs from that of the dominant society.

Sectarian collectivities typically emphasize primary, face-to-face relationships with fellow believers and high levels of commitment and loyalty to the group. Frequent interaction within this primary group, together with a distancing from nonbelievers' views, helps support members' worldviews in the face of real or perceived opposition from the "world." Even the most mundane actions are transformed to support the sectarian stance. For example, not only do some fundamentalist Christian groups eschew "worldly" media, they encourage members to immerse themselves in "appropriate" religious audio recordings and reading matter, even to the point of cooking with Christian cookbooks (Ammerman, 1987). The sectarian stance is oppositional to the dominant social institutions, as well; historically, sectarian groups have distanced themselves from their society's schools and universities, hospitals, trade unions, voluntary organizations, military, media, and so on (Wilson, 1989, 1993). Sectarian strategies in this stance toward "the world" include restricting members' social contacts, prohibiting marriage outside the group, running its own schools and otherwise protecting children from exposure to outside influences, and limiting reading and entertainments to approved content.

The relatively negative tension between a sectarian collectivity and its social environment is both a threat to the continued existence of the group and a potential basis for social change. Essentially, sectarian religious collectivities are a form of social dissent, exemplified by the Anabaptists of Reformation times, and Quakers and Christadelphians in eighteenth- and nineteenth-century America. Contemporary sectarian expressions include the Amish, the Jewish Hassidim, Jehovah's Witnesses, Seventh Day Adventists, and Pentecostal and Holiness groups. Some newer religious groups such as the Children of God (later called the Family of Love), Unification Church, and the Branch Davidians are organized as sectarian collectivities (see Barker, 1982; Richardson et al., 1979; Robbins and Bromley, 1991, 1993a, 1993b).

Denominational Stance. Religious collectivities that exist in a positive relationship with society and accept the legitimacy claims of other religious collectivities have a denominational stance. They are perhaps even more generally accepting of the larger society because, with their pluralistic approach, they have no intent or possibility of controlling society. Denominational collectivities interact with their larger social environment with the sense that the "world" is "okay" (see Berger, 1961). Most major U.S. religious groups, at least at the level of their national organization, exemplify this stance; denominational pluralism is also common among Christian and Jewish groups in Canada, Australia, and much of Europe.

Cultic Stance. The cultic stance is characterized by acceptance of the legitimacy claims of other groups but a relatively negative tension with the larger society. Like the denominational stance, cultic religious collectivities do not claim to have *the* truth and they are tolerant of other religious groups. Like sectarian collectivities, cultic groups are a form of social dissent; however, their dissent is likely to be less extreme because of their pluralistic stance. Historical examples include medieval healing cults and nineteenth-century Spiritualism. In contemporary societies, numerous religious and quasi-religious groups represent a cultic stance; for example, many groups that teach spiritual paths to healing maintain a cultic tolerance for most other religious groups, but dissent from the larger society's treatment of both spirituality and health (McGuire, 1988).

Using the Model. Our model illustrates the constant tension and dynamism in the process by which a group tries to effect its religious convictions. There is an ongoing dialectical process between the religious collectivity and its social environment. The religious group responds to changes in its social situation; it can also effect changes in that situation. The types delineated in Figure 5.1 are thus not fixed categories but are *moments* in a dialectical process.

Thus, when we use this conceptual model, we need to specify (1) societal context, (2) level of analysis (e.g., national, regional, or local), and (3) time period. In Figure 5.2, for example, we can classify the Roman Catholic organization as churchly relative to French society of the thirteenth century (point A^1). Relative to U.S. society of 1840, however, the Roman Catholic organization was more sectarian, because it neither accepted nor was fully accepted by society (point A^2). Much of nineteenth-century American Catholic devotional literature, preaching, and practices of popular piety reflected sectarian strategies in response to the perceived hostility of the Protestant majority (Taves, 1986). Later in the United States, especially after Vatican II, the Roman Catholic organization appears to have become increasingly denominational (point A^3), because it is more accepting of and accepted by society and is more tolerant of other groups' claims to legitimacy (see Seidler and Meyer, 1989).

Specifying the level of analysis is necessary because within a national group with a denominational stance may exist congregations that have a more sectarian stance. For example, within the Presbyterian Church U.S.A., which exhibits a denominational stance at the national level, are many congregations that are uncomfortable with the tolerance and liberal openness of the upper levels of the national and regional organizations. While still more denominational than, for instance, fundamentalist groups, these congregations would be more sectarian relative to their own national organizations. In Figure 5.3, B^1 indicates roughly where the denominational Presbyterian Church U.S.A. would be on the continua; B^2 suggests where a sectarian congregation within that group might be.

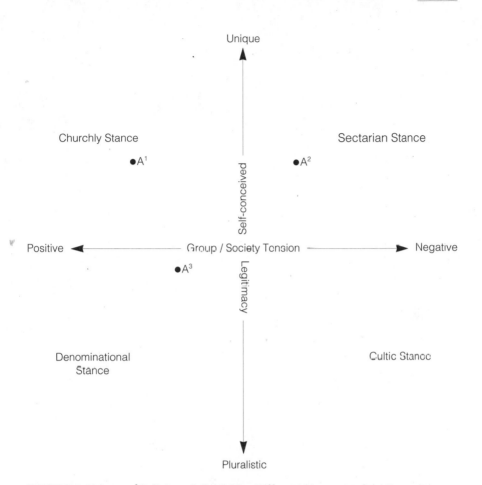

FIGURE 5.2 Stances of Religious Collectivities: Different Moments of the Roman Catholic Stance

This model is also useful for imagining the development of a schism within a religious collectivity. If, within a denominational religious organization, there is considerable support for a sectarian stance but also much support for a denominational stance, the pressure on the umbrella organization to change its stance may become very great. For example, within the Southern Baptist Conference there has been much pressure for a far more sectarian stance; indeed, in several years of conflict, the more sectarian wing of the group succeeded in gaining much control at the national level (see Farnsley, 1994). If the group eventually splits over these conflicts (especially regarding the crucial issue of authority), some congregations may identify with the denominational collectivity, others with the sectarian group. In Figure 5.4, C^1 shows roughly where the Southern Baptist Conference would be on the continua; C^2 and C^3 are where the respective splinter denominational and sectarian splinter groups might be, each a more homogeneous national unit than the previous Conference. Or the conflicting groups could stay within the national umbrella

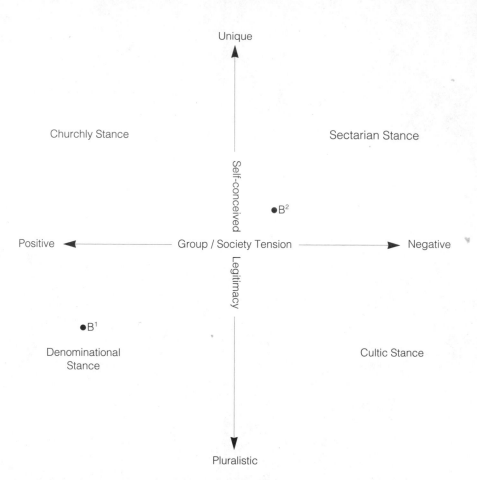

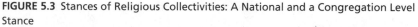

FIGURE 5.3 Stances of Religious Collectivities: A National and a Congregation Level Stance

group—which may or may not become more sectarian as a unit, albeit a highly heterogeneous collectivity.

These sketches show some of the interesting questions sociologists raise about religious collectivities and their dynamics. How do religiously shaped ideas of authority influence a group's tension with the larger society? How does society come to accept a religious collectivity as a "proper" expression of religion, thereby coming to ease the negative tension of that group with its larger social environment? How do divisions over accommodation versus boundary protection lead to internal conflicts and even schisms? How do different kinds of religious groups organize their members' participation and commitment to support a particular stance toward the rest of society and toward other religions? These features of religious collectivities are related to how they are formed and how they develop and change.

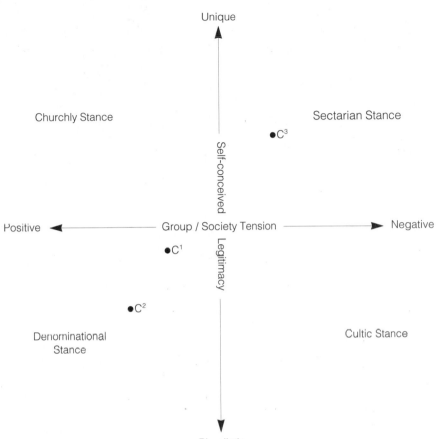

FIGURE 5.4 Stances of Religious Collectivities: A Sectarian/Denominational Schism

TYPES OF RELIGIOUS ORIENTATION

It is helpful to distinguish sectarian, cultic, denominational, and churchly *individual orientations* from the parallel patterns of *collectivity stance,* because several different orientations can occur within the same collectivity. The model of religious orientation types shown in Figure 5.5 is derived from the previous model of stances of religious collectivities. It is based on the characteristic commitment patterns of members. Although each type in Figure 5.5 corresponds to a type of group, members of a group often display differing individual orientations.

Types of Religious Orientation

The two key characteristics for conceptualizing individual orientations are (1) the extent to which the member's role as a religious person is segmented into a separate role or is expected to be diffused throughout every aspect of the person's life, and (2) the extent to which the individual judges self and others

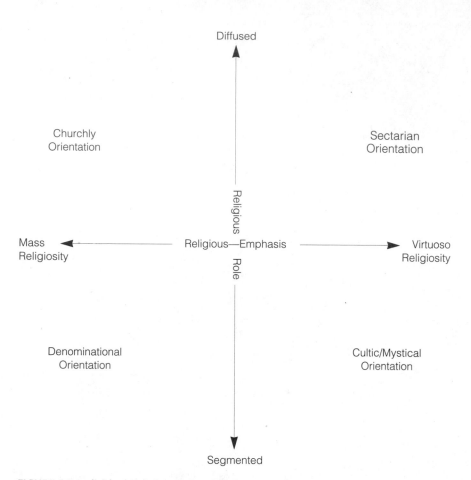

FIGURE 5.5 Individual Religious Orientations

according to standards of "mass" or "virtuoso" religiosity. These qualities are present in varying degrees, so we can speak of relatively "mild" or "strong" sectarian, cultic, denominational, or churchly orientations as represented on these two continua.

Figure 5.5 ties together a number of theoretical and empirical observations about sectarianism. Some theorists (Parsons et al., 1961:251) have suggested that one of the main characteristics of sectarianism is much less differentiation of members' religious role from other social roles (e.g., work or family). The sectarian orientation means that the religious role is the master role, pervading and subordinating all other aspects of the person's life. By contrast, a denominational orientation is more segmented and the religious role is separate from other aspects of life; this does not mean that the religious role is not important, but it has less influence in the total picture of daily life.

The second variable refers to a related normative aspect: the extent to which one judges oneself and others by standards of "virtuoso" religiosity.

Religious virtuosi are those who strive for religious perfection (however defined) and are not satisfied by the normal levels of religiosity of the masses (Weber, 1963:163–165). The person may define religious perfection in terms of some moral, ascetic, mystical, or even body-emotional ideal. Individual orientations are therefore guided by whether the religious role applies to all spheres of life and by whether it is judged by the norm of perfection.

Sectarian. The sectarian orientation is characterized by normative virtuosity and diffusion of the religious role (cf. Knudsen et al., 1978). The sectarian orientation toward perfection is part of the reason for the sect's negative tension with the "world." Sectarian dissent is a form of judgment of the imperfection of the rest of society. The sectarian orientation insists on the pervasiveness of the religious role, which is, ideally, the organizing principle of work, family, leisure activities, political stance, and personal life. This commitment to perfection and the diffuseness of the religious role explain why sectarianism is frequently characterized by total commitment (Snow and Machalek, 1983; Wilson, 1967:1–45, 1970; cf. nonreligious institutions described in Coser, 1974). The sectarian orientation is also more prone to religious extremism precisely because of its characteristic emphasis on applying a virtuoso religious norm in all spheres of social life; persons with a sectarian orientation are less tolerant of those who do not likewise strive for perfection according to the same norm.

Denominational. The key features of a denominational orientation are an acceptance of ordinary levels of religiosity and a religious role segmented from other aspects of life. This orientation is satisfactory for most members of a denomination because their beliefs and practices are less at odds with society than are those of the sect. Because a denominational stance considers its worldview as pluralistically legitimate, commitment tends to be more individualistic. Members feel free to select not only a "church of their choice" but also which elements of that religious view they wish to accept. The social acceptability of denominational membership, together with this individualism, make the characteristic orientation of denominationalism somewhat resemble commitment to other voluntary organizations such as Kiwanis and League of Women Voters. This does not mean that the religious content of denominationalism is not important, but commitment is much less intense or urgent than in sectarianism. Because of their characteristically segmented roles, persons with denominational orientations are more likely than sectarians to have close friends and valued commitments in other groups in the wider community; they are more likely to associate—in their other roles—with people who are religiously different from themselves.

Cultic/Mystical. The cultic/mystical orientation is also prevalent in contemporary religion, either by itself or in combination with denominational modes. It is characterized by separation of religious roles from other aspects of life, as in denominationalism. By contrast, however, a cultic orientation also involves seeking a higher level of spirituality, comparable to sectarian "perfection" except that the definition of perfection is less specific and more eclectic. Members' commitment is that of the seeker, permanently searching for particles of truth from a wide variety of sources. They may be committed less to the cultic group or religious movement itself than to a particular spiritual practice, such as a form of meditation or ritual chanting, central to that group (Wilson and Dobbelaere, 1994). Persons with a cultic orientation often draw on several different groups, either simultaneously or serially. A person deeply involved in Spiritualism may be simultaneously "into" Silva Mind Control and also be an active Methodist.

One prominent source of a cultic orientation is *mysticism*. Troeltsch's (1960) original typology named mysticism as a definitive type (comparable to sect), because he recognized the opposition and critical dissent that was intrinsic to this religious orientation. Because mysticism is based on an inner, immediate experience of the divine, it is highly individualistic and not amenable to control by religious authorities. Unlike sectarianism (which lays strong emphasis on the community of fellow believers and on the relationship between individuals), mysticism focuses on an unmediated relationship between the individual and the divine (however conceived). Mysticism is also a virtuoso pattern of religiosity and, implicitly, it is in opposition both to the formal dogmas and objective institutional features of organized religion and the low level of spirituality of the masses (see Turner, 1991). Mystical religiosity clearly is one special form of the cultic orientation, but not all cultic religiosity is mystical.

This individualism is a distinctive feature of cultic adherence because it contributes to the precariousness of cults as organizations. Cultic individualism means that no clear locus of authority exists beyond the individual member, and there is thus no way to handle "heresy" and therefore no clear boundaries between members and nonmembers. In the cultic mode, as in the sectarian, the individual finds some mutual support from fellow believers, but this group contact is less strong because of the characteristic pluralism and individualism (Campbell, 1977; Martin, 1965; Wallis, 1974, 1977).

Churchly. The major characteristics of the churchly orientation are acceptance of a "mass" standard of religiosity and a religious role diffused throughout everyday life. Religious commitment is very general, sustained by the whole fabric of society, and does not simply comprise the religious group itself. Religious activity may be considerable or negligible, but rarely does the individual consciously choose to be religious. Although church-type religious organizations do not exist in U.S. or Canadian society, individual churchly orientation is possible in certain subcultural settings (e.g., a Utah Mormon enclave).

Using the Model. The differentiation between types of religious collectivity and characteristic modes of individual orientation helps explain the diversity of orientations within a given group. Churches and denominations in particular, but also sects and cults, often have members with very heterogeneous orientations. This diversity is important in understanding the dynamics of religious groups. The interaction of these members often leads to important transformations in the collectivity itself. Such diversity, however, complicates the task of analyzing religious collectivities.

These distinctions of various religious orientations may help explain how it is that many contemporary Catholics are both adamant about remaining Catholic and, at the same time, highly critical of numerous church positions and policies (Hout and Greeley, 1987; Ray, 1994; Winter et al., 1994). One woman who was highly critical of Papal pronouncements, her bishop's insensitivity to the needs and voices of the laity, and the local parish's ineffective liturgies and programs, said, "It's my church, darn it, and they're not going to take it away from me" (quoted in Spickard and McGuire, 1994). As a result, she practiced a selective Catholicism, believing little that church officials preached and participating in only those parts of church life that she considered "true" to the core of the faith, yet considered herself a deeply committed Catholic. Some sociologists are surprised to hear such individualistic positions articulated by members of a religious organization that historically socialized its members in churchly or sectarian orientations. Yet, if we consider the extent to which post–Vatican II Catholicism—especially in the United States, where it is part of a pluralistic milieu— has promoted a denominational orientation, these members' individualistic and voluntaristic assertions of Catholic belief and commitment make sense. Their view is the epitome of a denominational individual orientation: "I have chosen my religion; I have chosen which elements of it I will believe and practice; I have chosen to which areas of my life my religion will apply, and—without extraordinary effort—those who are (perhaps mistakenly) in charge can't make me leave."

In one study, considerable diversity was found within the same denomination, with some members having highly sectarian patterns of involvement. Across four denominational groups (Congregationalists, Presbyterians, Disciples of Christ, and Baptists), sectarian members of one denomination also resembled sectarian members of other denominations in several important ways—especially socioeconomic status (Demerath, 1965).

The religious group must cope with the conflict produced by these differing orientations; but if it can balance the opposing impulses of sectarianism and denominationalism, the organization may emerge stronger. A parallel is the strength gained by medieval Roman Catholicism when it incorporated the sectarian impulse in the form of monasticism. Conflicting modes of orientation often exist in developing religious movements, too. Thus, in the early years of the Catholic Pentecostal movement (now called the Catholic Charismatic Renewal), sectarian and cultic orientations were both prevalent. Sectarian adherents tended to be highly involved in the prayer group and likely to view pentecostalism as the only way to be "really" a Christian. They were more likely than cultic adherents to consider their religious role as the master role in their lives. Cultic adherents, by contrast, were more likely to have segmental commitment to their prayer group and were less likely to view the movement as having the exclusive truth (Harrison, 1975; McGuire, 1982). These two orientations have produced considerable tension within the movement, with the cultic adherents decrying the others' elitism and authoritarianism and the sectarian adherents denouncing the lack of commitment and the independence of those with a more cultic orientation (cf. Neitz, 1986; Warner, 1988). The resolution of conflicts such as this is a significant part of the transformation of religious collectivities.

Conflicting Orientations and the Origin of Religious Movements

When individuals in a religious collectivity have individual orientations that differ from many other members, they may seek out small groups of others who have similar religious orientations. Joachim Wach (1944:173–186) referred to these little groups as *ecclesiolae in ecclesia* (i.e., little churches within the church). These groups consist of members of a religious group who are dissatisfied with the spiritual or moral tenor prevailing in the larger group and advocate a different kind of religiosity, often aiming at eventual conversion or reform of the entire collectivity. Their members press for a different tension between the collectivity and the "world" and are dissatisfied with the imperfection of the religious group. Wach distinguished three general kinds of *ecclesiolae,* which are remarkably descriptive of contemporary small religious groups.

Collegium Pietatis. The simplest form is only a step beyond individual protest: the *collegium pietatis* (i.e., an association of religious colleagues), which is essentially a cultic orientation within the larger collectivity. According to Wach's description, these collegia do not see their beliefs and practices as uniquely legitimate. They are loosely organized, stressing individual attitudes and seeking greater personal perfection. Historic examples of collegia include German pietism in the early part of the Reformation and the Oxford Group Movement in nineteenth-century Anglicanism. Similarly, Karaism (eighth and ninth century), Kabbalism (especially sixteenth and seventeenth century), and neo-Kabbalism (twentieth century) among Jews spawned groups like the collegia. Contemporary Christian examples of collegia include loosely organized prayer or fellowship groups within larger religious organizations. Societies related to some special method of spirituality or specific religious practice (e.g., a healing or devotional group) could also be collegia.

Fraternitas. Although common religious ideals and experiences are sufficient to draw people together as in collegia, many groups seek a closer association—a strong sense of community. The establishment of groups with some measure of communal life marks the *fraternitas* (i.e., brotherhood). Their communality, however, is integrated more by "spirit" than formal organization; thus, they are often in conflict with the larger collectivity. Such groups often have egalitarian concepts of fellowship. The fraternitas also represents a somewhat cultic orientation within the larger collectivity. Examples include early Moravians (eighteenth century), Methodist fellowship groups (especially nineteenth century), and the earliest years of the Bruderhof and kibbutzim (twentieth century). The "underground church" movement in American Roman Catholicism (1960s and 1970s), "basic ecclesial communities" and "house churches" among both Protestants and Catholics, as well as the havurah and schtibel movements in Judaism (1970s and 1980s) are contemporary *fraternitates* (see Bubis et al., 1983; McGuire, 1972, 1974; Prell, 1989).

The Order. Wach's third type of internal dissent is the *order,* characterized by greater stratification and specialization of functions and a more rigid conception of community. The order is essentially the organization of sectarian dissent *within* the larger collectivity. It expects the individual's total commitment. Members are bound in their adherence to the religious community and in their renunciation of competing ties. Monasticism is the most obvious and universal example of the order. Through the institution of monasticism, the larger collectivity absorbs and segregates sectarian protest. At the same time, the order may bring about reform in the larger collectivity, such as the reforms in medieval Roman Catholicism brought about by Cistercians (twelfth century). Monasticism is a feature not only of Christianity but also of Islam, Buddhism, Jainism, and Taoism; but in different cultures it has different meanings because the characteristic tension between church and social environment is different. There is often much tension between the order and the larger collectivity of which it forms a part. Even though the order may compromise its dissent, it represents a considerable source of independent authority within the collectivity. Contemporary examples of orders include modern descendants of medieval Roman Catholic monastic orders and nineteenth-century Anglican orders. Other twentieth-century examples include "covenant communities" among Catholic Pentecostals, and Protestant communal groups such as Iona in Scotland.

The order is an expression of the sectarian elite within a larger collectivity and is the ultimate social organization of virtuoso religiosity. The collegium and fraternitas similarly express virtuoso religiosity but in much less organized and homogeneous form. Part of the nature of their negative tension with "the world" is their dissatisfaction with ordinary religiosity. Even where religious virtuosi have not banded together in "little churches," there is a marked sectarian impulse among a religious elite (Berger, 1958; Fichter, 1951, 1954). The main difference between orientations of the order and the sect is the order's willingness to accept a dual standard of religiosity—one standard for the elite and one for the masses. Sectarian collectivities, by contrast, do not tolerate this distinction and see their standard as the only way (M. Hill, 1973a). Protestantism has historically tended to spawn groups taking a sectarian stance because it is more likely than Roman Catholicism or Eastern Orthodoxy to reject the dual standard (Martin, 1965:191–193).

The pressure to change brought by religious virtuosi is analytically different from the impetus to change in the challenge of a charismatic leader and followers. The charismatic leader proclaims a new message and new basis of authority. Religious virtuosi, by contrast, generally call for a return to the "true" tradition; authority is legitimated by reference to tradition. While this latter pressure for change is less threatening to the larger collectivity, it still has considerable potential for stimulating dramatic changes through something of a "revolution by tradition" (M. Hill, 1973a:3). Thus, dynamic processes within religious organizations themselves sometimes bring about major changes in the collectivity; and the existence of conflicting orientations within the same groups helps account for some of these transformations.

HOW RELIGIOUS COLLECTIVITIES DEVELOP AND CHANGE

An interesting subject in itself is how a religious group comes into being. What brings people together? What attracts them to the group? What repels them from other alternatives? What holds them together? What shapes their organization and collective direction? Even more important, however, is the relationship between the formation of the group and larger social processes. What social factors contribute to the formation of a new religious group? What factors promote or inhibit its success, and how does the developing religious group interact with its larger social environment? Developing religious groups have often had impact both on society and on the collectivities against which they were dissenting, so the dynamics of religious collectivities are linked with the process of social change. At the same time, social change has historically been an impetus for religious change, so religious dynamics indirectly reflect larger social processes.

The Formation of Religious Groups

Social factors, as well as theological ones, appear to be very important in the formation of new religious groups. Explanations of the formation of religious movements and new religious collectivities have generally focused on socioeconomic factors in the creation and mobilization of dissent. Historically, recruits to new sectarian religious movements have been disproportionately drawn from the lower classes (Aberle, 1962; Lanternari, 1963). Sectarian groups are vehicles for dissent from society as well as from the churches from which they break away. Persons who are relatively comfortable with society are unlikely to participate. On the other hand, the most destitute stratum, partly because of its characteristic fatalism and sheer lack of resources, is rarely involved. Furthermore, economic deprivation is not sufficient to explain all sectarian movements, because middle- and upper-class persons are also drawn to them. If deprivation is a factor in the formation of sects, the concept needs clarification to fit these diverse empirical situations.

Toward such clarification, Charles Glock (1973:210) proposed the notion of **relative deprivation,** which refers to "any and all of the ways that an individual or group may be or feel disadvantaged in comparison either to other individuals or groups or to an internalized set of standards." People's subjective assessment of their circumstances, rather than objective circumstances themselves, is the key criterion. Glock also notes that economic deprivation is only one of several ways that a person can feel deprived; a person can also feel social (i.e., status), organismic (i.e., physical), ethical (i.e., philosophical), or psychic (i.e., values) deprivation.

The existence of relative deprivation does not *cause* a movement of dissent, but it does create a pool of dissatisfied persons who, if mobilized, would express dissent through a political or religious movement. The deprivation model is helpful in conceptualizing the relationship between sources of dissatisfaction and a movement's social base. It does not, however, adequately explain why dissatisfied persons should choose a particular movement. Furthermore, in applying the model empirically, there is a tendency to reason from after-the-fact evidence. Some interpretations suggested that because a movement provided X, that provision constitutes evidence that recruits were deprived of X. Nevertheless, the deprivation model is useful in understanding not only sect formation but also subsequent transformations in the organization of the collectivity.[3]

Another approach to understanding the formation of religious movements is the emphasis on socioeconomic *change*. One study of historical millenarian movements suggested that the key factor was the ambiguity surrounding a disaster (natural or social) that produced the impetus for such movements (Barkun, 1974). Further evidence for this interpretation comes from analysis of a period of prolific movement development, the early nineteenth century in upper New York State. The revivals, experimentation, and development of new movements in this time may have been correlated with dramatic economic fluctuations, which had produced alternating rising expectations and disappointments (Barkun, 1986). A related interpretation suggests that religious and political movements arise in response to major changes in the world

---

3. A student of Glock subsequently used an extreme form of this reasoning to attempt a definitive explanation of religion. Accordingly, religion is defined as "a system of general compensators based on supernatural assumptions" (Stark and Bainbridge, 1979:121; see also Stark and Bainbridge, 1980; Bainbridge and Stark, 1984). Compensators refer to intangible promises that *substitute* for certain rewards that people desire, such as prosperity, success, and physical and mental health. From this definition, the authors derive a typology of church, denomination, sect, and cult, distinguished by the type of compensator each offers. This approach has been severely criticized for its *reductionism;* i.e., it tries to reduce religious belief and practice to overly simplistic this-worldly considerations: cost versus benefit, whether real reward or substitute compensation (Wallis and Bruce, 1984, 1986). Like relative deprivation theories, this approach tends to argue backward: This religion offers X; therefore, X must be a compensation for desired reward Y. It imputes to religious adherents motives and meanings that they would probably neither recognize nor affirm as theirs (McNamara, 1985). Critics further argue that the theory is not adequately supported by empirical evidence (Wallis and Bruce, 1984; Bibby and Weaver, 1985)—especially not by the kind of evidence produced by sound ethnographic research that listens to believers' accounts of their motives, meanings, and experiences.

economic order. The type of movement that develops depends, according to this theory, on whether its adherents are central or peripheral to emerging sources of economic power (Wuthnow, 1978, 1980).

All available evidence suggests that a strong relationship exists between social conditions and the development of sectarian and cultic expressions of dissent. None of the existing models alone is adequate to explain this relationship, though they are not necessarily wrong. What is needed is a sufficiently complex model that incorporates varied causes. Focusing on socioeconomic concomitants of religious movements, however, may lead to ignoring or downplaying members' personal motivations and aspirations in the group's formation. Religious aspirations themselves are important sources of religious movements. Religious virtuosi seeking to perfect their spiritual lives, cultic seekers looking for new techniques for spiritual experience, persons dissatisfied with the level of religiosity available to them in other collectivities—for these people, *religious* aspirations are foremost among their reasons for forming or joining a developing religious movement. Indeed, parent organizations such as a church often give impetus to such movements by raising members' aspirations. Vatican II and changes within the Roman Catholic church in years immediately following the Vatican Council may have raised religious expectations of church members to such a degree that they subsequently sought developing movements either within the church (e.g., Cursillo, Underground Church movement, and Charismatic Renewal) or outside it.

Movement Organizations

Just as our typology of religious collectivities can be viewed as a general typology of organizations in tension with their social environment, so too can the processes by which religious groups change be understood as general. Some useful insights into these processes come from the theory of movement organizations (see Beckford, 1975a; Lofland and Richardson, 1984). This approach explores how social movements organize themselves toward achieving their goals; how they recruit, consolidate, and deploy members; how they maintain their organization and arrange leadership. This approach has been fruitfully applied to a study of the Young Men's Christian Association (YMCA) (Zald, 1970), a Jesus movement commune (Richardson et al., 1979), Jehovah's Witnesses (Beckford, 1975b), Pentecostalism (Gerlach and Hine, 1970), Catholic Pentecostalism (Harrison, 1975), and Soka Gakkai (Snow, 1987).

Socioeconomic factors may provide the impetus for a religious movement, but aspects of the movement organization itself determine whether the movement will develop and spread. One study of the spread of Pentecostalism identified several movement organizational factors including (1) the relationship among local groups of the movement—flexibly linked networks of local groups promoted the movement's spread; (2) recruitment along the lines of preexisting class social relationships—family and friends; (3) emphasis on a clearly defined commitment act or experience; (4) an ideology offering a simple, easily communicated interpretive order, a sense of sharing in the control and rewards forthcoming, and a feeling of personal worth and power; and (5) emphasis on the perception of real or imagined opposition (Gerlach and Hine, 1968). The researchers found that this model applied not only to expressly religious groups but also to other dissenting movement organizations such as the Black Power movement (Gerlach and Hine, 1970).

Not all of these contributing factors are present or significant for all movements. Jehovah's Witnesses, a rapidly spreading movement, recruit very little among preexisting networks of social relationships (Beckford, 1975b). Some cults recruit almost totally anonymous persons. In 1975 two itinerant preachers, who proclaimed salvation through the aid of extraterrestrial beings with flying saucers, recruited over one hundred members (who gave up jobs, family ties, and material possessions) in four brief meetings, usually after less than six hours of contact with any member of the group (Balch and Taylor, 1977). Nevertheless, the approach is helpful for analyzing some of the ways in which the internal organization of a movement influences its outcome. It also suggests some of the ways (e.g., increasing the perception of opposition) that the group itself influences the tension between it and its social environment.

Organizational Transformations

Considering the dynamic forces built into our model of religious stances (Figure 5.1), it is not surprising that groups change continually. Forces for change exist within the organization itself, as well as in the relationship of the group to its social environment. The processes by which a religious group changes from one type of organization to another are sociologically important because they reflect both the impact of the group on society and the influence of society on the group.

Institutionalization: O'Dea. A movement cannot have any lasting impact on society without some measure of **institutionalization**—that is, the creation of some objective structure by which it tries to realize its values and goals. O'Dea (1961) delineates five inherent dilemmas that institutionalization often poses for any social movement, religious and nonreligious alike.[4] On the one hand, institutionalization enables the emerging movement to express and spread its perspective; on the other hand, it contains built-in tendencies toward stagnation and disorganization, separating members from the initial shared experiences of beliefs that held them together. Religious reform or revival movements are prototypes of this process, but other movements such as civil rights, environmentalism, communes, and political reform movements also wrestle with the dilemmas of institutionalization.

One such dilemma is the organizational contradiction of **mixed motivation.** According to O'Dea, the earliest stages of a movement are characterized by a singlemindedness of followers; either the cause or the charismatic leader unites members' motivations. As the movement stabilizes, other motives come

---

4. The following synopsis of five dilemmas from O'Dea (1961) is used with permission.

to the fore: desire for prestige, expression of leadership and teaching abilities, drive for power, search for security, respectability, economic advantages, and so on. The institutionalized movement is stronger precisely because it is able to mobilize the self-interested as well as the disinterested, but members' pursuit of vested interests often conflicts with some of the movement's original ideals. A movement that relies totally on members with "pure," disinterested motives has little chance for survival, especially after the departure of the charismatic leader who initially united the members. On the other hand, mixed motives frequently lead to compromise and corruption of the movement's ideals.

A second dilemma involves the movement's **symbol system**—its language, ceremonies, and physical symbols of central meanings. In the earliest stages of a movement, members develop symbols by which they attempt to share significant experiences. These special words, ceremonies, and objects express (however incompletely) what it means to be a part of the movement and its distinctive way of life. Without this objectification (i.e., transforming subjective experiences into concrete, shared images), there would be little basis of commonality among movement members. Shared symbols hold the members together and become part of the message they carry to prospective recruits. At the same time, however, there is an inherent tendency in the process of institutionalization for these symbols to become remote from the experiences and meanings that shaped them. They are cut off from the subjective experience of members who use the symbolic objects, group language, or ceremonies; they become ends in themselves or "things" to be manipulated for achieving ends. For example, the "kiss of peace" was an early Christian symbol of communal love, forgiveness, and unity. In many settings now, however, the "kiss" is merely a ritualistic handshake "performed" at a designated part of the worship service. Without symbolization, the central meanings of the movement cannot be transmitted or shared, but symbolization (i.e., objectification) contains the inherent risk that symbols will become alienated from the subjective life of the members.

Movements also face the third dilemma of **organizational elaboration of structures versus movement effectiveness.** As Weber (1947:358ff.) pointed out, the long-term impact of a movement hinges on transformation of bases of authority and leadership from a charismatic mode to either traditional or legal-rational structures. When a movement becomes established, there is a strong tendency for the organization to calcify around the memory of the early dynamism; its own tradition becomes the rationalization for why things should be done a certain way.

Early stages of movement organization involve simple structures such as the charismatic leader and followers or leader, core followers, and other followers. The transition to legal-rational structures is typically accompanied by the elaboration and standardization of procedures, the emergence of specialized statuses and roles, and the formalizing of communication among members. The early years of the Divine Light Mission in the United States were characterized by rapidly growing, loosely affiliated local ashrams (i.e., groups of devotees, usually living communally), united mainly by devotion to the ambiguous charismatic figure of Guru Maharaj Ji. As the DLM became increasingly structured and centralized, leadership and power focused in the movement's Denver headquarters. The guru's desire to consolidate his power and authority over the movement in the United States resulted in greater formalization: rules and regulations for ashram living, standards for recruited "candidates," and pressure toward certifying movement teachers (Pilarzyk, 1978:33–37). Although such division of labor and formalization of structures make the growing movement more efficient, the process may also make it less effective in the long run. The bureaucratic impersonalism and complexity may obstruct some of the movement's original goals. Members come to feel more peripheral to the movement, and their motivation becomes problematic.

A fourth internal dilemma is the **need for concrete definition versus legalism.** On the one hand, the movement's message must be translated into concrete terms in order to apply to people's everyday lives. On the other hand, the concretization that occurs as the movement becomes institutionalized promotes a legalistic approach to those very terms. The ethical insights of a religious movement might be translated into concrete moral rules (e.g., a norm against consuming alcohol); then, if a legalistic approach to this rule arises, members are reproached for breaking the "letter of the law" (e.g., using vanilla extract, which contains alcohol, in baking). Legalism reduces the insight of the ethical message to petty conformity.

The fifth dilemma of institutionalization is that of **power.** The developing movement needs to structure itself so that its message and meaning remain firm. It needs to protect itself from threats both within and without its boundaries. A certain amount of internal power gives the nascent movement strength against threats that would dissolve it. On the other hand, that same power may be asserted against deviant and dissenting members. Members' compliance may be assured by external pressures or coercion rather than by inner conviction and commitment. Similarly, desire for power and influence may motivate movement leaders to assert themselves in other institutional spheres, diverting the movement from its original message and goals. The direction taken by the leadership of the People's Temple illustrates one (albeit extreme) resolution of this dilemma. A number of its practices (e.g., the move from California to a remote part of Guyana and the requirement that members give up personal property) can be understood as efforts to overcome the inherent precariousness of charismatic leadership. They were strategies to consolidate power, similar in kind to strategies adopted by leaders of other nascent charismatic movements (see Johnson, 1979).

O'Dea (1961) describes these five contradictions as dilemmas to emphasize the extent to which neither direction (institutionalized or noninstitutionalized) is completely satisfactory. The very processes that would enable a social movement to realize its ideals contain inherent contradictions that simultaneously prevent the movement from accomplishing these goals. The concept of dilemmas, however, is somewhat misleading because rarely does a social movement make conscious choices to institutionalize. Nevertheless, O'Dea's analysis is useful for showing that frequently criticized features of institutionalized religion (or any other institutionalized social movement) are actually results of the ways in which that movement has organized itself for effectiveness and social impact.

All religious organizations, from the smallest informally organized sect to the largest bureaucratically organized church, can be analyzed for how they acquire and use the resources of ideas, people, and materials. Organizations can also be compared for how they socialize, recruit, and control their members. And they can be examined for their structural qualities: specialization, centralization, formalization, and authority relationships (Beckford, 1975a:34–92). Transformations in religious groups frequently involve changes in some of these organizational aspects. These organizational qualities are not definitive, but rather reflect the inherent tension between the religious group and its social environment. Thus, examination of how change occurs in different types of religious organizations focuses on the two key characteristics defining our model of organizational types. Transformations in religious groups can be analyzed in terms of two questions: (1) How does the group's sense of its monopolistic or pluralistic legitimacy change, and (2) what are some of the forces affecting the group's tension (negative or positive) with its social environment?

Transformations of Sectarian Collectivities. The dynamism of sectarian dissent makes the sect a potential force for social change. Yet successful sectarian movements seem to accommodate to society with great regularity, giving up much of their negative tension with society (Niebuhr, 1929). Weber noted that while charismatic leadership frequently stimulated social change, the process of trying to maintain and defend innovation typically resulted in routinization. **Routinization of charisma** is the process by which the dynamism of charismatic leadership is translated into the stability of traditional or bureaucratic organization. The very organization of a community of followers is thus part of this process of routinization (Weber, 1963:60, 61; see also Weber's essay "Sect, Church and Democracy," 1968:1204).

The change of sectarian groups to a more denominational stance is the prevalent direction of change in the United States. Other directions are also theoretically possible, however. Sectarian collectivities could adopt a churchly stance by accommodating to the larger society and consolidating their position to achieve a monopoly of legitimacy. Christianity began as a minority sect within Judaism, and in its spread to other cultural settings, it gradually consolidated its position, achieving an effective monopoly of legitimacy under Constantine.

Similarly, sectarian movements could change to a more cultic stance. Because this transformation involves giving up claim to unique legitimacy, the process is often the result of authority dissolution within the group. If a charismatic leader has successfully organized the group around his or her personal authority to present "the truth," the death or downfall of that leader might delegitimate that "truth." If the group comes to view those teachings as pluralistically legitimate, it is likely to become a cult (if it survives the leadership crisis at all).

A good example of this dissolution is the development of one flying saucer cult (Festinger et al., 1956). The group initially began with a cultic stance, one of several groups from which members drew their inspiration, but gradually became sectarian, especially as its role in the prophesied imminent end of the world was made clear by the increasingly authoritative pronouncements of group leaders. The failure of the prophecy undermined the authority of key leaders and eliminated a major basis for the group's distinctiveness—its position as the "elect" to be saved by flying saucers. Although failure of the prophecy did not destroy the group altogether, it gradually became more cult-like again, receding into the general cultic milieu.

Sectarian and cultic groups are organizationally more precarious than groups with denominational and churchly stances because dissent and deviance are difficult to maintain in the face of opposition from other religious groups and the larger society. Sects, in particular, are likely to clash because of their claim to unique legitimacy. On the other hand, history provides several examples of groups that sustained their sectarian dissent and never developed denominational patterns. The Old Order Amish have for many generations preserved their distinctive beliefs and way of life, largely by withdrawing into their own communities.

Such groups may be considered *established sects* (Yinger, 1970:266–273), groups engaged in sectarian dissent that have organized their internal and external social arrangements toward the perpetuation of their distinctiveness. Some internal arrangements that promote the success of social movements— the quality of leadership, commitment and recruitment processes, social control, and ideology—enable sectarian groups to become more stable. Arrangements for the relationship of the group to society also promote the establishment of groups with sectarian stances; like the Amish (see Kraybill, 1989), the group may withdraw physically from the rest of society and systematically control which elements of cultural change to allow "into" its everyday experience. Or it may migrate to a more favorable environment, like the Hutterites and Bruderhof (Peters, 1971; Zablocki, 1971). Or it may insulate itself by regulating members' contact with nonmembers and the rest of society, like the Hassidim (Poll, 1969; Shaffir, 1974), censoring or forbidding exposure to mass media, limiting occasions of visiting with nonmembers, or absorbing members so totally that outside influences cannot affect them. Such insulation may intensify the group's sectarianism, as in the case of the Exclusive Brethren (Wilson, 1967:287). Established sectarian groups have effectively developed boundaries—physical, ideological, or symbolic—between themselves and the threatening "world." Interestingly, pluralistic societies make it difficult for sectarian groups to establish these boundaries because such societies are more likely to be tolerant of the dissenting group and not intensify its sense of opposition. In a totalistic society, by contrast, social opposition can fortify the group's boundaries, indirectly promoting its survival.

At the same time, however, established sectarian groups appear to be somewhat more accommodated in some aspects than new sectarian movements. They may be less strident in their claims of uniquely legitimate authority, or they may develop a slightly more positive tension with "the world," perhaps relaxing part of their stand. While maintaining a measure of sectarian withdrawal, some Hutterite colonies have accepted modern technology, resulting in changes in community structure, work, and gender roles (Peter, 1987). A sectarian stance toward society depends partly on how society relates to the religious group. The dominant social or religious group may become more tolerant, allowing the sectarian group greater freedom to maintain its distinctive ways (Harper and Le Beau, 1993). The development of Mormonism into an established sectarian collectivity resulted partly from internal organization and partly from gradual tolerance by society. Interestingly, one of the most significant accommodations by Mormonism was socioeconomic: the abandonment of its early nineteenth-century working-class ideals of cooperation, common ownership, and classlessness. The religious group was transformed into a major financial force, a "theocratic corporation," espousing conservative politics, individualism, and free enterprise (Baer, 1988).

Sectarian groups become denominational by giving up their claim to exclusive legitimacy and by reducing their dissent, accommodating to society. Certain kinds of groups tend more to such changes than others. Those that express their dissent by withdrawing from society are less vulnerable to accommodation pressures than are dissident groups that aim to convert society (Wilson, 1967:26–27). Similarly, sectarian groups that dissent against the evils of society are more resistant to transformation than those emphasizing individual sin or anxiety (Yinger, 1970:272). Among sectarian groups that are trying to expand rapidly, a strong incentive toward the greater tolerance and social acceptability of the denominational stance is their wish to form coalitions and mergers with other like-minded groups. The growth of Calvary Chapel, while still a relatively sectarian movement, resulted largely from entire congregations merging into a more denomination-like affiliation (Richardson, 1993b).

The process of accommodation is not necessarily all on the part of the dissenters; the dominant religion or social group can change as well (B. Johnson, 1971). As the society or established church changes, the sectarian stance toward that larger group also changes, and the collectivity may be thrust toward greater or less negative tension with society.

The transformation of sectarian to denominational stance is not always in one direction. A group may move toward denominationalism, then have an internal "reform" and move back toward sectarianism. In the 1880s, the Society of Friends (i.e., Quakers) underwent a crisis in which some members criticized the group's growing compromise with "the world." There were both sectarian and denominational orientations within the group, which had gradually become prosperous and denomination-like. During the crisis, members returned to their historical roots and decided to eliminate a number of their more denomination-like practices (e.g., having ministers and birthright membership) (Isichei, 1967:181). Our model of types of religious collectivities consists of dynamic forces on two continua. Thus, we can imagine groups moving in either direction on either continuum: becoming more or less pluralistic and having more or less negative tension with society.

The prosperity of the Quakers illustrates another factor in transformation of the sectarian stance. Does the very success of a group make it difficult or impossible to retain original ideals and dissent? It is easier for group members to reject the things of "the world" if they are too poor to afford them. As the members become established and increasingly respectable in the community, there is pressure to relax restrictions and become more like their neighbors.

These changes are further promoted by the fact that children of the generation that founded the religious group have not had the distinctive experiences of their parents. The second generation's commitment to the group may be serious, but certain issues that were very important to their elders may not hold much significance for them. The *sabras* (i.e., subsequent generation) of the Israeli kibbutzim did not have the intense group experience of hardship and fellowship of the initial Zionist pioneers who settled on the land in the newly created Jewish homeland. The pioneers had forged a group vision of an ascetic communal existence, but subsequent generations of members did not have that unique experience as the basis of their commitment (Spiro, 1970:38–59).

A number of studies have suggested that the values and social interaction of some sectarian groups, while dissonant from those of the larger society, have the unintended effect of making members more prosperous and acceptable by the standards of society. Certain fundamentalist Christian movements encourage their members to work hard and reliably, to respect authority, and to refrain from drinking, smoking, gambling, and entertainments. Thus, group values increase the likelihood of members' earning steady wages while not spending that money—except on religious activities. The overall effect is that members are socialized into acceptable roles in society and are able to enjoy material comforts (Johnson, 1961); their sectarian dissent inadvertently brings dissidents "back" into the dominant society. For example, the Black Muslim movement changed from a violently sectarian separatist stance to a more denominational status within Sunni Islam. A major factor in the change was the increasingly middle-class character of the movement, brought about by the Black Muslim work ethic and morality (Mamiya, 1982).

This process does not occur in all groups, however. Sectarian collectivities that live communally withdrawn from society are better able to control the socialization of their members and to disperse material comforts over the entire religious community. Some, in fact, socialize their members into roles that are extremely dysfunctional in the larger society, making exit from the group very difficult (Beckford, 1978b). Nevertheless, material success is a built-in threat to a sectarian stance because it is more difficult to dissent from a society in which one has a stake and is enjoying some benefits.

Similarly, organizational success is a threat to stability. If a sectarian movement is successful in attracting many members and expanding into new geographical areas, it may grow more denomination-like. The group may become bureaucratized and rely on religious specialists—ministers, fundraisers, religious teachers, and so on. Some established sectarian collectivities (e.g., Jehovah's

Witnesses) demonstrate the possibility of retaining the dissenting stance under a complex organization. Nevertheless, organizational success brings structures that make it difficult to retain the initial sectarian enthusiasm and small-group intensity.

Transformations of Cultic Collectivities. Cultic groups are highly precarious organizational forms. Their characteristic pluralism together with their members' general individualistic mode of adherence make them more unstable than other types of religious collectivity. The typical transformation is from cultic to sectarian stances, but some cultic groups can stabilize. Spiritualism, now some 150 years old as a movement, has retained its cultic form. It is loosely organized around independent Spiritualist "mediums," some of whom are affiliated in a network of contacts but with little organizational control or authority (Nelson, 1969).

By contrast, a large number of groups that began with a cultic phase became sectarian. The Jesus People (Richardson, 1979), Christian Science (Wallis, 1973), the Bruderhof (Zablocki, 1971), Scientology (Wallis, 1977), and the Unification Church (Lofland, 1966) all began as cultic collectivities. The fragility of cultic groups as social organizations results from their indistinct and pluralistic doctrine, their problem of authority, and the individualistic, segmental mode of commitment characteristic of cultic adherents. In the transformation from cultic to sectarian stance, the key feature is the arrogation of authority; certain members successfully claim strong authority, thereby enabling them to clarify the boundaries of the group belief system and membership. The successful claim to strong authority gives leaders a basis for exercising social control in the group and for excluding persons who do not accept the newly consolidated belief system. Movement toward sectarianism may be a deliberate strategy of group leaders or simply the by-product of other organizational decisions (Wallis, 1974, 1977).

Other factors promoting this transformation, besides internal group factors, include factors relating to individual members and to the social environment. Interaction between the group and other elements of society sometimes promotes the transformation. Largely because of the material and moral encouragement of branches of certain established sectarian groups the Jesus movement's initial cultic independence was transformed. It became increasingly sectlike, consolidating its belief system along the lines of these outside groups (Richardson, 1979). One element of this belief system was a greater emphasis on authority in the religious community and in families.

Other transformations (i.e., to denominational or churchly stances) are theoretically possible but not common. Two groups that have made considerable changes toward a denomination-like accommodation with worldly success are 3HO and Vajradhatu, but their pronounced roots in Eastern religion (3HO is derived from a yogic/Sikh tradition, Vajradhatu from Tibetan Buddhism) leave these cultic movements far from denominational acceptance by the larger society (Khalsa, 1986). Similarly, Hare Krishna (Krishna Consciousness in America or ISKON) has both sectarian and cultic characteristics, and, in its first decades, its leadership moved between emphasizing sectlike or cult-like stances. After coming under attack from the anticult movement, however, leaders decided to try to become a "denomination of the Hindu church" (Rochford, 1985). While the group's religious freedom is more likely to be respected if it is identified as a Hindu ethnic church, U.S. religious pluralism does not as yet grant equal legitimacy to religions outside the Judeo-Christian traditions. Thus, no matter how accommodating these groups are, societal attitudes would impede acceptance of groups based on Eastern traditions as having full denominational status in the United States.

A more common pattern is for cultic groups to first undergo internal organization changes, often becoming more sectarian, and later edging toward being more denomination-like. Christian Science began as a cult but experienced the early consolidation of authority of Mary Baker Eddy. Its organizational form was somewhat sectlike for most of its history, though its middle-class members were often more cultic in their style of adherence. With increasing social acceptability, the group now appears to be moving toward the denomination end of the continuum.

Transformations of Denominational Collectivities. The denomination is the predominant organizational form of religious groups in the United States and Canada today. Widespread religious pluralism is, however, a relatively recent phenomenon, so there is little historical evidence concerning denominational changes. History provides several examples of groups arising as denominations and becoming established as denominations. The Reform movement among Jews was from its outset more tolerant and positively oriented toward society than its Orthodox predecessor (see Steinberg, 1965). The movement required organizational stability and popular acceptance before it could become an established denomination. Similarly, Congregationalism and Methodism were born as denominations rather than sects (Martin, 1965:4). Some of the mainline black denominations (e.g., the African Methodist Episcopal church) also began as denominations. Theirs was not a theological protest against the parent group but mainly a reorganization to allow blacks to achieve leadership and other important roles in their church (see Gravely, 1989).

A group that begins with a denominational stance but moves away from pluralism is likely to develop a more sectarian organizational form. The Salvation Army, for example, became sectarian after its more pluralistic beginnings (Robertson, 1967). Similarly, recent clashes among Southern Baptist leaders have resulted in greater sectarianization of that denomination (Ammerman, 1990). Just as sectarianization is a strategy for consolidating control in cultic collectivities, so too are denominational groups susceptible to sectarianization. But societal pluralism makes it highly unlikely that a denominational collectivity could achieve a sufficient monopoly of legitimacy to become churchlike.

Changes in society appear to be influencing some denominations toward a more sectarian stance. To the extent that society does not support or accept key beliefs of groups, they may become sectarian to protect those beliefs and practices. Societal changes are creating more negative tension between many established religious groups and society, thus encouraging sectarian stances (see Roof, 1978:215–227). For example, increasing sexual "permissiveness" in society presses denominational groups that do not accept this attitude to retreat toward a more defensive, negative posture.

At the same time, some denominational groups' willing accommodation of cultic patterns of spirituality within their folds may be strengthening the larger group, providing their congregations new vigor and attracting an enlarged pool of participants. There is some evidence that congregations of such diverse denominational groups as Unitarians and Episcopalians have been invigorated by incorporating spiritual practices such as mysticism and healing, borrowed by active members who are, at the same time, cultic adherents of such alternative movements as Tibetan Buddhism, Wicca, Native American spirituality, and Jain yoga and meditation (Davie, 1995a; Lee, 1995; Roof, 1993). This structural envelopment may turn the tolerance and openness of denominationalism into a strength, rather than a weakness.

Transformations of Churchly Collectivities. Similarly, changes in social setting create changes in modes of organization in churchly collectivities. If a group is unable to maintain its monolithic status, it cannot continue in its churchly stance. Churches that retain their claims to unique legitimacy are likely to take on sectarian characteristics. The Roman Catholic church of early twentieth-century immigrants in the United States maintained a highly sectarian stance to protect its members from the beliefs and practices of the predominantly Protestant culture. More recently, its greater tolerance of other religions has moved American Roman Catholicism toward a denominational form.

In Canada the Catholic church has more legal recognition by the state but tends to be more geographically concentrated and less culturally assimilated into mainstream values (Westhues, 1978). Thus, even in a society very similar to that of the United States, the Catholic church has a different organizational form. In England the transformation of the Church of England (i.e., Episcopalian) from its churchly form to a denominational form was primarily the result of increasing society-wide pluralism (Swatos, 1979).

Implications of Organizational Transformations

The types of religious collectivity characteristic of various periods or societies vary. Changes in religious collectivities often reflect responses to society. Religion may be undergoing some significant changes in its social form and location (as Chapter 8 will describe in more detail). An examination of transformations in religious collectivities can help us grasp some emerging patterns.

Sectarianism in Contemporary Society. Whether as a deliberate strategy or simple adaptation, increasing sectarianism may well be necessary for certain religious groups in contemporary society. The primary, close-knit, face-to-face relationship is an important part of the sectarian orientation because it supports the believer in the face of real or perceived opposition. The group helps maintain the believer's assurance of being part of the "one way" despite a hostile or unbelieving "world." The close-knit group provides support for a **cognitive minority**—a group of people whose worldview differs from that of the dominant society. Primary group relationships provide a structure within which the believers' distinctive worldview is plausible, whereas outside the group that worldview is disconfirmed (Berger, 1967:163).

To the extent that *any* religious group attempts to maintain its distinctiveness and applicability to all aspects of life, it runs counter to developments in society—especially the compartmentalizing of religion into the private sphere. In this sense, religious fundamentalisms and related movements are trying to reverse the differentiation of social institutions characteristic of modern societies. Even established religious groups may take on some sectarian qualities as they organize to protect their beliefs, practices, and values in the face of society's lack of support. The group must provide its own social support for its beliefs and way of life. Members become a cognitive minority and therefore structure their interactions with each other and the "outside" to protect their worldview (Berger, 1967:163ff.). This change toward sectarianism is, however, primarily a *reaction* to changes in society. One comparative study of strategies of a mainstream Orthodox and a Lubavitch Jewish congregation found that both groups had to adjust their teachings in response to issues posed by modernity, specifically individualism, pluralism, and feminism (Davidman, 1990a). Any group, regardless of how extreme or mainstream its beliefs, that seeks a different role for religion in society is thrust into a minority position. Even doctrinally "orthodox" groups find their posture shaped in response to societal forces.

A related reaction to the "modernizing," pluralistic impetus of "the world" takes the form of religious extremism. In the 1970s and 1980s, there was a marked increase in media attention to religious sectarian pressures in the political sphere. In the United States, the "New Christian Right" has attempted to mobilize sectarian dissent into political action. In Israel similar impulses have been expressed in the form of religious ultranationalism. The Shi'ite branch of Islam in Iran has been notably successful in gaining political clout to turn back the forces of "modernity." These developments are discussed further in Chapters 7 and 8.

The sectarian orientation thrives on a sense of opposition. Ironically, backlash movements often inadvertently contribute to the cohesion of the very groups they attack. Similarly, although the larger society may not appreciate the polarization, the homosexual rights groups and their "moral majority" opponents are often actually symbiotically linked. Sectarian groups often provoke their opposition, partly because of their sense of their own rightness and partly because they typically enjoy greater cohesion and sense of purpose when they feel their values and goals are under attack.

Privatized Religiosity in Contemporary Society. Other types of religious collectivity are well suited to privatized religiosity. Denominations that focus on religiosity in the private sphere (i.e., family, community, parish)—and cults, with their characteristic individualism and eclectic approach to beliefs and practices—fit very well into their privatized slot. Neither claims exclusive legitimacy, and neither expects its beliefs and practices to govern behavior in all aspects of social life. A member of an Eastern meditation group may find in it greater interior peace and joy, relief from stress and tensions in everyday life, and more satisfying ways of relating to others. At the same time, the person may be an effective corporate manager whose religious beliefs and practices do not challenge or guide behavior in the business world. Belonging to a cultic religious group can perform an adaptive function, providing meaning and a sense of belonging without conflicting with members' roles in a highly differentiated society (Robbins et al., 1975).

In this context, we can examine the significance of emerging religious movements in contemporary Western society. The last three decades have witnessed the emergence of several new religious movements and the rebirth and transformation of some old ones. Although most of these movements are clearly still religious minorities, they have attracted significant numbers of people. In 1976 the Gallup Poll found that about 12 percent of those polled were engaged in one or more of certain recent movements (i.e., mysticism, Oriental religions, yoga, Transcendental Meditation, and the charismatic renewal). Cultic movements, TM and yoga, accounted for 4 percent and 3 percent of the response, respectively. Mysticism (which could be practiced in combination with most other religious perspectives, including traditional religions) accounted for 2 percent, as did the charismatic renewal, which includes members of many traditional religions such as Episcopalianism, Lutheranism, and Roman Catholicism. A further 1 percent were involved in an Eastern religion. Projecting from this sample, these figures suggest that approximately 19 million Americans were involved in these various religious movements (Gallup Poll, 1976).

The sheer number and diversity of new religious movements make it difficult to generalize about these developments. Some groups (e.g., Zen Buddhism) are based on elements borrowed directly from traditional Oriental religions. Others (e.g., Soka Gakkai and the Unification Church) are syncretic religions that emerged in the Orient and took on further dimensions in their American translation. Some emerging religions such as the Jesus People, the charismatic renewal, and neo-Evangelicalism are adaptations of older Christian movements. Other movements such as Arica and Kerista took their inspiration, as well as some of their language, from the Human Potential movement—an amalgam of religious and psychotherapeutic beliefs and practices.

The following Extended Application suggests a sociological interpretation of these emerging religious movements.

EXTENDED APPLICATION: EMERGING RELIGIOUS MOVEMENTS

The image of contemporary religious movements that have caught the attention of the press, public, and social scientists is of a diverse, colorful, strange, or exotic assortment of religious groups. The picture includes saffron-robed, head-shaven youths on street corners singing "Hare Krishna." It includes middle-aged women "channeling" messages from 35,000-year-old spiritual guides. Further images include flashy televangelists, door-to-door witnessing, Eastern gurus, suburbanite witchcraft covens, sales of crystals and incense, and tape cassettes with new hymns or religious messages by group leaders. In the late 1960s and early 1970s, some observers identified new religious forms as part of the youth culture or the counterculture, but by the late 1970s into the 1990s, it became clear that emerging religious movements were appealing to middle-aged (and older), economically comfortable, educated persons as well. Getting beyond the sensationalistic treatment that such movements have received in newspaper or television coverage is difficult. While some of the emerging religious fervor is the product of media hype or crass commercialism, much of it represents genuine religious experience and commitment.

Analyzing religious movements and established groups is more than a mere exercise in typologizing. The foremost purpose of categorizing religious collectivities is to conceptualize a situation and then to ask: Of what larger phenomenon is this an example? The types of collectivities characteristic of emerging religious movements are related to their cultural setting. Indeed, the types of social organization and characteristic religious orientations developed by these emerging religious movements are particularly well suited to the social structural place of religion in modern society. The types of collectivities characteristic of these new religious approaches are linked with peculiarly modern forms of individual-to-society connections (as discussed further in Chapter 8; see also Beckford, 1984; McGuire, 1985).

The Variety of Emerging Religious Movements

Writing around the turn of the century, Durkheim (1965:475) made some predictions about the future of religion that ring surprisingly true today:

> If we find a little difficulty today in imagining what these feasts and ceremonies of the future could consist in, it is because we are going through a stage of transition and moral mediocrity. The great things of the past which filled our fathers with enthusiasm do not excite the same ardour in us, either because they have come into common usage to such an extent that we are unconscious of them or else because they no longer answer to our actual aspirations; but as yet there is nothing to replace them. . . . In a word, the old gods are growing old or already dead, and others are not yet born. . . . It is life itself, and not a dead past which can produce a living cult.

The new religious movements appear anomalous because they have emerged precisely at a time when religion, as this society has traditionally known it, seems weakest. While theologians were exploring the meaning of the "death of God" and church administrators were trying to cope with membership slippages and general drops in attendance—and while social scientists were debating the thesis of "secularization" (see Chapter 8)—new religious movements emerged, and many of them flourished. Some of these movements are particularly interesting because they appealed to societal sectors that seemed least likely to be attracted to conventional religious movements—young, educated, cosmopolitan, comfortably middle-class persons.

The apparent newness and uniqueness of these developments are, however, misleading. New religious movements have been emerging throughout America's history, in part the result of U.S. religious and cultural pluralism. The diversity of sects that characterized America's early religious pluralism brought one Puritan minister to call the sect-filled commonwealth of Rhode Island "the sewer of New England." Later periods of great religious fervor were the Great Awakening (1730s) and the Second Great Awakening (1790s onward). The latter continued for some years and stimulated pioneer revivalism in the Western expansion well into the nineteenth century. The early nineteenth century saw many religious and communal experiments such as Ralph Waldo Emerson's Transcendental Fellowship (which spawned the Brook Farm commune) and John Humphrey Noyes's Perfectionists (out of which the Oneida commune was born). Millenarian groups such as Seventh Day Adventists and, later, Jehovah's Witnesses were founded and became especially active as the turn of the century approached. A related movement, Fundamentalism, arose within Protestantism and caused considerable controversy. The Holiness movement, out of which Pentecostalism emerged, also brought about conflict in some churches. A very different late Victorian development was the rise of Spiritualism and related cults such as "animal magnetism" (i.e., hypnosis) and various healing groups, including New Thought and Christian Science. A number of nonofficial religions flourished, though the actual size of their followings is unknown; many of these groups drew on occult traditions of Europe such as Gnosticism, kabbalism, astrology, Rosicrucianism, and so on (Ahlstrom, 1972, 1978; Albanese, 1991).

Recent new religious movements are, from this perspective, simply the latest in a long history of eruptions of religious fervor. Even the more exotic new religions emerged from a preexisting social base. The two main sources of new religious movements are the Judeo-Christian heritage and the more amorphous nonofficial religious area or "cultic milieu" (Ellwood, 1978). The Judeo-Christian strain is a fertile source of new religions because of its built-in tendency for cycles of renewal, reform, and schism. In recent times, there is an inherent potential for religious retrenchment to organize itself in sectarian forms, especially as the dominant culture in Western societies becomes less consistent with conservative Jewish/Christian ideologies.

Another segment of new religious movements emerged from the already present but unorganized stratum of nonofficial religion. Most ideas and many ritual practices of numerous new religions were already present in the cultic milieu, but the new movements shaped the ideas and adherents into an organized form. Some new religious movements claim to be totally new; others emphasize that they are older even than historical religions such as Christianity. For example, many new religions borrow from Eastern religions such as Zen and Tibetan Buddhism, various Hindu forms of yoga and meditation, or Sikh, Sufi, and Taoist traditions. Other emerging movements borrow heavily from Celtic lore, Native American religions, and shamanism. Other sources in inspiration include various psychotherapeutic group movements (often called the Human Potential movement), the "Green" movement (especially its spiritual "deep ecology" version), and various identity movements (particularly feminist groups, but also movements asserting social identity alternatives for racial and ethnic minorities, for persons with handicaps or chronic illnesses, for homosexuals, etc.). Nevertheless, as a social phenomenon even the more exotic new religions have their roots in identifiable clusters of prior beliefs and practices (Albanese, 1990, 1991; Greil, 1993; Lewis and Melton, 1992; MacDonald, 1995; McGuire, 1993; Melton et al., 1990).

Nonofficial religion tends to spawn cultic collectivities (as defined earlier in this chapter), though some of these are eventually transformed through changes in the authority pattern. One main reason for this tendency is that nonofficial religion itself is seldom a coherent, authoritative, organized belief system; adherents are often loosely affiliated, drawn together only by vaguely similar perspectives or parallel beliefs. Groups taking a cultic stance often emerge from the amorphous background of nonofficial religion, they organize around a teacher or a technique and are inherently fragile because cultic adherence is highly individualistic and eclectic. Thus, they typically dissolve back into the diffuse milieu from which they came, and members find other focuses for their beliefs and practices within the larger nonofficial religious setting. From the perspective of official religion then, with its emphasis on organizational permanence and numbers of adherents, cultic religious movements are insignificant. This view, however, overlooks the considerable staying power of general ideas supporting cultic dissent and of the numerous adherents who subscribe simultaneously to official and nonofficial religion (Ellwood, 1978).

Sectarian and Cultic Religion in Modern Society

An examination of two contemporary religious groups exemplifies some of the characteristics of modern sectarian and cultic religiosity in the United States.[5] The Women of Truth prayer group is an interdenominational fellowship of about sixty women who practice charismatic (or pentecostal) Christianity. The Meditation Circle is a group of approximately twenty men and

---

5. This information is based on my own research, fully reported in *Ritual Healing in Suburban America*, 1988; these names of groups are fictitious, and these descriptions are composites drawn from several similar groups in order to protect the anonymity of respondents.

thirty-five women who meet regularly for meditation and a variety of other spiritual practices. When we began the study, both groups had been in existence for several years, and both were steadily attracting new members.

Members of the Women of Truth were more homogeneous in age (about 35–55), education level (most had some college education, but only about a third had graduated and none had advanced education), marital status (all were married or widowed), and social class situation (virtually all were homemakers whose husbands earned middle-class incomes in business or lower-professional occupations). The prayer group met as a whole once a week, usually in a church to which some members belonged. Most activities were scheduled during daytime hours when members had the fewest obligations to their families. Most members also attended meetings of two or more subgroups that also met weekly for Bible study, visiting the sick, praying for each other, or for mutual support (e.g., a new mothers' group and a widows' group).

The Meditation Circle had a greater age range (about 30–60, with the average age a little lower than Women of Truth); members were generally better educated (nearly all had or were studying for college degrees, and about half had advanced degrees, as well). Fewer than half of the members were presently married; the rest were mainly never married or divorced. All but two of the members were employed out of the home. The income range was also greater, with three members who worked in the fine and performing arts earning near-poverty incomes and some established professionals (architect, lawyer, psychotherapist) earning enough to be in the upper-fifth income bracket in the region. The entire Circle gathered every Wednesday evening in a rented carriage house, and about half the group also met on Sunday mornings. Some members participated several times a week in other group-sponsored meditations, healing circles, classes, workshops, and social gatherings.

The sectarian orientation of the Women of Truth is particularly evident in their preoccupation with maintaining the boundaries of their truth. They emphasize a strong sense of right versus wrong, teaching constantly about sin, repentance, confession, forgiveness, and salvation. They reaffirm strict traditional moral norms regarding gender roles, marriage and divorce, sexual behavior and procreation, work, and childrearing. They interpret their daily lives as arenas of a war against Satan, a battle of powers. For example, one lengthy informal discussion shortly before Christmas was devoted to preventing members from buying for their children "dangerous" toys and games, like Dungeons and Dragons and fantasy books. Prayer meetings typically include prayers to protect members from the influence of Satan.

Like other pentecostals, the Women of Truth emphasize knowing God through experience, rather than just knowing about God. Personal experience (e.g., praying in tongues) and group experiences (e.g., hearing a prophecy—believed to be God speaking directly to the group through one of its members) are of primary importance. The belief system holds that the Holy Spirit is powerful and that this power is available to the individual through the charisms. Accordingly, the Holy Spirit empowers the individual for better prayer, personal transformation, fighting Evil, and personal healing. Similarly, the group is empowered in its battle with Evil and in its prayer, praise, and healing. Thus, religious experience is seen as both validating the group's belief system and empowering its members in their personal lives and relationships with others.

Another sectarian characteristic is the group's strong authoritarianism. Accordingly, there is only one truth; it is not up to each member to discover but is spoken authoritatively by pentecostal leaders, and the authority structure of the group is God-given. Specifically, the authority structure places women in submission to men, so the officers of the prayer group must have their husbands' permission to serve and their business meetings must be held under the supervision of a (male) pastor.

Within the context of the group, members found considerable support for integrating their pentecostal beliefs into their daily lives. One woman witnessed to how her husband's job entailed frequent moves from one city to another, and how Christ led her to a group like this in each new city to keep her close to the Christian way of life and to give her new friends. Another woman received prayers and moral support as she described the anguish of trying to stay married to an abusive man who had not yet "found the Lord." Yet another member described the "healing" of her relationship with her teenage daughter, who had stopped running away from home and had not dropped out of school. Prayer meetings were full of everyday concerns: children and husbands, relatives and friends, neighbors and pets, illnesses and personal problems, housework, churchwork, yardwork, family vacations, routine leisure activities. Supported by the close network of the prayer group, these women tried to apply their faith to all aspects of their lives. At the same time, the unity of the small, close-knit group of fellow believers served to protect members from the evils of "the world," shoring up members' defenses against those aspects of modern society their religion decries.

The Meditation Circle was, by contrast, highly eclectic in its beliefs and practices. Individual members were free to choose among many modes of spiritual development, including those not taught within this circle. The group had begun with a combination of Jain Yoga and Meditation, together with other meditations borrowed from Psychosynthesis and Arica, and other practices borrowed from Tibetan Buddhism and Reiki. Individual members had tried virtually every "New Age" or "Human Potential" movement introduced in that region: rebirthing, crystal healing, Shiatsu, neurolinguistic programming, Cornucopia, Reichian therapy, T'ai Chi, Transcendental Meditation, and so on.

No single path was considered right for all persons at all times. Rather, the group emphasized choosing change and developing one's self. One woman explained:

> Self-awareness and taking responsibility go together hand-in-hand. If I walk around unaware, then I feel like bad things just "happen" to me. I'm a victim! The more I understand myself and my situation, then the more

control I have in my life. I realize that I have choices—like I chose not to accept a promotion if it meant sixteen hour days, . . . or like choosing to be an adult and not a little girl to my mother for the first time in thirty-seven years. . . . And I can choose my responses to situations and to other people. Like I don't have to get caught up in stress or in anger or whatever.

. . . But there's no single right choice that's for everybody all the time. A lot of paths are good paths. Like, I'm a vegetarian these last few years, but that doesn't mean you aren't making the right choice for you if you aren't vegetarian. And maybe ten years from now, [being] vegetarian won't be what I need then. My responsibility is to choose which paths are best for me for now. . . . We keep growing by using that responsibility for ourselves.

Thus, there was little or no focus on fixed (much less traditional) moral norms, guilt, or sin. Rather, the good life was one of continual growth in understanding of self and others, greater awareness of one's body and emotions, greater balance and harmony in one's relationships to inner self, to others, and with the cosmos. While the "sacred" was conceived to be greater than the individual, it was tapped by going within oneself. Thus, the realm of the sacred was relatively imminent, subjective, and not fixed in one place, belief system, or religious object.

The group's ideal was a self that could be flexible and adaptable without being determined or overwhelmed by others. Group rituals served to free the self from learned constraints and negative effects of guilt and to open new potentials for choice and change. They proffered energizing experiences without expecting permanent commitment for the direction of that energy.

Group members were very loosely committed to the group and its practices. The practices changed continually, generating little long-term commitment. Members cared about each other as individuals and often went out of their way to support or help each other, but the group itself was not an end, much less the only possible source of spiritual belonging or meaning. Furthermore, many members belonged simultaneously to other groups (e.g., mainline denominational churches), which fulfilled some of their needs; through their jobs, professional ties, and other networks, they participated in wide, relatively cosmopolitan supports for other aspects of their identities.

Implications of New Religious Movements

The emergence of numerous diverse religious movements at this point in history may have come as a surprise to many observers, especially those predicting the demise of religion. New religious movements have emerged, as we have seen, from a broad background and long history of similar movements. Nevertheless, in interpreting this latest eruption of religious movements, we need to ask why they are emerging now and why these particular types of movements are finding such wide appeal.

One approach suggests that new religions develop in response to crises in world economic order and that the type of religious movement varies according to the location of adherents relative to the bases of economic power (Wuthnow, 1978). Another possible interpretation views new movements (in modern Western societies, at least) as responses to the disintegrating old bases of moral or political order. This interpretation explains dualistic movements such as the charismatic renewal and modern Christian evangelicalism as protests against the relativism and permissiveness of U.S. culture; other new movements (e.g., Human Potential and imported Oriental religions) turn the moral relativism into a virtue (Robbins et al., 1978; see also Tipton, 1982b).

Another approach suggests that these movements represent two kinds of responses to the demands of modern societies on individuals. Cultlike groups offer the ability to control and change one's social "face"—to manipulate one's social identities—while reaffirming a hidden-core self (Bird, 1978, 1979; Dreitzel, 1981). Sectlike groups address the same problem of self by consolidating all social identities into a single central, religiously defined self and by strengthening the institutions (e.g., religious group and family) that control and support that self.

One interpretation of the upswing of interest groups that, like the Meditation Circle, take a cultic stance is that they are possibly fulfilling Durkheim's prophecies about the shape of religion in modern pluralistic society: the "cult of man." Durkheim (1969:26) stated:

> As societies become more voluminous and spread over vaster territories, their traditions and practices . . . are compelled to maintain a state of plasticity and instability which no longer offers adequate resistance to individual variations. These latter, being less well contained, develop more freely and multiply in number; that is, everyone increasingly follows his own path.

Elsewhere Durkheim (1951:336) predicted that in complex societies religion would focus on the enhancement of the human personality because this would be all that members of such diverse societies would have in common. Traditional religions locate the sacred outside the individual, reflecting (according to Durkheim) the self-consciousness of the whole social group. Although the "cult of man" would express and dramatize social relations, sacred power in such religion would be located within each individual (Westley, 1978:139). The idealization of individual humanity lends itself to cult organization, with its characteristic pluralism and individualism.

Although all of the new religious movements assert alternate meaning systems and practices, their opposition to the dominant culture and religiosity is somewhat limited. Cultic groups, with their characteristic pluralism, are particularly accepting of many dominant culture values. They propose an alternative worldview and practices that emphasize neglected aspects of reality, but this alternative conveniently enhances adherents' chances for success or, at least, happiness in the context of the dominant society (Bird, 1978:181; Heelas, 1991, 1992). Sectarian groups, while encouraging members to see a gap between their way and the "ways of the world," are often similarly accommodated to the values of the dominant culture. Members of pentecostal prayer groups frequently witnessed that God had blessed them with success on the job or with material goods as evidence of God's power. The tension of contemporary religious movements with the rest of church and society appears to be segmented or privatized, and the apparent impotence of such protest may be related to wider issues, as discussed in Chapter 8.

This tension of new sects and cults with the society raises the issue of their long-term impact. Religious movements are potential sources of social transformation (explicated further in Chapter 7). Are the new religious movements transformative? Or do they serve merely to integrate their adherents into the dominant society? Many of the movements themselves explain their privatistic situation as a positive feature. They assert that real social change will emerge through their movement by the transformation of individuals. The organizational features of sectarian and cultic religion, such as exemplified by the Women of Truth and the Meditation Circle, are particularly well suited to the social structure of modern societies in which religion is privatized (see Chapter 8). The sectarian forms preserve religious purity by shoring up the supports of belief (usually in a transcendent sacred reality) by a close-knit enclave of fellow believers, protected from the threats of "the world." The cultic forms turn the individualism and freedom of choice, characteristic of the modern world, into a religious virtue and an advantage in the spiritual realm, consciously developing changeable selves in tune with a more subjective, imminent sacred reality.

Nevertheless, social movements such as these have in the past often influenced the society indirectly. Their ideas filter into the mainstream culture, and they serve as experimental models for the rest of society. The idea of holistic health and treatment of illness has filtered (in a watered-down form) from religious movements to mainstream medical schools. Whether such influences are ultimately transformative cannot yet be known. Are the new religious movements bringing about a "new age" or ushering in the millennium? Or are they ways of soothing such unpleasant aspects of modern life as loss of freedom and control in the public sphere and doubts about self and interpersonal relations in the private sphere? These implications suggest why an understanding of new religious movements and their impact is important for gaining an understanding of the society as a whole.

SUMMARY

Religious collectivities and their variety have been the subject of much analysis and typologizing. The dynamic model presented in this chapter emphasizes the ever-changing form of religious collectivities. The church–sect typology is a general developmental model for any organization defining itself as uniquely legitimate and existing in a state of positive (i.e., churchly) or negative (i.e., sectarian) tension with society. By contrast, the cultic and denominational stances do not claim unique legitimacy.

This model distinguishes between types of religious collectivity and types of religious orientation of members. The two main criteria for defining typical orientations are (1) the extent to which the person's religious role is diffused throughout all aspects of life and (2) the extent to which the person adopts standards of virtuoso religiosity for judging self and others. Diverse orientations can exist within the same collectivity. The tension between two or more different religious orientations sometimes produces change within the organization or stimulates formation of a new group.

Religious collectivities typically change as they interact with their social environments; organize themselves toward achieving their goals; recruit, consolidate, and deploy members; and arrange their organization and leadership. As a new movement becomes institutionalized, it faces several dilemmas. The very processes that would enable the movement to realize its goals simultaneously change the movement and reduce its chances of attaining those goals. Emerging religious movements illustrate this model of religious collectivities and orientations.

RECOMMENDED READINGS

Edited Collections

David G. Bromley and Jeffrey K. Hadden, eds. *The Handbook on Cults and Sects in America*. Vols. 3A and 3B of *Religion and the Social Order*. Greenwich, Conn.: JAI Press, 1993.

Thomas Robbins and Dick Anthony, eds. *In Gods We Trust: New Patterns of Religious Pluralism in America*. New Brunswick, N.J.: Transaction, 1990.

Case Studies

Nancy Tatom Ammerman. *Baptist Battles: Social Change and Religious Conflict in the Southern Baptist Convention*. New Brunswick, N.J.: Rutgers University Press, 1990. A richly contextualized analysis of ongoing conflicts within one of the largest U.S. denominational groups.

Phillip Charles Lucas. *The Odyssey of a New Religion: The Holy Order of MANS from New Age to Orthodoxy*. Bloomington, Ind.: Indiana University Press, 1995. The fascinating transition of a religious movement from cultic synthesis of diverse New Age and older nonofficial religious paths to becoming an order within the Eastern Orthodox church.

Armand L. Mauss. *The Angel and the Beehive: The Mormon Struggle with Assimi-*

lation. Urbana: University of Illinois Press, 1994. Illustrates the ongoing tension in a large, established religious movement between retaining some of its dissenting tension with "the world" and accommodating to the larger society in which it has become a prosperous participant.

R. Stephen Warner. *New Wine in Old Wineskins: Evangelicals and Liberals in a Small-Town Church*. Berkeley: University of California Press, 1988. A highly readable ethnography of a single Presbyterian congregation as it underwent significant internal changes, growth, and divisions; this book received the 1989 Distinguished Book Award of the Society for the Scientific Study of Religion.

Fiction

Umberto Eco. *The Name of the Rose*. New York: Harcourt Brace Jovanovich, 1980. Set in a fourteenth-century Italian monastery, this novel is full of mystery, historical and philosophical webs, occult and esoteric allusions, theology and religious history, semiotic riches, and, above all, the love of words—spoken words, written words, read words, experienced words.

Frank Herbert. *Dune* (1965); *Dune Messiah* (1969); *Children of Dune* (1976) all published by Berkley Books, N.Y. *God Emperor of Dune* (1981), *Heretics of Dune* (1984), and *Chapterhouse Dune* (1985) are published by G. P. Putnam & Sons, N.Y. This science fiction series includes many excellent illustrations of religious movement formation, mobilization, charisma, routinization, sectarianization, and so on.

6

◦◍◦

Religion, Social Cohesion, and Conflict

Religion contributes to both social conflict and cohesion. Some people find it surprising or objectionable that religion should be a force for conflict as well as cohesion. In many respects, however, the aspect of conflict is merely the obverse of social cohesion; some conflict is an integral part of what holds groups together. Religion's importance as an expression of a group's unity also makes it significant as an expression of that group's conflict with another group.

We must keep a neutral conception of conflict and cohesion. A tendency exists in our society to think of conflict, in the abstract, as "bad"; cohesion, in the abstract, seems "good." Yet when we examine some concrete instances of cohesion and conflict, we see that we evaluate the *content,* not the process, as good or bad. Was Moses's confrontation of the pharoah necessarily bad because it was conflict?

In this chapter, we first examine the ways in which religion reflects or contributes to the cohesion of a social group. Particularly problematic is the issue of whether religion contributes to the integration of complex societies such as the United States. Then we analyze the relationship between religion and social conflict, focusing on those aspects of religion and society in general that contribute to or reflect social cleavages. Finally, we apply these understandings to the case of "the troubles" in Northern Ireland, where religion is an important factor in civil strife.

RELIGION AND SOCIAL COHESION

Blest be the tie that binds
Our hearts in Christian love.
The fellowship of kindred minds
Is like to that above.

(Hymn, John Fawcett, 1782)

The theme of social cohesion is central to sociology. What makes society possible? What integrates separate members into a larger whole, the identifiable entity we call "society"? Society is more than an aggregate of people who happen to share a certain time and space. Although made up of individuals, society is not reducible to individual beliefs, values, and behavior. Social norms and traditions existed before the individual and have a force that is external to the individual. In a society with a norm against cannibalism, for example, persons known to violate the norm will be punished regardless of whether or not they agree with the norm. Another evidence of the external quality of society is the process of socialization, in which the child is confronted with the given expectations, language, and knowledge of that society.

What, then, is the nature of the unity of society that gives it this powerful quality? How is the individual linked to the larger society? How does the society gain the commitment and cooperation of its members? According to integration theories, societal cohesion and stability are assured by the functioning of institutions (e.g., religion, education, and the family) that represent the larger social reality to the individual and enable the individual to accept personally that definition of reality.

Religion is one important contributing factor in societal integration. Religious symbols can represent the unity of the social group, and religious rituals can enact that unity, allowing the individual to participate symbolically in the larger unity they represent. The Christian ritual of communion is not only a commemoration of a historical event in the life of Jesus but is also a representation of participation in the unity ("communion") of believers. Also, to the extent that a religion imparts important values and norms to its members, it contributes to their consensus on moral issues. Especially significant is religion's ability to motivate believers' commitment and even sacrifice to the group's purposes. By referring to a sphere that transcends everyday life, religion encourages individuals to seek the good of the group rather than their own interests. This consensus and commitment of members is largely a positive effect, but religion also wields powerful negative sanctions for noncooperation. Again, by reference to a transcendent realm, religious sanctions are more potent than earthly punishments.

Integration theories of society stress the equilibrium and harmony of the group. They show the ways that religion helps to maintain equilibrium whenever events threaten it. For example, funeral rituals help a group regain its social balance and morale after the death of a member (Malinowski, 1948:18–24). Religious healing is often a way of reintegrating the deviant member into the group (McGuire, 1988). Changes of social status (e.g., marriage, adulthood, or adoption) are integrated into existing status arrangements through religious symbols and rituals (Van Gennep, 1960). The balance of political and economic power is often ritually expressed and confirmed (Keitzer, 1988). Sometimes religious symbolism and ritual action overcome sources of group cleavages and conflict. In a contemporary Mexican village, the celebration of the fiesta of the Virgin of Guadalupe served to express ethnic cleavages (Maya and mestizo) within the community but then to overcome them by affirming the Virgin's protective relationship to all Mexicans (O'Connor, 1989).

Most examples of religion contributing unambiguously to social cohesion are from relatively simple, homogeneous societies. If religion is coextensive with society, its contribution to social cohesion is generally clear. Many societies, however, are not coextensive with a single religion. What is the basis of cohesion in a society such as the United States, in which many competing religions exist and a relatively large number of persons participate in no religious group? What is the function of religion for cohesion where conflicting societies presumably share the same religion (e.g., two warring Christian nations)? How can integration theories explain situations in which a religion arises in a society to conflict with the established ways of that society? Although religion does contribute to social cohesion in complex heterogeneous societies, its role is not very clear cut. In such societies, the basis of society-wide integration is problematic.

Religion as an Expression of Social Cohesion: Durkheim

One approach to social cohesion and religion holds that religion is the *expression* of social forces and social ideals. This perspective emphasizes that *wherever there is social cohesion, it is expressed religiously.* The classical statement of this theoretical approach is Durkheim's, and his insights are sufficiently important to explore in some detail. Throughout Durkheim's works—on the division of labor and on deviance, education, and religion—the theme of social cohesion is central (Bellah, 1973). The relationship of the individual to the larger society was relatively unproblematic for Durkheim because he identified the larger social group as the source of individuation. His theory of religion illustrates this resolution.

Durkheim based his discussion of religion on anthropological evidence about the beliefs and practices of Australian aborigines because he believed that their religion would illustrate the most elementary form. Although some of the material he used has been discounted by later anthropological investigations, Durkheim's general theory of religion is still useful for understanding some aspects of religion, particularly in relatively homogeneous societies. According to Durkheim (1965:22), religion is in its very essence *social.* Religious rites are collective behavior, relating the individual to the larger social group. And religious beliefs are **collective representations**—group-held meanings expressing something important about the group itself. Durkheim may have overstated the social aspect of religion, as Malinowski (1948:65) later pointed out, because subjective and even purely individual religious experiences figure importantly in virtually all religions. The experiences sought by the Christian mystic or a participant in a Sioux initiation rite, for example, are highly individualistic, though their meaning is derived from group-held beliefs and imagery. Furthermore, Malinowski noted, not all times of collective effervescence are religious, nor are all religious gatherings necessarily unifying (see Geertz, 1957). Bringing people together for a periodic religious ritual creates the potential for friction, especially in times of stress or hunger. Nevertheless, Malinowski agreed with Durkheim that religion provides a basis of moral cohesion to the social group.

Durkheim observed that a sense of *force* was central to primitive religions. The totemic principle that represented abstract force to the Australian tribes was comparable with other primitive religions' awe of a force (e.g., *mana, orenda,* and *wakan* in Polynesian and North American tribes). Durkheim emphasized that religious force was not an illusion. Although the symbols for expressing this force are imperfect, the force that people experience is real—*it is society.* Durkheim described religion as a system of shared meanings by which individuals represent to themselves their society and their relations to that society. Thus, religious meanings are metaphorical representations of the social group, and participation in religious ritual is experience of the transcendent force of society itself (1965:257).

Thus, Durkheim resolved the individual-to-society relationship by presenting individuals as transcending themselves in communion with the greater reality—society itself. But Durkheim added that this force is not entirely outside the individual; it must also become an integral part of the individual's being because society cannot exist except through individual consciousness. By this twofold relationship, religion assures the commitment of society's members and empowers them to act accordingly. Thus, religious beliefs are idealizations by which society represents itself to its members. Religious rites renew these representations by rekindling the group's consciousness of its unity. At the same time, they strengthen individuals' commitment to the group's expectations and goals (Durkheim, 1965:especially Book 3, Chapter 4).

How do collective representations and collective rituals link the individual to the larger social group? Durkheim's emphasis on language and ritual suggests two ways of understanding this individual-to-society link. First, language and other symbol systems (e.g., religious symbols) depend on shared meanings, and meaning requires a shared reality. By exploring how language and other symbol systems articulate a group's reality, sociologists may come to a better understanding of how people subjectively share that reality (see Fabian, 1974:249–272). Second, studies of ritual can point to how the individual is related to the larger society. Participation in certain religious rituals appears to reduce the sense of boundaries between participants, producing an experience of unity (see Douglas, 1966, 1970; Kertzer, 1988; Spickard, 1991; Turner, 1969, 1974a, 1974b). Both language and ritual articulate the unity of the group and serve to separate that group from others.

Durkheim's emphasis on the significance of ritual for both the individual and the collectivity points to some interesting problems of describing religion in modern society (see Pickering, 1984:442–456). If social cohesion is expressed through tribal collective rituals and meanings, how does it function in a society whose members are from numerous different ethnic backgrounds? Several observers have noted the close link between religion and ethnicity among immigrant peoples. Religious groups provided important resources to newly arrived immigrants, especially during the great waves of immigration in the nineteenth and twentieth centuries. They provided informal networks of association through which the immigrant could obtain a job, help with housing, and other mutual assistance. They also provided protection from the dominant society, keeping alive the old ways, educating children in "safe" environments where their backgrounds would not be held against them, and providing mutual protection from the hostility of those who did not accept the immigrants. When the historical need for these functions has become muted or disappears, the ethnoreligious community may still be a source of mutual aid, friendship, and a sense of belonging. For these individuals, the ethnoreligious group as a set of relationships—not the fact of ethnicity itself—provides a stable source of belonging.

Herberg analyzed the relationship between religion and ethnic group identity in the 1950s, specifically asking why people in the United States tended to identify themselves in terms of one of three religious "communities": Protestant, Catholic, or Jew. He suggested that as the immigrant became assimilated into the dominant U.S. culture, ethnic ties superseded localistic ties that characterized one's self-identity and self-location in the old country. The ethnic church was an important expression of these new ties. According to Herberg, however, ethnic differences are residual and disappearing. As assimilation proceeds even further, religious identity assumes even greater significance—not as actual affiliation with a particular religious group, participation in church activities, or even affirmation of the group's belief system—but rather as a basis of identification and social location. According to Herberg, self-identification in ethnic terms was not altogether satisfactory because it implied incomplete integration into U.S. life. By contrast, religion in the United States is an acceptable way for people to differentiate themselves and thus becomes a way for people to define and locate themselves in the larger society. For this reason, according to Herberg's interpretation, people will identify themselves with one of the three legitimate religious "communities"—Protestant, Catholic, or Jewish—even though they do not necessarily practice that religion or believe its tenets (Herberg, 1960:6–64).

Herberg's analysis was based on earlier waves of European immigrants who fit these three "communities." Indeed, it was in response to those waves of immigration that the United States gradually changed from being essentially a Protestant empire to considering itself a Judeo-Christian land. After new immigration laws took effect in 1968, however, both the volume and the sources of new immigrants changed dramatically. Between 1960 and 1989, a significant proportion of immigrants (37 percent) came from Asia; another 39 percent came from various parts of Latin America, and a much smaller but significant portion came from the Middle East (Kivisto, 1993). How applicable will Herberg's thesis be to immigrant groups whose religious traditions are completely foreign to the Judeo-Christian communities?

Herberg's thesis might thus apply to the high rates of church participation of Korean Christians, most of whom were Christians before emigrating from Korea (Hurh and Kim, 1990). Communities centered on Vietnamese or Hmong folk religions, however, would *not* readily be accepted as legitimate bases of separate identity within U.S. culture. What little evidence we have suggests that Southeast Asian refugees, resettled in the United States between 1975 and 1983, often change affiliation to Western religions (a Utah study found 36 percent had "converted"). Whether in Western or Asian churches or temples, however, the immigrants remained in ethnic congregations, which appear to be especially important in dealing with the drastic adjustments these refugees must undergo (Lewis et al., 1988). Thus, ethnic group identity appears to be linked with religion for this new wave of immigrants, even though neither their native religions nor their racial-ethnic identities are as likely as those of earlier European immigrants to be readily accepted as legitimate by the larger society.

Herberg's prediction of ethnicity's disappearance as a basis of personal identity in a pluralistic society may have been premature (as the discussion of ethnic self-identity in Chapter 3 suggests). As a basis of group identity, ethnoreligious ties are highly salient for many groups in the United States. Creating a sense of solidarity among a fragmented minority has important political implications for groups lacking a voice in their cities, denominations, and nation. "Latino" group identity, for example, is extremely difficult to forge given the wide range of cultural, geographic, and social class backgrounds of the approximately 25 million Americans the term was invented to unify (see Cadena, 1995). Catholicism may have initially been a source of group identity for immigrant Latinos, but many Latinos do not find the Catholic church in the U.S. to be a strong source of belonging; indeed, many have become Protestant, but many more simply do not identify with a Catholicism so foreign to their own cultural experience. Some also feel that the U.S. Catholic church fails to represent their interests in important political and economic issues as adequately as it represented earlier Irish, German, or Eastern European immigrant minorities (Stevens-Arroyo, 1994). Indeed, becoming Protestant may help immigrant Latinos become acceptably "American," while providing some of the socioeconomic supports of an ethnoreligious community (Greeley, 1991). Thus religion is potentially linked, in sometimes complex ways, with the social cohesion *and* conflict of culturally pluralistic peoples within a nation.

Moving from primitive religions to world and civil religions, Durkheim (1965:474, 475) asserted that all societies need regular events to reaffirm their shared meanings and central ideas. He saw no basic difference between specifically religious commemorations such as Passover or Christmas and civic rituals such as Independence Day or Thanksgiving. Thus, Durkheim implies that in modern nations (e.g., his own country, France) religious representations and rituals may comprise the civic religion of the national collectivity.

Is it possible for a highly differentiated, heterogeneous society (such as the United States, Canada, and much of Europe) to have a religious expression of its solidarity? Is there a unifying representation of the nation to inform and reflect diverse peoples' sense of "who we are as a nation" and "what kind of nation do we want to be"? What happens when, in the same country, there are conflicting religious representations of the nation? How is religion, then, linked with the nation?

Religious Representations of the Nation

The civil religion thesis is important because it proposes that a religious form exists for the unity of even highly differentiated, heterogeneous societies. **Civil religion** is "any set of beliefs and rituals, related to the past, present, and/or future of a people ('nation') which are understood in some transcendental fashion" (Hammond, 1976:171). Civil religion is the expression of the cohesion of the nation. It transcends denominational, ethnic, and religious boundaries. It includes rituals by which members commemorate significant national events and renew their commitment to the society. Such rituals and representations are religious in that they often represent the nation—the people—as a higher and more valuable reality than mere (i.e., human) social contract and convention. They may also stir religious fervor and sentiments regarding the national collectivity. This concept appears to apply to many features of U.S. religion as it is linked with America as a nation.

American Civil Religion. Many civil ceremonies have a marked religious quality. Memorial Day, Fourth of July, presidential inaugurations, all celebrate national values and national unity (Bellah, 1967; Cherry, 1969, 1992; Lerner, 1937; Warner, 1953; Wuthnow, 1994). There are national shrines such as the memorials in Washington, D.C., the Capitol itself, the birthplaces of key presidents, war memorials, and other "special" places. It is not their age or even historical significance but their ability to symbolize the transcendence of the nation as a "people" that inspires awe and reverence. A visitor to Independence Hall said, "Just standing here sends chills down my spine." National shrines are "sacred," in Durkheim's sense of the word. Similarly, there are sacred objects of the civil religion—especially the flag. Interestingly, the Bible is probably also a sacred object in civil religion, not because of its content but because it signifies an appeal to God as the ultimate arbiter of truth and justice. The extent to which these ceremonies, shrines, and objects are set apart as sacred can be seen in the intensity of outrage at inappropriate behavior or "desecration." Some people were arrested during the 1960s for wearing or displaying a copy of the U.S. flag improperly (e.g., on the seat of their pants), and in the 1990s various legislators introduced bills to criminalize the burning of the U.S. flag in symbolic protests.

Civil religion also has its myths and saints. Lincoln is a historical figure who particularly symbolizes the civil religion. His actions and speeches contributed to the articulation of that religion in a time of crisis, and his life from his humble birth to his martyrdom typifies values of the civil religion. Other "saints" include key presidents (Washington, Jefferson, Wilson, Franklin D. Roosevelt, Kennedy), folk heroes (Davy Crockett, Charles A. Lindbergh), and military heroes (MacArthur, Eisenhower, Theodore Roosevelt). Similarly, there are stories that exemplify valued traits (e.g., the Horatio Alger rags-to-riches genre) and images (the frontier). Socially important myths include the American Dream—the land of plenty—unlimited social mobility, economic consumption, and achievement. Although these shrines, saints, and ceremonies are not religious in the same sense as, for example, Greek Orthodox shrines, saints, and ceremonies, they are still set apart as special and not to be profaned. They are an important element of nonofficial religion (as described in Chapter 4) and exist alongside—separate, yet related to—official church religion.

If civil religion is the expression of the integration of a nation, we might expect its especially powerful articulation at the resolution of a conflict. Just as tribal rituals "heal" internal strife and celebrate the unity of the tribe, so too does the rhetoric of inaugural speeches and court pronouncements represent the resolution of conflict and appeal to the overarching integration of the group (Hammond, 1974). The symbolism of civil religion is also very evident when the nation is believed to be threatened by an enemy. During wartime, members' commitments and sacrifices are given special significance. Even one's vegetable garden becomes a symbol of patriotic effort—a "victory garden."

According to Robert Bellah (1967), American civil religion is related to biblical religion, yet is distinctively American. Biblical symbolism has prominent themes (e.g., chosen people, promised land, new Jerusalem, death and rebirth). On the other hand, the civil religion is genuinely American and parallels the biblical religions, not replacing them. Civil religion and Christianity, accordingly, are clearly divided in function: Civil religion is appropriate to actions in the official public sphere, and Christianity and other religions are granted full liberty in the sphere of personal piety and voluntary social action. This division of spheres of relevance is particularly important for countries such as the United States, where religious pluralism is both a valued feature of sociopolitical life and a barrier to achieving a unified perspective for decision making. By having a civil religion for the public sphere and a diversity of particular religions in the private sphere, the social structure has cohesion with the sense of individual freedom of choice. The success of this division is, however, problematic in U.S. society.

What emerges from sociological descriptions of American civil religion is a picture of diverse—even conflicting—values associated with what is central to the people. Bellah's version implies a single, clear-cut, yet ever-developing ideological stance. But is the American civil religion a unified entity? And if not, does it have the capacity to be unifying?

One explanation of diversity in American civil religion is the distinction between "priestly" and "prophetic" versions of civil religion (Marty, 1974; Wuthnow, 1988, 1994). The **priestly version** of American civil religion celebrates the greatness of the nation, its achievements and superiority. The **prophetic version** calls the nation's attention to its offenses against the idealizations for which it stands. Both versions are very much a part of American thought and rhetoric, but they are clearly in conflict. During the 1960s and 1970s, these versions were used to justify opposing stances on U.S. engagement in Vietnam. One set of bumper stickers during that period proclaimed, "America—Love it or leave it!" Another set of bumper stickers retorted, "America—Change it or lose it!" Are these expressions of the *same* civil religion?

A prophetic version of American civil religion reminds the nation of higher ideals that it must strive to meet. Drawing on this version of civil religion, a group of African-American clergy organized a public challenge to their city's economic development plan, claiming a moral agenda to legitimate their involvement on economic issues. They achieved legitimacy for their dissent by claiming moral "high ground"—not only by reference to their particular religious status and values (e.g., the Catholic Bishops' pastoral letter outlining Catholic values on the economy)—but also by reference to civil religious values such as inclusivity and justice drawn from this prophetic version (Williams and Demerath, 1991).

By contrast, the priestly version of American civil religion frequently devolves to nationalistic sentiments and beliefs or to identification of God's will with the aims of "our kind of people." In 1992 President Bush used civil religious imagery to legitimate the U.S. involvement in the war against Iraq in the Persian Gulf. He told a group of Christian radio and television station officials, "I want to thank you for helping America, as Christ ordained, to be a light unto the world" (quoted in Rosenthal, 1992). A priestly version of American civil religion has historically been used to legitimate intolerance, as illustrated by the U.S. history of agitation and discrimination against Asian-Americans since 1850, culminating in the internment of American Japanese during World War II (H. Hill, 1973).

Even the most vulgar forms of American civil religion have considerable appeal and motivating power, exemplified in the prosecutor's speech to the jury at the trial of some communist labor organizers in 1929:

> Do you believe in the flag of your country, floating in the breeze, kissing the sunlight, singing the song of freedom? Do you believe in North Carolina? Do you believe in good roads, the good roads of North Carolina on which the heaven-bannered hosts could walk as far as San Francisco? . . . Gastonia—into which the union organizers came, fiends incarnate, stripped of their hoofs and horns, bearing guns instead of pitchforks. . . . They came into peaceful, contented Gastonia, with its flowers, birds, and churches . . . sweeping like a cyclone and tornado to sink damnable fangs into the heart and lifeblood of my community. . . . They [the people of Gastonia] stood it till the great God looked down from the very battlements of heaven and broke the chains and traces of their patience and caused them to call the officers to the lot and stop the infernal scenes that came sweeping down from the wild plains of Soviet Russia into the peaceful community of Gastonia, bringing bloodshed and death, creeping like the hellish serpent into the Garden of Eden (quoted in Pope, 1942:303–304).

Civil Religion as Legitimating Myth. If, in practice, there is no compelling, unifying civil religion in the United States, perhaps we might better conceptualize these religious sentiments and rituals as competing legitimating myths (Hammond, 1994). **Legitimating myths** are stories out of which people live and which they use to justify their values, actions, and identity. Thus, images of America such as being "the Chosen People" or God's vehicle for realizing a millenarian utopia serve both to inform believers' ideas of who they are as a people and to legitimate their nation's collective identity and actions (Moorhead, 1994). Nineteenth-century Populism, for example, drew extensively on civil religious ideas, symbols, meanings, and values to inform and legitimate its ideology, using it as a cultural resource in mobilization of a political movement (Williams and Alexander, 1994). A legitimating myth can be simultaneously a subjectively meaningful source of meaning and belonging and also an ideology. Some of the linkages between religion and ideology, discussed further in Chapter 7, also apply to national legitimating myths.

As legitimating myths, religious representations of the nation are still potentially very powerful. They are cultural resources upon which citizens may draw, both for personal meaning and for mobilizing collective sentiment.

Some civil religious themes may recede in usefulness as resources. Some themes may be creatively used and combined in new and unforeseen ways; the application of civil religious themes to the environmental movement is a good example (Albanese, 1992).

If we think of civil religion as a cultural resource available for selective use rather than as a fixed institutional entity, it may help make sense of how this same civil religion could be a significant source of cultural *conflict* (Wuthnow, 1994). If Americans are in conflict over basic notions of "what it means to be one of us" and "what kind of a people do we want to be," opposing civil religious sentiments are likely to be stirred in debates about abortion, capital punishment, immigration, civil rights, family values, and economic justice, among others. Rather than a taken-for-granted foundation, the social solidarity of modern societies becomes the site of contestation; those engaged in these conflicts actively work to construct and maintain their vision of solidarity (Beckford, 1994).

It may well be that in advanced industrial societies, civil religious elements—like components of individuals' religions—have become somewhat loosened from their institutional coherence, as discussed further in Chapter 8. If this hypothesis is accurate, it suggests that civil religion loses much of its capacity to accomplish solidarity, especially a cohesion that transcends divisions of ethnicity, region, and particular religions. What symbolic power remains may be highly susceptible to political manipulation and commercialization. Many public rituals involve citizens not as voluntarily committed and active participants, but rather as spectators and consumers. Is the quality of civil religious commitment not changed when a Fourth of July celebration is an elaborately staged, professionally orchestrated spectacle, parading U.S. technology, serenaded by a professional band amplified through an elaborate sound system, highlighted by expert lighting technicians and fireworks choreographers, underwritten by corporate contributions, and promoted as a "Kodak moment"? (Wuthnow, 1994).

Why has the religious basis of collective solidarity become an important problematic theme for the social sciences at this time? One reason is the worldwide prominence of religious factors in the splintering of nation-states, such as the former Yugoslavia. Another is the changed location of religion in modern societies (discussed in Chapter 8).

Religion and Nation. The priestly form of civil religion is similar to nationalism—the ideological expression and building of individual identification with the nation. A useful distinction at this point is between state building and nation building. **State building** refers to developing an authoritative, utilitarian organization for expediently conducting a country's internal and external business. **Nation building** refers to developing a country's sense of solidarity and identity as a people (Eisenstadt and Rokkan, 1973). An example of this analytical distinction is the Quebecois of Canada. Although their primary identification may be regional (i.e., to their province), most Canadians also identify with the Canadian nation, as well as participate in the Canadian state. The Quebecois, however, do not feel part of the Canadian national identity or solidarity, even though they are subject to the authority of the Canadian state. Their sense of themselves as a people is identified with entirely different symbols, myths, folk heroes, history, and national holidays from that of the rest of the country (Kim, 1993). Indeed, many Quebec residents who voted against separation from the Canadian state did so for economic and political reasons, rather than any sense of national identification with Canada. Existing civil religions in Canada are not nation building.

Many contemporary nation-states came into existence arbitrarily as the result of colonial expansion, division of former colonial territories, or political mergers imposed upon an area by a ruling power. The transformation of these political units (states) into nations is not automatic, and indeed depends upon several cultural factors, including religion (Anderson, 1991).

Civil religion is clearly an element in this process. It can shape a national vision, sacralizing the ideals and "destiny" of a people. It can give national solidarity and identity a religious quality, enabling peoples of diverse tribal, regional, ethnic, and religious groups to come together in a central, unifying cultural experience. Yet, where cultural nationhood is not firmly grounded, where people are divided by religious, ethnic, and linguistic ties, civil religion cannot readily bring people together. Any civil religion arising from tribalistic particularism cannot be a basis of social cohesion for a nation integrating diverse peoples. In the war-torn former Yugoslavia, for example, the legitimating myths supporting Serbian identity as a "people" illustrate this tribalism: They depict Serbian Christians battling Turkish Muslims more than 600 years ago, folk heroes of resistance to Turkish rulers, Serbs displaced by Croats, rural and mountain people resisting urbanization and cosmopolitan attitudes (Kifner, 1994). Such symbols do not have the potential to unite a nation that includes many people of several other ethnic, language, religious, and cultural backgrounds. The problem of tribalistic particularism is discussed further in the second part of the chapter, because such civil religions—as national legitimating myths—are potent sources of social conflict, as exemplified by the situation of Northern Ireland (see Extended Application).

Several factors affect the potency of a religious tie with nationhood. One obvious factor is the degree of religious *homogeneity* in the nation. If a people are already of a single particular religion, that religion can (but does not necessarily) serve as the link with nationhood. Where historically that single religion has been the sole available means of expressing nationality against foreign domination, that religion is especially likely to develop as a potent form of civil religion as well (D. Martin, 1978:107; see also Ramet, 1989). This religion–nationalism relationship is exemplified by the situation of Palestine, Cyprus, Poland, and Ireland.

Because religion has considerable potential for legitimating (or delegitimating) the ruling authority of the state, leaders and would-be leaders often try to shape and interpret the religion-to-nationhood linkage in a way that would favor their own power and legitimacy. Sensing the growing influence of opposition Shi'ite fundamentalist leaders in the 1960s and 1970s, the Mohammad Shah of Iran tried to create a civil religion under the control of the state in order to undermine the traditional religious institutions. He attempted to persuade the people that he was the rightful guardian of the faith. He began to replace religious courts with state-sponsored "Houses of Justice," religious educational institutions with state-controlled ones, and mosques under the control of his religious opponents with state-sponsored ones. The Shah's religious establishment had little effect on the loyalties of most Iranians; indeed, many were incensed, believing that the Shah's civil religion was a perversion of their Islamic tradition (Voll, 1982:292–296). The failure of the civil religion contributed to the decline of the legitimacy of the monarchy and its eventual overthrow by revolutionary religious fundamentalists.

Often, representatives of particular religions are highly critical of civil religions, especially when the particular religion seeks *hegemony* (i.e., dominant influence) in the state. Some critics decry the dilution or adaptation of the true (particular) faith and its symbols. Others criticize any accommodation with religious perspectives other than their own. In the United States, much of the conflict over prayer in public schools is due to the tension over the state's pluralistic accommodation of non-Christian or nonreligious minorities. Among the most outspoken proponents of prayer in the schools are those who seek a conservative Christian hegemony in the nation. Such tensions over the "proper" religious national vision illustrate that there may actually be conflicting or competing civil religions. They have also on occasion tried to yoke elements of the civic faith, such as flag worship, to their political agenda by claiming to be uniquely the "true" patriots.

The development of civil religion in Israel also reflects the influence of particular religious perspectives in trying to shape the image of nationhood. Israel exemplifies a situation in which there are multiple competing national visions with their respective civil religions. Between 1919 and 1945, the Israeli civil religion could be characterized as "Zionist-socialism," uniting around a blend of socialist-worker beliefs and symbols together with an image of the state of Israel as the culmination of Jewish history. "Statism" was the second phase, between 1948 and 1956. It was characterized by an emphasis on the state of Israel itself as a centralized political reality and as a symbol of the collective Jewish people. The statist model, however, did not develop adequate symbols and rituals to be effective. Since 1956 a new civil religion has developed; it focuses more on Judaism and Jewish tradition as a basis for collective Israeli identity and on Israel's connection to worldwide Jewry. Thus, Judaism and Jewishness have become increasingly central to Israeliness. The explicit linkage of the particular religion and national identity has been effectively reasserted (Liebman and Don-Yehiya, 1983; Liebman, 1993). Different national images and their relative emphasis on traditional Judaism have profound influence on the problem of integrating the society, treatment of indigenous (Muslim) Palestinians within the country, and on formulating a viable diplomatic policy relative to Israel's Arab neighbors (Aran, 1991; Lustick, 1988; O'Brien, 1987; Ruether and Ruether, 1989; Sprinzak, 1993).

The comparison of America's and other nations' legitimating myths is useful for an understanding of the religious link between the citizen and nation. It is also related to church–state and religion–power issues, discussed further in Chapter 8.

The civil religion thesis is an important sociological concept. It explains certain aspects of American religiosity that are not related to particular religions. It provides a hypothesis for understanding expressions of national unity in a heterogeneous, highly differentiated society. American civil religion, as described by Bellah and Robertson, is related to the apparent weakness of particular religious institutions in the public sphere. This suggests that the development of a separate civil religion may be related to processes of modernization and secularization, discussed further in Chapter 8.

Finally, the civil religion thesis proposes a basis for the relationship of the individual to the larger modern society. It allows, in theory, the societal needs of cohesion and commitment of members and the individual needs of identity and belonging to be met by the same social processes. To what extent American civil religion really accomplishes this linkage remains unclear. We need to know to what extent people identify their interests and sense of belonging with the nation as a whole. Or, by contrast, to what extent do they locate their interests and community in particular segments of the society that are in conflict with other segments? The relationship of the individual to society is a critical issue in understanding modern societies. If a civil religion can effect a special kind of individual-to-society connection, it forms an important clue to that understanding (Bellah, 1978; Robertson, 1978). And if a civil religion is unable to achieve that connection, the reasons for this inability are also useful for our understanding of modern society.

RELIGION AND SOCIAL CONFLICT

Mine eyes have seen the glory of the coming of the Lord.
He hath trampled out the vintage where the Grapes of Wrath are stored.
He hath loos'd the fateful lightning of His terrible swift sword.
His truth is marching on.
Glory, Glory, Hallelujah . . .
His truth is marching on.

("Battle Hymn of the Republic," Julia Ward Howe, 1861)

This hymn, which has as its context both particular (Protestant) religions and American civil religion, illustrates the capacity of religion to inspire and reflect social conflict. In this section, we consider factors that make religion a powerful basis or reflection of social cleavages.

Conflict as the Obverse of Cohesion

We must keep in mind that cleavage and conflict are, in many respects, merely the "other side of the coin" of cohesion and consensus. We tend to think of conflict as a breach in sociation, but as Georg Simmel (1955:18) reminds us, conflict is one form of sociation. Simmel emphasized that "a certain amount of discord, inner divergence, and outer controversy, is organically tied up with the very elements that ultimately hold the group together."

Thus, our discussion of religious expression of social cohesion is necessarily related to a consideration of conflict as well. For example, the Doukhobors are a religious sect in Canada and the United States whose values and way of life often conflict greatly with those of the dominant society. Members have often clashed with educational, social welfare, and police authorities; for example, in 1953, Canadian Doukhobors protested educational authorities' toughened stand of public schooling for their children by burning schools and making nude pilgrimages (Woodcock and Avakumonic, 1977). Their strong cohesion as a group sets them off from non-Doukhobors. At the same time, their experience of opposition from the rest of society increases their group's cohesion and commitment. So cohesion at one level of association can produce conflict on another level, and conflict from the outside can contribute to internal cohesion.

Religion has been historically related to conflict at several levels. Perhaps the most obvious has been conflict *between* religious groups, especially when religious boundaries are coextensive with political boundaries. Religion played an important part in the reconquest of Spain by "the Christian monarchs" (end of the fifteenth century). This event was accompanied by powerful "us-against-them" sentiments, resulting both in expulsion of the Moors (i.e., Moslems) and suppression of indigenous Jews. Conflict between religious groups within modern nations tends to be more subtle (e.g., much of the anti-Semitism in America); but when religious boundaries are coextensive with other boundaries (e.g., social class, race, or ethnicity), open conflict can erupt.

Another level of conflict arises *within* a religious group. Conflicts during the early part of the Protestant Reformation exemplify this type, as protesting groups were defined as splinter groups within the Roman Catholic church. Similarly, many current religious groups came into being as offshoots from larger religious bodies with whom they had some quarrel. The history of U.S. and Canadian Protestantism is the record of many hundreds of such internal conflicts.

Sectarian religious groups also exemplify another level of conflict—between a religious group and the *larger society*. Sometimes the conflict results in reprisals by the larger society, as when courts override family or educational arrangements of the Mormons, Amish, and Doukhobors and jail Quaker conscientious objectors. And sometimes sectarians express conflict only by condemning and withdrawing from the "ways of the world."

Social Sources of Conflict

Social Cleavages. Some religious cleavages have their sources in the organization of society. As noted in Chapter 2, religious belonging is one basis of self-identification. A strong sense of belonging (e.g., to a national, religious, kinship, or ethnic group) entails a sense of barrier between members of that group and those outside it. Because religion is one important basis for group identification in society, it is a potential line of cleavage (Coleman, 1956:44, 45). A closer look at religious conflict, however, suggests that the situation is more complex. Often religious boundaries overlap with other lines of cleavage such as social class, race or ethnicity, political or national allegiance, and so on. What appears to be a religious struggle may be also an ethnic or social class conflict.

When religious divisions are coextensive with other lines of cleavage, it is difficult to distinguish the exact role of religion in conflict. In 1844, Protestants and Roman Catholics in Philadelphia engaged in armed combat. Protestants were angered that the Catholic bishop had persuaded the school board to excuse Catholic children from religious instruction, which was then a standard part of the public school curriculum. Mass meetings were held to attack the change, and a Protestant crowd marched into a Catholic neighborhood. Street fighting and general rioting resulted, and Protestant mobs set fire to several houses and Catholic churches. As violence mounted, the governor sent the militia, against which the Protestant mob fought with its own cannon and muskets. Thousands of Catholics fled the city (Shannon, 1963:44).

On the surface, this historical event appears to be a simple case of religious conflict, but the divisions between Roman Catholics and Protestants were also cleavage lines of ethnicity, economic interest, politics, and neighborhood. Catholics involved were almost all Irish, and many were recent immigrants. Protestants were largely part of a rising political stream of anti-immigrant fervor, later culminating in the "Know-Nothing" Party (1854). Economic factors included the competition of immigrants with WASP "natives" for jobs. While the extent of anti-Catholic prejudice at that time should not be understated, it is nevertheless difficult to distinguish elements of religious conflict from other sources of divisiveness.

Because religion is often coextensive with other lines of cleavage, it is frequently used as a way of expressing other divisions. A South African can use loyalty to the Dutch Reformed church to express many other loyalties: racial (white, as opposed to black), ethnic (Afrikaner/Dutch, as opposed to English), language (Afrikaans and rabidly anti-English), and political (Afrikaner Nationalist Party) (see Moodie, 1978, and parallels with the Lebanese situation discussed in Labaki, 1988, and the Northern Irish conflict described in further detail in this chapter). Overlap of cleavage lines also means, however, that sometimes religious conflict coincides with other perhaps more fundamental cleavages. In India, for example, state policy regarding education (and by extension economic opportunity) must deal with competing or conflicting demands of groups, defined by coinciding religious–ethnic–language identities (e.g., Aryan Hindus using Sanskrit, North Indian Muslims speaking Urdu, Sinhalese Buddhists using Pali, and Punjabi Sikhs writing Gurumukhi). When these groups conflict over an educational policy, religion is a relevant factor, but it is inextricably intermeshed with ethnic, linguistic, and often economic and regional divisions as well (Oomen, 1989). Because of this kind of linkage with other important sources of group identity, religion has often been a mask or even an overt legitimation for political and economic conflict.

Both the rhetoric and inspiration of religious conflict can cover aspirations for political power or economic gain. Right-wing radio evangelists of the 1930s, Protestant minister Gerald L. K. Smith and Roman Catholic priest Charles E. Coughlin, organized a politico-religious campaign. Virulent anti-communism, anti-Semitism, and nativism were central to their message. Coughlin attracted a radio audience of approximately 10 million weekly listeners, who contributed so much mail and money that he needed 145 clerks. Together Smith and Coughlin organized the Union Party for the 1936 presidential elections. After the defeat of their presidential candidate, Coughlin emphasized anti-Semitism even more, primarily as a symbol of his attack on the country's economic and money system. Although Coughlin received some active church opposition and very little official Roman Catholic support, this right-wing coalition had special appeal to discontented urban Catholics, the elderly, and rural poor (largely Protestant). The political ambitions of the coalition were served by the anti-Semitic and anti-"subversive" messages proclaimed with religious teachings over the radio (see Bennett, 1969; Lipset, 1963; Ribuffo, 1983). In this example, a religious cleavage between Christians and Jews was exploited to further political ends.

Religion and Nationalism. Today religion often feeds nationalism or nationalist aspirations. Occasionally it is captured by them. Ethnic conflict in the former Yugoslavia, for example, is organized more along religious than along other lines. This is not because the Serbs, Croats, and Bosnians are particularly religious people, but because their formal religious allegiances have become the only salient division between them. Serbian identity is historically tied to Eastern Orthodoxy, Croatian identity to Roman Catholicism, and Bosnian identity to Islam. Despite some protestations to the contrary, they share a language, a culture, and a gene pool, but they do not share a national sense of self. Their "ethnicity" is constructed from their communal identity, rather than the other way around (see Bauman, 1992; O'Brien, 1993). The Serbs traditionally saw themselves as a vanguard in the historic fight against Islam: the guardians of Europe's eastern gates against the invading Turks. The Croats look west, rather than east—identifying themselves with Rome, not Constantinople. Both continue to look down on Bosnian Muslims as traitors and collaborators—Slavs who converted to Islam during the long Ottoman occupation centuries ago (Ignatieff, 1993).

In this situation religion has become a marker for other social factors that underlie the conflict. Some of these are economic: Poorer provinces resented the advantages given the richer ones when Yugoslavia was united (Botev, 1994). Others are political: Certain politicians found that they could gain power by appealing to nationalist passions (Silber and Little, 1994). But the cultural factors are most important. Serbs, Croatians, and Bosnians imagine themselves each to be "nations," each defined in opposition to the others.

Religious differences did not create those nations, but they did mark them off from one another.[1] Once marked, the communities could be mobilized by politicians on all sides.

Religion does not always figure in the creation of such "imagined communities" (Anderson, 1991), but it is a potent resource when it is so used. Sociologists have identified some of the many other factors involved (see Calhoun, 1993; Greenfeld, 1993; Hobsbawm, 1992). Religion is often an important factor in its own right, however: from the nationalism of the French Revolution (Bell, 1995) to the Northern Ireland "troubles" described further in this chapter. Religion can become a "sacred marker" that accentuates and stands for the totality of divisions between people (Hanf, 1994). When conflicting groups treat their religions as cultural *possessions,* the value of which rests not in religious practice itself, but rather in defending the cultural possession against others, religion is particularly likely to serve as such a marker (Tambiah, 1993). Even in modern societies, religion seems to be as salient as ever for dividing people from one another.

The Marxian Perspective. Marxian interpretations hold that religious conflict is merely the expression of fundamental economic relationships. Accordingly, the dominant class attempts to impose ideas (including religious ideas) that legitimate its interests. Or it may exploit existing religious divisions among subordinate peoples to prevent them from realizing their true class interests. As we will note in Chapter 7, religion can contribute to legitimation of the existing social system and also express real conflicts in that system (Turner, 1977).

According to Marxian theory, religious dissent is often an expression of economic dissatisfaction, but its religious nature prevents dissenters from fully realizing their economic class interests. Religious conflicts have often divided members of the working class when their true interests, according to Marx, would be best served by uniting against the common enemy, the ruling class. In the late nineteenth century, some 100,000 members of the American Protective Association pledged not to vote for, strike with, or hire Catholics if non-Catholics were available (Ribuffo, 1988). Such movements attracted economically threatened working-class persons who might have otherwise benefited from uniting in political and labor movements with Catholics who shared their class interests.

Part of the reason why Marx's predictions of increasing class conflict have not eventuated in many modern societies is that their members are subject to

---

1. Because imagined communities want to believe that their basis for unity is an essential or "pure" quality, they often deny the syncretism (of religion, cultural heritage, language, and genetic makeup) that has inevitably already occurred. For example, Greek nationalism promotes the myth of a racially and culturally direct line back to classical Greek civilization, despite the obvious intervening influences of other cultures (especially those introduced over centuries by the Ottomans). Nationalist rhetoric then claims for Eastern Orthodoxy a pure link with Hellenism, despite its cultural adaptations and syncretic absorption of popular (non-Hellenic) religious practices (Kokosalakis, 1987). These elements—religion, language, cultural tradition, geography, "race"—are woven together to create an imagined community, with obvious political uses (Stewart, 1994).

many **cross pressures** other than social class interest, thus making religion a less volatile source of conflict too. Cross pressures refer to the conflicting loyalties that individuals feel when they identify with several different roles and reference groups (Coleman, 1956:46). An example of cross pressures is the conflicting loyalties felt by a black woman who manages the local branch of a bank, lives in an integrated, middle-class suburb with pleasant Jewish and Irish Catholic neighbors, and belongs to a local Baptist church. When an issue arises in which different attachments conflict (e.g., whether tax monies should be used to fund abortions for poor women), she may acutely feel the cross pressures. Her loyalties to poor blacks or to women may urge her to support funding the abortions; her identification with her religion or the privileged classes may press for the opposite stance.

In many societies, this kind of cross pressure is absent or minimal because all major meaningful kinds of personal identification coincide. If the dominant class is of a totally different race, religion, and cultural background from the subordinate class (as in many colonial societies), the potential for social conflict is great because there are few conflicting loyalties within the person. In the United States, lines of cleavage—religious, ethnic, racial, economic, residential, regional, political—are seldom coextensive. Although real status inequalities exist (e.g., whites and Protestants are considerably overrepresented among the political and economic elites), geographical mobility, relative economic mobility, and mass communications have contributed to blurring the lines of cleavage. The existence of these cross pressures within individuals means that the likelihood of concerted conflict along any *one* line of cleavage (e.g., social class or religion) is reduced (Coleman, 1956:47).

Because religious cleavage sometimes masks other cleavages does not mean, however, that there are not real religious interests involved. Protection of religious interests frequently results in conflict. The struggle for *religious liberty* — the freedom to hold and practice the religion of one's choice—has been a recurrent cause of political conflict. Violent suppression of the Huguenots (i.e., French Protestants) resulted in continued conflict from their inception in the early 1500s until 1789. The Huguenots persisted in asserting their right to practice their religion, and the conflict included civil wars, mass emigration, massacres, and torture. To many people today, religious intolerance seems impolite and, at worst, unjust. We find it difficult to understand why religion would be the basis of such strong antagonism.

Conflict over Bases of Authority. One of the key reasons why religion can result in such extreme conflict is that it not only reflects lines of cleavage in society, but it can also challenge the bases of legitimacy by which authority is exercised. The French rulers believed (probably accurately) that the Protestant worldview constituted a challenge to the basis of their authority because that authority was justified largely by the Roman Catholic worldview. Religious ideas and interests have historically been significant forces in establishing authority.

Authority, in contrast with power relations, requires that subjects consider it legitimate. **Legitimacy** means the social recognition of an authority's claims to be taken seriously, and it implies negative social sanctions for failure to comply with authoritative commands. Conflict is implicit in authority relationships because they involve dominance and subordination. Yet they are based on more than pure power. Other considerations (e.g., scientific or religious knowledge) may also be sources of legitimacy of authority. There is constant potential for conflict between authorities who base their claims on different sources of legitimacy. When religion is related to civil disobedience (e.g., the actions of the Berrigan brothers in the 1960s and 1970s), religion is largely the basis of challenging the civil authority's source of legitimacy. This kind of civil disobedience is saying, in effect, "Your rules and practices may be right according to your authority, but our actions appeal to a higher basis of authority."

The place of religion in societal authority relationships is complex because religion is both a *basis* of legitimacy and, often, the *content* of authoritative pronouncements; thus, the pope pronounces religious messages and bases his claim to speak authoritatively on religious tradition. Religion has historically been an important source of conflict because it offered both a basis for legitimacy of authority and a set of ideas around which conflict centered. The apparent decrease in religion's role as a source of authority in modern society is due partly to its loss of legitimacy—its capacity to compel people to take its claims seriously.

Sources in the Nature of Religion

The nature of religion and religious groups also contributes to social conflict. In many respects, this capacity for promoting conflict is simply the obverse of religion's ability to engender social cohesion. Religion is one way of expressing the unity of the "in-group"—"our people." This in-versus-out dichotomy, however, applies to conflict both outside and within the group.

Conflict with Outsiders: Boundaries. This distinction between in-group and out-group implies distancing oneself from outsiders. Just as religious rituals celebrate the identity and unity of the group, they simultaneously maintain the boundary between that group and outsiders. Not only do religious groups protect their external boundaries, but they are also concerned with internal purification—another potential source of conflict (Douglas, 1966, 1970). The we–they dichotomy is both a structural and cognitive framework that includes how people think about themselves and others. This cognitive boundary is based on the fact that "we" have shared central experiences that "they" have not. In-group language embodies these shared experiences and further distinguishes "us" from "them." "Born-again" Christians consider their religious experience an important distinction between themselves and others, and their ways of witnessing to their special experience of being "born again" symbolize this difference.

Religion figures importantly in the socialization of children in most societies and thus becomes part of the "we–they" cognitive framework from an early age. Developing a sense of religious belonging, children come to think of themselves as part of the in-group and share their group's way of perceiving the rest of society. This cognitive framework may also include the perception of others as enemies or inferior to members of one's in-group. The values and attitudes learned in one religious group often vary considerably from those taught children in another group. Thus, religion's significance in socialization enhances its potential for divisiveness (Coleman, 1956:54, 55; Gorsuch and Aleshire, 1974; Wach, 1944:36).

Conflict with Outsiders: Particularism. Certain religious perspectives appear to promote conflict. Particularist worldviews encourage intolerance and prejudice toward the out-group. Religious **particularism** is the viewing of one's own religious group as the only legitimate religion (Glock and Stark, 1966:20). The clash of worldviews in a complex society is resolved in a number of possible stances: "Our way is totally right, theirs is totally wrong"; "Their way is good, our way is better"; "Our ways are both right but appropriate for different people"; "Their way and ours are essentially the same, and apparent differences are only incidental matters." The belief systems of some religions include particularist judgment of nonbelievers. For example, a study of Canadian racist extremists found that social class and education levels were not closely correlated with involvement in extreme racist activism; rather, the most striking correlation was active membership in certain fundamentalist Christian religions (Barrett, 1987). A U.S. study also found a consistent correlation between fundamentalism and the *general* tendency to discriminate against others, as well as actual discrimination toward specific targets, such as women, blacks, and homosexuals (McFarland, 1989).

Indeed, religious particularism seems to require a sense of opposition; one's own religion is seen as triumphant *over* some other. The in-group needs an out-group against which it can compare itself (Glock and Stark, 1966:29; Raab, 1964).[2] Religious worldviews that involve a particularistic stance toward the rest of humankind hold greater likelihood for promoting conflict than less triumphalistic worldviews.

Groups with particularistic worldviews are often able to mobilize the efforts of their members precisely because of this sense of opposition. Particularism enhances militance for one's beliefs. The strong element of religious particularism in Islam promoted its early missionary expansion throughout the Middle East, west to North Africa and Spain, east toward India and Southeast

---

2. For example, some years ago I was talking with a boy of about 12 who had recently been confirmed in his church. Responding to my query about what that ritual meant to him, he said, "It means being willing to fight and even die for Christ." I asked whom he thought he had to fight, and he replied, "I don't know—I guess the Jews." Nascent anti-Semitism perhaps, but probably only a reflection of the boy's sense that his Christianity had to triumph over some other group. The Jews may have been the only non-Christian group of which he had ever heard.

Asia, and north toward Eurasia. This particularism, combined with religiously legitimated militance, was also embodied in the *jihad* (i.e., holy war)—a recurrent feature of Islamic conflict. Contemporary conflicts in Afghanistan, Egypt, and Palestine have been viewed as jihads. The idea of missionary expansion is foreign to most non-Western religions because relatively few religions combine the particularism of having "the truth" with the mandate to convert the entire out-group. Interesting parallels with Christianity and Islam, however, are the worldviews of modern totalistic movements such as nazism, communism, and Maoism.

Internal Conflict in the Religious Group: Deviance and Control. This same in-group versus out-group dichotomy applies to the relationship of the group toward its own members who are defined as deviant or heretical. Again, conflict is the obverse of cohesion. The way in which a group treats a deviant member involves some form of conflict by which the group exerts its control. **Deviance** is behavior that is contrary to norms of conduct or expectations of a social group. Because it is the group that sets norms and identifies individual instances of deviant behavior, analysis of deviance describes the group as much as the deviant member (Becker, 1963). If a group sets a norm (e.g., against gambling) and labels a member as deviating from this norm, the group is both punishing the gambler deviant and proclaiming its own identity as a nongambling people.

Especially in small, closely knit societies or religious groups, the group's social control over deviant members can be powerful. Mennonites sometimes "shun" deviant members. Family and neighbors refuse any social interaction, sometimes for years, until the deviant repents of a "sin" or recants a "heresy." Because religion is often a source of the social group's norms and values, it is frequently important in the social response to deviance. The actions of the deviant member are seen as not only hurtful to the group but a violation of things that the group holds sacred.

Durkheim (1965:Book 3) emphasized that "piacular rites" (e.g., expiation for wrongdoings and cleansing of impurities) are just as expressive of the group's core unity and force as "positive rites" (e.g., communion). By ritually reincorporating deviant members who repent, both individual and group are strengthened in their solidarity around the norm. Threat of deviance from within is potentially more disruptive than opposition from outsiders because insiders "ought to know better"; outsiders can be dismissed as uninformed and ignorant of the "truth." Insiders must be taken more seriously as "one of us." A fellow member who goes against an important group norm becomes an affront to the essential unity of the entire group.

Deviance may contribute to group solidarity even when the deviant member is not repentant. In uniting against the deviant member, the group is strengthened. Even more than external opposition, internal conflict with the deviant member sharpens the group's sense of its boundaries and norms (see Durkheim, 1938:65–75; Mead, 1918). Collective rejection of the person accused of adultery reminds the entire group how much it should abhor adultery. Sometimes norms may be unclear or changing, and collective treatment of the deviant member may actually articulate the norm. In recent years, for example, numerous community or religious groups have attempted to proscribe or punish homosexual practices. Some have gone so far as to stigmatize AIDS patients due to the association of their disease with homosexuality. One study found that intolerance (e.g., prohibition of children with AIDS from schools) toward AIDS patients is correlated with factors of lower levels of education and self-esteem, political conservatism, and religious fundamentalism (Johnson, 1987). These actions are not merely a reflection of antagonism toward homosexuals; more importantly, they are groups' attempts to assert their norms in a situation where societal norms are unclear or changing. By uniting against what they define as deviance, they are confirming their own norms for themselves.

Religious groups with particularistic worldviews appear especially intolerant of deviance. Their certainty of their own total rightness increases their condemnation of the wrongness they perceive of any other beliefs and practices. Particularistic religious groups give deviant members more reason to fear group sanctions. If you have been socialized in a religion that you believe is uniquely true—the only path to salvation—you are not likely to consider leaving the group. You are likely to try very hard to follow its rules for behavior, accept its punishments for infringement, and hope never to be forced to leave. If you accept the group's claims to be the only true religion, expulsion (i.e., excommunication) is the worst possible punishment. In uniting against the deviant, members of particularistic religious groups gain both a sense of solidarity and a sense of their own moral rightness; they are triumphant over external and internal opposition. The core of righteous members of such groups often develops purist and elitist characteristics, which further promote the likelihood of conflict.

Internal Conflict: Issues of Authority and Heresy. Internal strife in religious groups can develop over nonreligious interests too, including socioeconomic issues, leadership and power, and other social cleavages within the group. For example, in the late fourteenth and early fifteenth centuries, the Catholic church experienced a major schism resulting from a purely political split rather than theological dispute. The same group of cardinals, acting apparently in good faith, elected two rival popes; subsequently, various kings and princes from most regions of Europe sided with one or the other pope to legitimate their own political claims. The subsequent resolution of the schism was due to political rather than any religious rapprochement (see Blasi, 1989).

Nevertheless, conflict over religious ideas and practices frequently centers on the issue of **authority,** which implies control both of the religious organization (however small) and over the articulation of central beliefs and practices. The separation of Eastern ("Orthodox") and Roman Catholicism was fundamentally a conflict over authority. Ostensibly, the schism was over theological issues such as whether Christ was "of the same substance" as God the Father or was "similar" to the Father. Behind these theological issues, however, were deeper sources of conflict: Greek philosophy and polity versus Roman philosophy and law. In particular, the split represented the Eastern patriarchs' dissatisfaction with the consolidation of church power and authority in the hands of the single Western patriarch (i.e., the pope in Rome). The specific theological issues, important as they were to conflicting parties, did not cause the split; the issue of authority in the church was the central source of conflict (Niebuhr, 1929:111–117; "Orthodox Eastern Church," 1943, 16:938–939).

Similarly, the main division between the dominant form of Islam (Sunni) and Shi'ism (the most powerful of the challenging Islamic sects) occurred over the issue of the *imam,* the legitimate successor to the Prophet, Muhammad. According to Shi'ite tradition, there is a line of succession to the imamate from the Prophet through his cousin Ali to their contemporary leaders; the imam is thus the rightful, divinely ordained leader of Muslims (Voll, 1982:7–31).

Other sources of conflict over authority are religious revelations. Because religious experience and ways of knowing are so intensely private, there is always the potential for believers to receive revelations or come to interpretations that differ from official ones. Revelations, prophecies, insights, and new interpretations of scripture and tradition have all been significant sources of intragroup conflict in most major religions.

When members assert a belief or practice that differs in some important way from those authoritatively established, it is often defined as not mere deviance but **heresy.** This kind of dissent implicitly challenges the existing authority structure of the group. It suggests that the entire group should consider a different basis for its core beliefs and practices. The labeling and condemnation of heresy is the official leadership's assertion of its authority in the face of the challenge. The vigor with which labeled heresies have historically been prosecuted illustrates the seriousness of their challenge to established religious authority.

Conflict over authority is exemplified by the issue raised in the Roman Catholic church of *Americanism,* which Pope Leo XIII condemned as heretical in 1899. *Americanism* generally referred to the idea that religion should be adapted to the individual cultures in which it is practiced; specifically it meant that the American culture required a different brand of Catholicism than European cultures. The origin of this perspective is attributed to missionaries to the Americans in the mid-1800s. They sought to adapt their missionary appeals to the peculiar characteristics of American culture, especially its democratic and pluralistic organization. Latent in the papal condemnation of Americanism, however, was the assertion of one particular system of church authority (Rome/pope/curia-centered) over another system (national/bishops/collegia-centered). The possibility that any national church authorities might develop separate strains of teaching authority was perceived as a direct challenge to the existing system of authority (Cross, 1958; McAvoy, 1957).

The label of *Americanism* continued to suppress much internal dissent in the American Roman Catholic church in the first half of the twentieth century. These same issues of authority were central in the deliberations of the

First and Second Vatican Councils (1869 and 1962). The politics of the Americanist "crisis" have not been eliminated and continue to recur in clashes between Vatican authorities and some Catholic scholars and clergy (see Kurtz, 1986). The Roman Catholic church provides clear-cut examples of the definition of heresy because it has developed specific measures for defining and dealing with it. But Protestant history also involves numerous instances of heresy, schism, and clash over authority. While the particularist worldview of many Christian groups promotes this kind of internal conflict, non-Christian religions (e.g., Hinduism, Islam, and Shinto) have also known many schisms and internal divisions.

Internal conflict over heresy is a dynamic process. It can mobilize members of the group for action against the heresy and indirectly promote internal changes, revising its teachings, shifting organizational arrangements, or developing other innovations [e.g., see a case study of the 1953 Boston heresy case (Pepper, 1988)]. Or the group can absorb the alternative beliefs and practices, changing in the direction of alternative authority. Even in the suppression of heresy, the group impels adherents of the heresy to form an alternative social movement. The history of Christian sects illustrates this dynamism.

Religion is thus an important factor in social conflict, both within the group and with outsiders. This potential for conflict results both from qualities of religion and religious groups themselves and from the nature of the society as a whole. Nevertheless, the aspect of conflict is basically the obverse of social cohesion; a certain amount of conflict is part of the very structure that holds groups together. And because religion is one important way by which groups express their unity, it is also a significant factor in conflict. This dual relationship is illustrated in the Extended Application section of this chapter: an examination of recent conflict in Northern Ireland.

EXTENDED APPLICATION: THE CONFLICT IN NORTHERN IRELAND

Recent civil strife in Northern Ireland illustrates religion's capacity to promote both cohesion and conflict. The history of divisions in Northern Ireland is long, complex, and confusing to outsiders. This brief essay focuses on the religious dimension of this struggle, recognizing that other important historical, political, economic, and social factors are also involved.

Religious Conflict?

One receives the impression from television and newspaper coverage that the strife in Northern Ireland is purely religious, with two antagonistic camps— Roman Catholic and Protestant—pitted against each other in senseless, deadly violence. Between 1970 and 1994, the sectarian strife claimed 3,171 lives, more than half of whom were civilian noncombatants, and injured 36,807 persons (Darnton, 1995; see Smith and Chambers, 1991, for the demography of these deaths and injuries). Neighborhoods and business districts have been bombed or burned, and a generation of children has grown to adulthood knowing no other way of life than intermittent warfare in their streets.

The image is puzzling to most Americans, who see the antagonists as so similar to each other. Unlike the U.S. national experience with conflict over segregation and discrimination, the opposing sides in Northern Ireland appear to share the same racial stock, language, and social class. These apparent similarities are, however, based on only a casual glance. Participants themselves apply finer distinctions; they are quick to identify someone as "one of us" or "one of them" on the basis of simple items of information—name, address, or school attended.

Nevertheless, the only differences apparent at a superficial level of analysis center around religion. Many Americans find it difficult to believe that religion is so important to anyone as to be worth fighting over. Others note the relative ease with which both Catholic and Protestant Irish immigrants adjusted to the heterogeneous and pluralistic religious scene in America (MacEoin, 1974b:1, 2). To participants in the Northern Irish conflict, however, the salience of religion as a genuine source of antagonism is unquestionable. In 1970 a British Broadcasting Corporation (BBC) interviewer asked a Belfast Protestant, "What do you have against the Roman Catholics?"—to which the response was, "Are you daft? Why, their religion of course" (cited in Rose, 1971:247). Americans and other outsiders ask: Is the conflict in Northern Ireland indeed a religious one?

This analysis will suggest that the answer is yes, but with qualifications. The conflict is not over theology or doctrine. It does not center on Protestantism or Catholicism as most adherents of these religions in the United States, Canada, or elsewhere in Europe know them. The conflict in Northern Ireland could perhaps best be understood as the outcome of two mutually exclusive religious representations of the nation: one "Protestant" (strongly identified with "Orangeism") and the other "Catholic" (strongly identified with "Republicanism"). Although the concept of civil religion is controversial, it appears useful in explaining the Northern Irish situation of religiously focused political strife.

Conflicting Civil Religions

The role a civil religion takes in a society varies according to the relationship that exists between the particular religions and the civil society. In the United States, civil religious rituals and symbols can represent the cohesion of persons of diverse particular religions (e.g., Episcopalian, Baptist, Catholic, Mormon), in large part because no particular religion has dominance in the state. Separation of church and state makes possible (and, some sociologists would argue, necessary) a civil religion that *transcends* particular religions.

Northern Ireland, by contrast, represents a situation in which religion (Protestant) is not separated from the state. Religion has been a significant consideration in political decision making, in applying social policy, in the actions of police and other officials, in the curriculum of national schools, and in the content of mass media. Relatively few persons in the country doubt that religion should be linked with the state; there is, however, profound dissension over *which* religion should shape the nation.

Both Protestant and Catholic perspectives in Northern Ireland entail visions of nationhood. Whatever their political thrust, both national visions are also intensely religious in both content and style. Each group builds its own version of national identity imbued with religious significance, and each group engenders a strong sense of "us" against "them."

While particular religions (i.e., Roman Catholic and various Protestant groups) contributed to the creation of these opposing visions, they cannot control them or how opposing factions use them.[3] Thus, religious leaders in both camps who attempt to quell the violence typically have little power over the strong sentiments set into motion in the name of religion. Indeed, adherents of opposing factions are sometimes strongly critical of church leaders who promote interfaith tolerance and ecumenical cooperation.

The opposing civil religions of Northern Ireland illustrate that social cohesion is often the obverse of social conflict. Each group is held together largely by its sense of opposition to the other group. Both groups—especially the Protestants—are thrust into enclaves of tribal togetherness out of fear of the other. Reformation and Counter-Reformation symbols and myths that are several hundred years old shape these fears and lock believers into nationalistic ideologies that cannot accommodate full religious pluralism.

The Social Context

Northern Ireland was created in the 1920s after rebellion against Britain, when twenty-six counties of Ireland became a self-governing dominion (eventually recognized as an independent republic). The other six counties, an area of 5,200 square miles in the northeast corner of the island, continued their union with Britain—that is, the United Kingdom. The part of the island retained by Britain included most of the industry and was economically and strategically important. Partition of the island in the 1920s encompassed as much area as possible in the British-held part without upsetting Protestant electoral dominance.

Roman Catholicism is the largest single denomination in Northern Ireland, accounting for roughly 38 to 45 percent of the 1.6 million population (the accuracy of the 1981 and 1991 decennial censuses were severely compromised by the large proportion who refused to answer the census question on religion) (Morrow, 1995). The remainder of the populace is Protestant, particularly Presbyterian and Church of Ireland (i.e., Episcopalian). The Repub-

---

3. Further corroboration of the distinction between particular and civil religions in Northern Ireland is found in survey data that show that commitment to particular Protestant religion is only weakly related to "Protestant" political attitudes about the conflict; commitment to Catholic religion is unrelated to "Catholic" political stances (McAllister, 1982).

lic of Ireland, by contrast, is predominantly Catholic, so that on the island as a whole Catholics outnumber Protestants approximately three to one (MacEoin, 1987; Prokesch, 1990).

Irish Protestants and Catholics tend to be more fundamentalist and theologically conservative than their counterparts in the United States and England. Both groups exemplify religious particularism, the belief that one's group is the only legitimate religion. The self-righteousness and sense of opposition characteristic of particularistic worldviews make both groups strong sources of intolerance. Religious polarization in Northern Ireland has produced a social and psychological split in which there are no neutrals; even unbelievers are identified as Protestant or Catholic unbelievers. One of the most important products of this religious polarization is that the public does not believe (indeed, cannot conceive of) groups that claim to be nonsectarian (Beach, 1977).

All aspects of life in Northern Ireland are divided by these religious poles. There are separate Protestant and Catholic neighborhoods, playgrounds, schools, social clubs, charitable organizations, political parties, youth activities, sports, newspapers, and cultural events. As suggested earlier in this chapter, religion is especially likely to be a source of conflict between groups when it is coextensive with other important sources of identity. In Northern Ireland, religion is coextensive with almost all other significant social divisions.

Much of the social segregation results from systematic discrimination against Catholics, especially since the 1920s partition. To protect their power and economic advantages, Protestants severely discriminated against Catholics in employment, housing, civil service, public board appointments, electoral districting, and representation. By restricting housing for Catholics to certain neighborhoods and then gerrymandering the boundaries of voting districts, Protestants were able to maintain strong majorities in the councils of communities where they were numerically the minority. Largely due to the polarization and violence of the last twenty-five years, many communities became more segregated after these policies were ended; by 1991, about half of Northern Ireland residents lived in areas that were more than 90 percent Catholic or 90 percent Protestant (Murray, 1995).

But not all segregation of Catholics was imposed by Protestants. Segregated Catholic schools are supported by the Catholic church hierarchy, partly in order to retain greater control over the socialization of Catholic youths and to protest the (Protestant) religious content of national school instruction (see Menendez, 1973). More recently, the bishops have resisted the spread of integrated schools, arguing that excluding religious education in the schools would lead to secularism (Fulton, 1991); their emphasis in Northern Ireland has been on achieving equity in building and funding Catholic national schools, while promoting tolerance within segregated schools (Davis, 1994; Murray, 1995; Smith, 1995).

The net effect of such total segregation is that few of the cross pressures that could reduce prejudice and conflict operate in Northern Irish social life. Polarization makes it difficult, if not impossible, to socialize with persons outside one's enclave. "They" are not real persons; "they" are described only by group myths, not by personal contact. Indeed, polarization has produced strong social controls. In-groups punish members who overstep boundaries and are too friendly to the out-group. Members of both enclaves have been beaten, tarred and feathered, and even killed for crossing the social boundaries. Sometimes religious sanctions are brought to bear on people who do not stay in their religious enclave. One Catholic bishop refused to confirm children who were not sent to Catholic schools. Some Protestant ministers use their pulpits to denounce fellow Protestants who accept ecumenism.

These current polarizations are the result of a long history, which figures significantly in the myths, symbols, and legends of the two civil religions. This history, however briefly presented here, is critical for understanding contemporary sentiments.

Some Historical Background

Although the history of England's involvement in Ireland may be traced to the twelfth-century Norman conquest, events of the seventeenth century were especially critical in shaping contemporary problems. In the sixteenth and seventeenth centuries, the chief objective of England's Irish policy was to prevent Ireland from becoming a center of English rebels or a stepping-stone for continental enemies. Thus, under the Tudor monarchs, the English began a system of plantations in Ireland, substituting loyal English settlers for potentially disloyal Irish or Old English landholders. The plantation system was particularly important in shaping the religious composition of Northern Ireland (Beckett, 1966:38–63).

Plantation and Insurrection. In the seventeenth century, the policy of plantation was extended with vigor and increasingly religious overtones. The English wished to disempower the native Irish and Old English landholders (predominantly Roman Catholic) and substitute new English and Scottish settlers (Protestant). The most extensive and successful plantation was in the province of Ulster (northern Ireland). Large areas of Ulster (i.e., counties Armagh, Cavan, Derry, Donegal, Fermanagh, and Tyrone) were confiscated and given to immigrant English and Scottish landholders. Since the seventeenth century, there has been a major social class distinction between the English (generally Church of England, called "Protestant") settlers and the Scottish (Presbyterians, Covenanters, etc., called "Dissenters"). Some Scottish-Irish later identified with the native Irish because of their own previous experience of discrimination from the English. In today's Northern Ireland, however, both Anglicans and Dissenters generally view their Protestantism as "common ground" against Catholicism. Some of these seventeenth-century settlers removed the Irish tenants from their land and replaced them with British tenants, while many kept their Irish tenants. Only about 2,000 British families lived on these Ulster plantations by 1628. Thus, the large-scale removal of native Irish that the government intended did not occur then (Beckett, 1966:64–74; Clarke, 1967).

Discontent of the native Irish festered, and in 1641 insurrection broke out, especially throughout the north of Ireland. Ulster natives (mostly Catholics) attacked the Protestant colonists and seized many towns and fortifications. The Ulster insurrection had special significance to the English, who were then involved in a civil war between Royalists (largely Catholics and Anglicans) and Parliamentarians (mostly Puritans and Dissenters). Exaggerated tales of the 1641 massacres served as fuel for the English reconquest of Ireland. In 1649 Oliver Cromwell (a Puritan) landed in Ireland with a force of 12,000 soldiers and began a campaign so ruthless that 350 years later he is still identified as perhaps the most hated symbol of English oppression. Cromwell saw his mission as not only to quell a royalist uprising but also to bring divine revenge for the 1641 massacres. Reporting on his conquest of Drogheda, after which some 2,000 townspeople were put to the sword, Cromwell wrote, "I am persuaded that this is a righteous judgment of God upon those barbarous wretches, who have imbrued their hands in so much innocent blood" (quoted in Beckett, 1966:79, 80).

The settlement following this reconquest solidified English dominance. The settlement forced all "disloyal" landlords to forfeit their lands, and new "loyal" settlers took their places. Catholics were particularly affected by this policy because the English parliament felt that they were, by definition, disloyal to English interests. In 1641 the majority of Irish landlords had been Catholic; after the Cromwellian settlement, the majority were Protestant (Beckett, 1966:74–81). Catholics lost virtually all political power because representation in parliament was based on landholding.

A subsequent revolution in England further polarized the Irish situation. The English monarch James II (a Catholic) was supplanted by a Dutch (Protestant) ruler, William of Orange (after whom the Orange Order of Northern Ireland is named). Irish Catholics sided with James II, and he brought his army to Ireland, hoping to reconquer England from that base. The Protestant colonists of Ireland had sided with William of Orange, who then brought an expeditionary force to Ireland. James took Dublin and laid seige to Londonderry (i.e., Derry), the main source of Orange resistance, but the city held out for fifteen weeks until reinforcements came. In 1690 the two armies met in a decisive battle at the River Boyne, and James was beaten on July 12. William's victory not only secured his position as king of England but also established Protestant supremacy in Ireland (Beckett, 1966:90–95; Simms, 1967).

These events, which occurred some 350 years ago, are still enshrined in the civil religions of Protestants and Catholics in Northern Ireland. The Catholic civil religion celebrates the heroes of the revolutionary uprising, and the Protestant civil religion commemorates the Orange defenders. In contemporary Derry, an annual (and sometimes violent) Protestant celebration parades the walls of the old city and proclaims to the Catholics living below the walls the seventeenth-century slogan "No Surrender" (see vivid descriptions in MacEoin, 1974b:87, 227, 232). In Belfast July 12 is a major Protestant holiday with bonfires, parades, speeches, and a mock battle to commemorate William's victory at the Boyne. The defeat is remembered in the Catholic civil religion as merely one of a long string of Protestant and English acts of oppression and injustice (the myths and images of this civil religion figure significantly in Irish literature, especially of the era immediately preceding the 1916 uprising; see Thompson, 1967). Thus, the history of "our nation" is remembered differently by the two groups. Events centuries past are the basis for two completely opposed sets of myths, legends, and heroes.

The Deepening Split. Events following 1690 further deepened the rift between Protestants and Catholics. Catholics were effectively excluded from Parliament, and a number of "penal laws" excluded Catholics (the great population majority of the island) from Parliament, army and militia, positions in municipal corporations, all civil service, and the legal profession. Laws of inheritance and land tenure were changed so that it was virtually impossible for Catholic landholders to leave land to a Catholic heir. The laws forbade Catholics from sending their children abroad for education, and Catholics were forced to pay tithes to the (Protestant) Church of Ireland. These laws, in effect for over 100 years, had the desired result of suppressing Catholic political and economic power. The first widespread nationalist movement, arising in the first half of the nineteenth century, focused on the issue of "Catholic emancipation," especially from strictures on political representation and voting. This history laid the basis for the distinctively Roman Catholic character of modern Irish nationalism (Beckett, 1966:96–144).

Meanwhile, the economy and political situation in Ulster (the north) developed further apart from the rest of Ireland. Ulster included the only developed industrial sector in the country, and Ulster business owners and industrialists feared nationalist separation from Britain—their primary source of markets and supplies. Ulster was also the locus of the only large Protestant enclave, which felt threatened by the prospect of a Catholic nationalist movement in power. Some English politicians deliberately manipulated these fears in their fight against Irish home rule (i.e., allowing an independent Irish parliament). They "played the *Orange card*"—that is, encouraged religious fears and sectarian strife, advising violent resistance if the British Parliament should pass a home-rule bill. Lord Randolph Churchill visited Belfast in 1886 to encourage Protestant fears of home rule and left the slogans "Home Rule Is Rome Rule" and "Ulster Will Fight; Ulster Will Be Right"—mottos kept alive in contemporary Protestant civil religion in Northern Ireland (Beckett, 1966:146–157).

In the nineteenth century, both Protestant and Catholic national visions grew. Sometimes they were embodied in organizations and movements, but more generally they developed as an attitude or part of a worldview of the respective communities. The development of these nationalisms illustrates the distinction between *particular* religions (i.e., Roman Catholic, Presbyterian, and Church of Ireland) and Catholic and Protestant *civil* religions. The Roman Catholic hierarchy frequently opposed the nascent Irish nationalism, even though the movement was heavily Catholic. As the twentieth century approached, church leaders found their interests more closely allied to those of the establishment. They were particularly unhappy with the socialist strain of the pro-independence labor movement. Thus, while the Catholic hierarchy encouraged a general antagonism toward Protestants, it discouraged the growing Catholic nationalism. Similarly, especially in the North, the Orange Order and related Protestant groups developed their power and beliefs independent of particular Protestant churches, yet with their approval. Both Protestant and Catholic religious leaders generally encouraged distrust toward each other, but Protestant and Catholic civil religions developed independently of the churches.

The nature of the Irish Revolution (1916–1921) itself laid the grounds for the present conflict. The fighting was primarily between two relatively small, irresponsible armed forces, neither of which was controlled by its government. The British government was unable to exercise control over its special force, the "Black and Tans," and the Irish Republican Army (IRA) was not responsible to the Dail (the newly established Irish independent parliament). Never actually militarily "won," the revolutionary war did prepare both sides for a compromise: the division of the island into a politically independent part and a unionist part (see Beckett, 1966:157–166). The significance of irresponsible paramilitary forces in that conflict laid the groundwork for contemporary paramilitary violence in Northern Ireland; and the IRA (Republican nationalist) and Ulster Volunteer Force (Orangeist) are offspring of the armed forces of the Irish Revolution.

Since Partition. The 1921 treaty that followed the Irish Revolution established partition of the country. Northern Ireland consisted of six counties, which together had a Protestant majority of approximately 65 percent, but the Catholic minority was large and numerically dominant in several localities. These six counties had a parliament in Belfast and remained in "union" with England—thus the political tag "Unionist," retained even today. The other twenty-six counties (overwhelmingly Catholic) were given an independent dominion parliament in Dublin that eventually proclaimed the country a republic. The border was arranged to keep as much economically advantageous area for Northern Ireland as possible while retaining Protestant political dominance. Slogans of Unionists during border disputes (1922–1925) figure in their contemporary civil religion: "Not an inch" and "What we have we hold" (McCracken, 1967).

The fears of Protestant Unionists in Northern Ireland erupted in violence during the treaty years. Armed mobs in Belfast viciously attacked Catholic neighborhoods, forcing Catholics to depend on the IRA (i.e., the branch of the revolutionary army that is now an illegal paramilitary secret society) to protect them. Often, Protestant fears were deliberately exploited by (Protestant) employers to prevent Protestant and Catholic workers from uniting in labor disputes. With unemployment at 30 percent during the 1930s depression in Belfast, Catholic and Protestant workers organized a bipartisan demonstration. In response, Orange Order leaders stepped up their appeals to religious sectarianism, suggesting that the dearth of employment was all the more reason to discriminate against Catholics. The grand master of the Belfast Orange Lodge (and subsequent member of the Northern Ireland Senate) said in 1933:

> It is time Protestant employers of Northern Ireland realized that whenever a Roman Catholic is brought into their employment it means one Protestant vote less. It is our duty to pass the word along from this great demonstration, and I suggest the slogan should be: "Protestants, employ Protestants!" (quoted in MacEoin, 1974b:66).

The social and political pattern that emerged from increasing sectarianization severely discriminated against the Catholic minority in housing, jobs, law enforcement, and elections. Proportional representation in Northern Ireland's Parliament was an early casualty of the Protestant struggle to retain power. Abolition of proportional representation guaranteed that minority parties could not coalesce to counterbalance the Unionist party. The Northern Ireland government also approved a systematic gerrymandering of electoral districts so that Catholic voters were isolated and underrepresented (see McCracken, 1967). Immediately prior to the present strife in Northern Ireland, gerrymandering resulted in Derry's 20,000 Catholic voters being able to elect only 8 councillors, compared to 12 councillors elected by only 10,000 Protestants. Similarly, in Omagh, Catholics comprised 61 percent of the population but elected only 9 of the town's 21 councillors; in Dungannon, Armagh, Enniskillen, and other large towns, Catholics were systematically deprived of a fair proportion of power in county and town government (London Sunday Times Insight Team, 1972:34, 35; MacEoin, 1974b:58–59).

Catholics were also discriminated against in housing. This feature was politically significant because until 1969—when the British government forced the Northern Irish government to change its electoral practices—only householders and their wives could vote. Thus, policies that prevented Catholics from becoming householders also kept down their voting strength. City and county councils have controlled the allocation of much housing, apartments and single-family dwellings alike. These councils gave strong preference to Protestants in public housing, and they further restricted Catholics by reducing the number of new houses built in their neighborhoods. In predominantly Catholic Derry, the Protestant-controlled city council built only 136 houses between 1958 and 1966 and none after 1966. Partly as a result, over 1,000 single-family housing units in the city were occupied by more than one family and sometimes by seven or eight. Over 1,500 families (almost all Catholic) were on the waiting list for housing, with an average of ten years waiting time (London Sunday Times Insight Team, 1972:35–37; MacEoin, 1974b:59).

Widespread job discrimination was the result of preferential treatment of Protestants in both private and public employment. Belfast's largest employer, the shipyards, had in 1971 only 400 Catholic employees out of 10,000 (London Sunday Times Insight Team, 1972:36). Northern Irish public officials systematically granted more and superior jobs to Protestants. In 1961 Catholics accounted for only 13 of 209 officers in professional and technical grades of

Northern Irish civil service, and there was only 1 Catholic out of 53 people at
the top administrative grade of civil service. Catholics were also greatly underrepresented on the twenty-two public boards in charge of such services as
housing, tourism, hospitals, and electricity. Disproportions created since partition have not been substantially altered by reforms introduced after the British
resumed Direct Rule in 1972. In 1973 Protestants retained 95 percent of the
477 top civil service positions (MacEoin, 1974b:67–69).

The religious segregation of entire occupations and industries also meant
that Protestants had an effective stranglehold on large areas of vital services.
When in 1974 the British got several government groups to agree to share
some political power with Catholics, the plan was destroyed by a brief strike
of Protestant workers. These workers controlled entire occupations so strate-
gic that their strike paralyzed the country. Protestant extremists have held
demonstrations and threatened similar work actions to destroy the 1985
English-Irish pact on Northern Ireland.

Despite the widespread discrimination, however, the Protestants are not
overall much better off economically than Catholics. Because both groups are
largely working class, the conflict is not class conflict; Protestants of all classes
unite against Catholics of all classes (cf. O'Brien, 1974; Rose, 1971). The economic discrimination has direct political implications, however. There is
higher unemployment among the Catholic working class than among the
Protestant working class, thus encouraging workers to emigrate (usually to
England) to find jobs. The higher emigration rate of Catholics had, for many
years, counterbalanced their higher birth rate, preventing them from becoming the population majority (see London Sunday Times Insight Team,
1972:29–31); in the economic recession of the 1970s and 1980s, however, decreased employment opportunities in England reduced incentives to emigrate
and exacerbated economic problems in the Catholic areas of Northern
Ireland.

Protestants also controlled "law and order." Before the British intervention, the two main police forces in Northern Ireland were the RUC (Royal
Ulster Constabulary) and the B-Specials (Ulster Special Constabulary). The
RUC was established at partition and was supposed to be bipartisan. Catholics,
however, were never proportionately represented on the force and, when re-
cent strife broke out in the 1960s, were less than 10 percent. The B-Specials
was a militia, almost exclusively Protestant. Indeed, the prime minister of
Northern Ireland boasted in 1922 that "it is also from the ranks of the Loyal
Orange Institution that our splendid B-Specials have come" (quoted in
MacEoin, 1974b:62). Catholics had a reasonable basis for their suspicions that
police forces used their power against the interests of the Catholic community.
There is evidence that in some conflicts, the B-Specials openly joined Protestant mobs against Catholic civilians (see London Sunday Times Insight Team,
1972:132–142).

Besides possessing an arsenal (e.g., armored cars and automatic weapons),
both forces were allowed discretionary powers by the Special Powers Act (introduced as a "temporary" measure in 1922 and eventually made permanent).

This act, a source of considerable embarrassment to the British government, abridged civil freedoms by allowing any police officer to search, arrest, and imprison suspects without warrant, charge, or trial. A 1935 British inquiry concluded that "the Northern Ireland government has used Special Powers toward securing the domination of one particular political faction and at the same time toward curtailing the lawful activities of its opponents" (cited in MacEoin, 1974b:61–62).

Grave dissatisfaction with these features of Northern Irish life led to the 1964 formation of the Northern Ireland Civil Rights Association (NICRA), patterned after the American civil rights movement. Its demands were modest. They were (1) "one-man-one-vote" in local elections, (2) removal of gerrymandered boundaries, (3) laws against discrimination by local government and provision of machinery to deal with complaints, (4) allocation of public housing on a points system, (5) repeal of the Special Powers Act, and (6) disbanding of the B-Specials (London Sunday Times Insight Team, 1972:49). The confrontations that resulted from the generally nonviolent demonstrations of NICRA occurred largely because the Northern Ireland government refused to recognize the legitimacy of the movement's grievances and equated it with Republicanism—the movement for national independence from Britain. Protestant enclaves similarly saw the civil rights demonstrations as "them" organized against "us." The stage was set for a long period of sectarian violence: Protestants versus Catholics and both Protestants and Catholics versus the British.

After three decades of conflict, several of the problems addressed by the Northern Irish civil rights movement continue to be major sources of tension and of serious social problems. One of the most pressing of these issues is inequality of socioeconomic opportunity. Comparing socioeconomic status of Catholics and Protestants in 1968 and 1978, one study found very little improvement in the relative status of Catholics (Moxon-Browne, 1983). Outright discrimination, together with workers' fears of hostility and unfair practices in the workplace, have perpetuated historical patterns of greatly disproportionate Catholic unemployment and underemployment. Despite nearly comparable attainment of educational qualifications, Catholics continue to be virtually absent from entire sectors of the economy, including better-paid manufacturing jobs, law enforcement, banking, insurance, and big business. In the expanding public service sectors, Catholic employees tend to be females, working in education, health services, and clerical jobs; except for police and prison work, public sector jobs for males decreased in the 1980s, providing little or no opportunity for Catholics (Eversley, 1989; Smith and Chambers, 1991).

The Northern Irish economy, already disadvantaged relative to England and other parts of Europe, has further declined; were it not for massive subventions from Britain and the European Economic Community, the standard of living in Northern Ireland would be comparable to countries like Mexico (Smith and Chambers, 1991). The role of the British Government in the conflict is further exacerbated by the fact that it subsidizes a very high proportion of what employment exists in Northern Ireland. Among other things, the British Government provides—directly or indirectly—nearly 40 percent of all jobs in the province, compared to 25 percent in Britain overall (Stevenson, 1994b). Government statistics show the Northern Ireland unemployment rate to be 13.2 percent, compared to 9.8 percent for all of Britain. Male unemployment among Northern Ireland Catholics is particularly severe. The 1991 census found that, compared with 12.7 percent of Protestant males, 28.4 percent of Catholic males were unemployed. In some Catholic urban ghettos, however, unemployment affects more than two-thirds of all working age males (Stevenson, 1994a). Furthermore, employed Catholics were greatly overrepresented in low-paid jobs with little job security. Interestingly, the strife itself may have artificially raised rates of employment, especially among Protestant males, through the provision of thousands of jobs in the various police forces and prisons (Rowthorn and Wayne, 1988). Not until 1990 was a tough equal opportunity act put into effect in Northern Ireland, and it remains to be seen how effective this legislation will be.

In the area of housing, major changes occurred since the early civil rights protests. To eliminate the discrimination in public housing, the newly created Northern Ireland Housing Executive assumed responsibility for building, managing, and allocating public housing (about 40 percent of all new housing is in this public sector). By 1985 most of the worst housing had been replaced, and overcrowding was greatly reduced. At the same time, however, neighborhood segregation had increased. Many previously integrated neighborhoods became all Protestant or all Catholic, as families who were in the minority in each neighborhood left in fear or were actively driven out. Due to considerable government expenditure in Northern Ireland compared to the rest of Britain, public housing was greatly improved and made more accessible, although there is some disparity still in waiting time for housing in Catholic neighborhoods. There is also still a higher proportion of Catholics living in substandard private-sector housing (Melaugh, 1995).

Police actions in the same twenty years have, if anything, generally increased Catholics' sense of injustice. Throughout the 1970s the various police forces in Northern Ireland were expanded, while the proportion of Catholics on the forces remained extremely small (usually less than 5 percent). Simultaneously, the scope of police powers was expanded and civil liberties were curtailed. Police detained several thousand people without ever bringing charges, much less conducting a hearing or trial. Amnesty International investigated conditions in Northern Ireland in 1977 and found evidence of frequent maltreatment of suspects and political prisoners. These police actions were perceived as discriminatory against Catholics. A survey conducted shortly after the Amnesty findings were reported in the media found that some 63 percent of Protestants, but only 20 percent of Catholics, believed that reports of prisoner maltreatment were unsubstantiated (Moxon-Browne, 1983:147). Although in the 1980s levels of street violence were lower and police were better disciplined than in the 1960s and 1970s, the failure to integrate Catholics into their ranks proportionate to their numbers in the communities leaves police forces open to plausible suspicion of biases and outright partisanship in the administration of justice.

Public perceptions of the causes of the violence in Northern Ireland are divided. A much larger proportion of Catholics than Protestants view socioeconomic factors such as discrimination, unemployment, inferior housing, and education as significant causes of the "Troubles." Furthermore, actual socioeconomic deprivation is linked with support for extremist positions (Smith and Chambers, 1991; see also Bell, 1990; Whyte, 1990). This capsule history of strife in Northern Ireland illustrates how two groups, defined by their religions, became so utterly segregated in their associations that cross pressures that could have prevented polarization never developed. The combination of this segregation with the religious particularism of each group makes the situation especially volatile. Indeed, if anything, the years of strife have further crystallized the separation of the two groups (Moxon-Browne, 1983:124–136). Systematic political and economic suppression of a large religious minority of natives by a colonial power made the resulting conflict reflect nationalistic as well as religious sentiments.

Now let us examine some of the chief characteristics of the opposing civil religions of Catholics and Protestants.

Catholic Civil Religion

The national vision identified as Catholic in Northern Ireland is one of a united republic, encompassing all counties of the island. Thus, it is necessary to view the conflict in the context of the religious and political situation of the Republic of Ireland as well. Although some Catholics in Northern Ireland do not favor union with the Republic of Ireland because of its lower standard of living (especially its social welfare benefits), they are nonetheless likely to identify themselves as Irish and feel a strong tie to the Irish cultural heritage (Beckett, 1966:175). A more militant version of this civil religion draws on the centuries-old Irish revolutionary myth; this version envisions the (Catholic) natives rising up to overthrow the oppressive (Protestant) colonial power. This militant nationalism does not accept the validity of the settlement that partitioned Ireland. A 1968 survey showed that 33 percent of the Catholics in Northern Ireland approved "on balance" the constitution of Northern Ireland, a proportion probably reduced by subsequent polarization. The study concluded that Northern Ireland is a state that is governed without consent (Rose, 1971:189). A later survey (after various intervening constitutional changes had been tried unsuccessfully) found that about 25 percent of Catholics still favored a United Ireland as the "most workable and acceptable solution" (Moxon-Browne, 1983:103). Refusal to accept the Northern Ireland government's right to rule further feeds Protestant citizens' notions that Catholics are disloyal or traitorous.

An important factor in the Catholic civil religion of Northern Ireland is its readiness to accept the church–state relations of the Republic of Ireland as normative. Thus, Protestant fears are fed by what they see as the effects of the Catholic church on political and social policies in the neighboring Republic of Ireland. Unlike Spain and several other predominantly Catholic countries, Ireland's Catholic church has retained much of its dominance as a public religion and has made few allowances for pluralism and incorporation of minorities (see Casanova, 1994, for useful analytical concepts; see also Fulton, 1991).

Although the influence of the Catholic church in the Republic is indirect, it is still potent (see Fulton, 1991; MacEoin, 1987, 1974a, 1974b; Morrow, 1995; O'Brien, 1995; Schmitt, 1973; Whyte, 1980). The Roman Catholic church is the established church in the Republic of Ireland (Eire), including in its membership 95 percent of the Republic's 3.5 million population. The 1937 constitution of Ireland declared, until it was abridged in 1972, "a special position of the Holy Catholic Apostolic and Roman Church as the guardian of the Faith professed by the great majority of the Citizens" (Article 44 quoted in Whyte, 1980:24–61). Although the Protestant minority in the Republic is seldom discriminated against, the Catholic church has influenced legislation and social policy with which Protestants are unhappy. The Catholic church's considerable influence in national (i.e., public) schools affects Protestants, especially in areas where they are not numerous enough to support alternative schools. Although public referenda in the 1990s relaxed laws somewhat (Clarity, 1994), church-influenced legislation in areas of "private" morality (e.g., divorce, contraception, abortion, and censorship) is similarly seen as unfairly restrictive by many Protestants. Thus even though Northern Ireland is politically distinct from the Republic of Ireland, both its Protestant and Catholic nationalisms refer directly to beliefs and practices of the Republic.

The salient imagery and ritual in the Northern Irish Catholic civil religion are symbols of independent Ireland: the flag ("tricolor"), the Irish language, national anthem and ballads of independence, celebration of heroes, and events of revolution. These symbols are always near the surface of interactions with the opposing group. Even nonsectarian events (e.g., civil rights demonstrations) are often the occasion for invoking the symbols of this civil religion; despite organizers' plans, demonstrators sometimes break into singing nationalistic ballads and hymns or unfurl a tricolor flag. These are natural expressions: Catholic militance on behalf of their rights has long been equated with nationalism. Catholic civil religion in Northern Ireland likewise often celebrates—with even more vigor than in the Republic—the heroes and events of Irish revolution: Emmet, Pearse, Connolly, Tone, Father Murphy, the Easter Insurrection, the Fenian Rebellion, Catholic Emancipation, Land League, Gaelic League, United Irishmen. All these elements mesh together in a national vision.

This nationalism is both anti-British and anticolonial (i.e., against Protestant "settlers"). The British "peace-keeping" forces in Northern Ireland have been embattled from both sides and are intensely disliked by both Protestants and Catholics; both groups believe the troops to be siding with their opposition. Even moves by the British government to pressure the Protestant powers in Northern Ireland to recognize some Catholic rights are distrusted by the Catholic community. Britain has a long history of political maneuvers resulting in detriment to the Irish.

The Catholic national vision in Northern Ireland is generally more toler-ant of Protestants than the Protestant version is of Catholics. This acceptance is largely because Protestants, too, were prominent in the revolutionary her-itage of Ireland. Protestants such as Wolfe Tone and Robert Emmet, inspired by the ideals of the French Revolution, led Irish revolts and are celebrated as martyrs in the civil religion of Catholic nationalists.

Protestant Civil Religion

The Protestant civil religion of Northern Ireland is, by contrast, virulently anti-Catholic. Both a 1968 survey and a 1978 comparison found Northern Irish Protestant antipathy to Catholics and Catholicism to be considerably greater than Catholic antipathy toward Protestants and Protestantism (Rose, 1971:256; Moxon-Browne, 1983:94–95, 124–136). The religious form of Protestantism as a civil religion in Northern Ireland is particularly evident in the power and influence of the Orange Order. Orangeism was, until the mid-1970s, essentially the state religion of Northern Ireland. The Unionist Party, which has dominated the government since its inception, has close, overt links with the Orange Order and is almost exclusively Protestant. Lord Craigavon, prime minister of Northern Ireland from 1921 to 1940, proclaimed, "I am an Orangeman first and a politician and a member of this parliament after-wards. . . . All I boast is that we are a Protestant parliament and a Protestant state" (cited in MacEoin, 1974b:53). More explicit about the link between the controlling party (i.e., Unionist) and Orangeism is the opinion of Brian Faulkner, prime minister of Northern Ireland before Britain stepped in to rule the province directly in 1972. Faulkner stated:

> There is no alternative to the invincible combination of the Orange Order and the Unionist Party. . . . The Unionist Party relies upon the Orange Order and likewise we in the Order trust the party. That vital faith must never be jeopardized by either partner (cited in MacEoin, 1974b:243).

The Orange Order, founded in 1795, has been especially strong in North-ern Ireland in this century. The grand master of the Belfast Orange Lodge (the largest of the several lodges in the country) described the Order as "basi-cally religious and only coincidentally political, a fellowship of all who em-brace the Reformed faith, founded to safeguard the interests of the Protestant people against the aggressions of the church of Rome which historically claims to have power over princes" (cited in MacEoin, 1974b:31). The salvation the-ology of Orangeism requires a sense of opposition: It fights for salvation not only from sin but also from Catholicism. The grand master further stated, "We have to fight the pretensions of the church of Rome. . . . I am convinced that it has never stopped its efforts to obscure the gospel, and I believe we must contend for the way of salvation through Christ."

The Orange Order has maintained nearly monolithic control over Protes-tant opinion. It includes in its membership one-third of Protestant men in

Northern Ireland, and its strength is estimated at 100,000. The Order includes all social classes, Church of Ireland and Dissenters alike. Indeed, historically it was the main vehicle for establishing Ulster Protestant unity despite significant religious and social class differences, especially between members of the Church of Ireland and the more fundamentalist Free Presbyterians (Bruce, 1987). Together with active women's and youth auxiliaries, it involves more than 90 percent of the Protestant community in some places (MacEoin, 1974b:62). There is some indication that popular support for the Orange Order has waned in recent years, and its active overlap with Protestant paramilitary groups may not be as widespread as earlier in the present conflict (Morrow, 1995; see also Buckley and Kenney, 1995).

The anti-Catholic stance of Orangeism is directly relevant to the Protestant national vision in Northern Ireland; a fundamental premise of that vision is that Catholics must not be allowed power. When some Catholics attempted to work within the Northern Irish political system in the 1960s, Unionist leaders tried to prevent their participation on grounds of their religion. One of the foremost Unionist leaders stated:

> I would draw your attention to the words "civil and religious liberty." The liberty we know is the liberty of the Protestant religion. . . . It is difficult to see how a Roman Catholic, with the vast difference in our religious outlook, could be either acceptable within the Unionist party as a member, or bring himself unconditionally to support its ideals. Furthermore to this, an Orangeman is pledged to resist by all lawful means the Ascendancy of the Church of Rome (cited by London Sunday Times Insight Team, 1972:37).

According to Northern Irish Protestant civil religion, Catholicism is incompatible with the Protestant conception of democracy, and the Protestant nationalist vision is, by definition, opposed to any Catholic power. One DUP (Democratic Unionist Party, the most extreme Unionist group) member of Assembly, interviewed in 1986, said:

> The Christian beliefs held by the vast majority of the DUP clearly influence their political decisions. We believe that Ulster's future is best secured in a society where the gospel message can be preached freely, and that is not under Rome. Where the gospel light in many little towns throughout the Irish Republic has died, it has gone out because of the control of Rome. . . . I'm in politics to preserve this province so that the gospel can be preached freely and the Christian message on any street corner and that cannot be done in a Catholic dominated state . . ." (quoted in MacIver, 1989:367–368).

This anti-Catholic stance is celebrated in the fiery speeches of Orange parades and commemoration of their history, Protestant heroes, and legends. One popular Protestant tune, called "Croppies (i.e., Catholics) Lie Down," goes:

> Poor Croppies, ye know that your sentence was come,
> When you heard the dread sound of the Protestant drum.

In memory of William we hoist his flag,
And soon the bright Orange put down the Green rag.

Protestant group identity is defined mainly by what it is *not:* Northern Ireland Protestants are adamantly not Catholic. Despite the fact that they are a privileged majority, the Protestant enclave experiences itself as "under siege" (Buckley and Kenney, 1995; Darnton, 1995). Parallel to the legitimating myths of Catholic nationalist sentiments, Protestant collective representations try to legitimate a separate "Ulster identity." Their legitimating myths include a Protestant reading of history and the claim to be a separate ethnic group. For example, in the 1980s the myth of the Cruthin—supposedly the original inhabitants before the invasion of the Gaelic peoples—was promoted to legitimate the political idea of an Ulster ethnic identity for the Scots-Irish, whose claims on the land would, thus, precede those of the "native" Irish (Davis, 1994). This myth does not appear to be widely held, but it illustrates the extent to which religious and ethnic group identities are very much "imagined communities" (Anderson, 1991). Constructing separate identities in places like Northern Ireland is especially difficult, because the cultural heritage of one group (e.g., Protestant farmers) is largely the same as that of the other group of the same social class and vicinity. Thus, the cultural segregation that has developed has to be actively cultivated, and separate group identities have to be continually carved out and maintained (Buckley and Kenney, 1995).

Although this national vision is inimical to the prospect of union with the Republic of Ireland, it also includes strong anti-British sentiment. Northern Irish Protestants are keenly aware of their inferior status in the British class structure and resent being treated as colonials. The militant version of the Protestant civil religion holds that loyalty to the British Crown is contingent upon the British enforcing the Protestant ascendancy of the seventeenth-century "Revolution settlement" (MacEoin, 1974b:40, 285). Some observers suggest that the British intervention may have unintentionally exacerbated the contemporary conflict, especially by creating conditions for an upsurge of Protestant populism (Bew and Patterson, 1983).

Nationalism Without Nation Building

The Protestant and Catholic national visions of Northern Ireland appear to fit our model of civil religion, albeit in a different form from that of civil religion in the United States. Both Protestant and Catholic versions are mutually exclusive images of the nation. Both entail extensive myths, legends, rituals, and symbols of a long history of growing separation of peoples. Both civil religions entail imagery of the "chosen people." The Protestant version defines themselves as God's chosen people struggling to maintain the truth in a land of paganism and idolatry, whereas the Catholic version describes the Gaelic peoples as the chosen children of God, oppressed by the Protestants. Commenting on this Old Testament imagery, one astute observer remarked, "One could say that Ireland was inhabited, not really by Protestants and Catholics, but by two sets of imaginary Jews" (O'Brien, 1974:288).

The efforts of nonsectarian political groups, concerned clergy on either side, and various ecumenical peace groups to defuse the religious antagonisms in Northern Ireland are typically thwarted because the conflict is less between particular religions than between two civil religions. The differences in belief and practice between the particular religions—Protestant and Catholic—are important but are only a small part of this conflict. The division is essentially over national identity and vision, which are strongly shaped by a long history of experiences and cultural differences between the two groups.

As suggested earlier in this chapter, civil religion in many countries has considerable potential for modern nation building, forging unity from diverse peoples of a state and bridging barriers of language, particular religion, tribalism, and regionalism. In the case of Northern Ireland, however, neither civil religion has this potential. The state governs without consensus, and the people operate from opposing senses of national identity. Another model of civil religion or of church–state relations would serve the interests of nation building better, but the issue of what nation to build is still a political issue.

SUMMARY

Religion is a source of both social cohesion and social conflict. It contributes to the cohesion of a group and, at the same time, expresses its unity. Durkheim suggests that the moral unity of the group itself is the source of the sense of religious force or power experienced by participants. Thus, religious rituals incorporate members into the group by reminding them of the meanings and obligations of being "one of us." The civil religion thesis is an attempt to understand the expressions of moral unity of modern societies. It suggests that the nation may have a religious expression distinct from the particular religions of the people.

Religion contributes to social conflict because it is one basis of social cleavage. In societies where other loyalties cut across religious ties, religion is less likely to be a focus of serious conflict. Nevertheless, religion's capability for defining the boundaries of "us" against "them" makes it an inherently potential source of conflict. Particularist worldviews especially tend to intolerance and conflict. Religious groups are also subject to internal conflict over issues of deviance and control, authority and heresy.

The aspects of religion that promote both conflict and cohesion are illustrated by the strife in Northern Ireland. While particular Protestant and Roman Catholic religions have historically contributed to the active bigotry and intolerance there, a broader religious division—that between opposing civil religions with national visions, one Catholic and one Protestant—may also exist.

RECOMMENDED READINGS

Articles

Robert N. Bellah. "Religion and Legitimation in the American Republic." *Society* 15 (4), 1978:16–23.

Robert N. Bellah. "Civil Religion in America." *Daedalus* 96, 1967:1–21; reprinted in McNamara (1984) reader and in Richey and Jones ed., 1974.

Theodor Hanf. "The Sacred Marker: Religion, Communalism and Nationalism." *Social Compass* 41 (1), 1994:9–20.

Rhys H. Williams and N. J. Demerath, III. "Religion and Political Process in an American City." *American Sociological Review* 56 (August, 1991):417–431.

Books

Anthony D. Buckley and Mary Catherine Kenney. *Negotiating Identity: Rhetoric, Metaphor, and Social Drama in Northern Ireland*. Washington: Smithsonian Institution Press, 1995. With a theoretically sophisticated appreciation of the social construction of identities, this book is a highly readable ethnography of the everyday practices by which rural Northern Irish Protestants accomplish separate identities from those of their Catholic neighbors with whom they share a largely common culture.

David Chidester. *Shots in the Streets: Violence and Religion in South Africa*. Boston: Beacon, 1991. Although South Africa has begun to move politically beyond its decades-old policy of apartheid, this analysis of the complex ways that religion has been linked with cohesion and conflict will continue to be useful for understanding the problems that country will face in the future, as well as those it faced in the immediate past.

Emile Durkheim. *The Elementary Forms of Religious Life*. Tr. K. Fields. New York: Free Press, 1995 [1915]. Durkheim's classical study uses illustrations from the religion of the Arunta of Australia to explicate his theory of the social foundations of religious beliefs and practices. This new translation is far superior to the older English edition.

Will Herberg. *Protestant-Catholic-Jew*. Garden City, N.Y.: Doubleday, 1960.

Herberg's description of the "American way of life" as a quasi-religious belief system and his descriptions of Protestant, Catholic, and Jewish history in America are especially useful. While somewhat dated, this essay on religion in American life during the 1950s is still provocative.

David Kertzer. *Ritual, Politics and Power*. New Haven, Conn.: Yale University Press, 1988. A lively analysis of how ritual promotes cohesion and conflict, liberally illustrated with examples from many different cultures and historical periods.

Charles S. Liebman and Eliezar Don-Yehiya. *Civil Religion in Israel: Traditional Judaism and Political Culture in the Jewish State*. Berkeley: University of California Press, 1983. This volume gives both an excellent contrast of civil religion in another sociopolitical setting and a helpful basis for understanding contemporary Israeli political action.

Stanley Tambiah. *Buddhism Betrayed? Religion, Politics and Violence in Sri Lanka*. Chicago: University of Chicago Press, 1992. How does a religious group that has historically promoted peace and renunciation of worldly power and gain become entangled in nationalistic politics, violence, and war? Without oversimplifying the complexities of this small country's postcolonial development, Tambiah explores the role of Buddhism in its social cohesion and conflict.

7

The Impact of Religion on Social Change

The process of social change highlights the relationship between religion and other aspects of the social system, especially the economic and political spheres. This chapter examines some of the key ideas about the interrelationship between religion and social change. Because the issue of religion and social change is so complex, it is useful to separate the ways in which religion supports the status quo and inhibits change from the ways that it promotes social change. This division is, however, largely analytic: We can speak of aspects of religion that promote or inhibit social change, but in most cases both are occurring simultaneously. We will first examine the ways in which religion supports the status quo and inhibits change, then follow by examining how religion promotes social change. The concluding analysis poses the questions: Under what conditions does religion have the greatest impact on society, and under what conditions is religion's influence likely to inhibit or promote change? Discussion of these questions is followed by an Extended Application to the case of African-American religion.

FACTORS IN SOCIAL CHANGE

The processes involved in social change are complex. It is difficult to isolate a single aspect such as religion in a chain of events that result in social change. As the example of millenarianism (discussed in Chapter 2) shows, a religious movement that arises in response to social change may itself help bring about social change.[1] A further complexity is the meaning of religion. Do we mean religious ideas, religious personages, religious movements, or religious organizations? All these aspects of religion need to be considered; yet in any given situation, the relationships are often very complex.

We need to remember that change itself is neither necessarily good nor bad. **Social change** refers to any alteration in the social arrangements of a group or society. Of particular sociological interest is change that results in basic structural rearrangements (e.g., a new basis for social stratification or a change in a group's fundamental mode of decision making). Although the process of change itself is neutral, we often evaluate specific changes and potential changes by culturally established criteria. For example, we might consider the development of modern technology to be good or bad, but the criteria by which we make this judgment are not inherent in the process of change itself.

The effect of religion on a social development is not necessarily intentional. A religious idea or movement may become transformed into something very different from what its originators intended, and the influence of religion is often indirect. The Society of Friends (i.e., Quakers) became one

---

1. In framing the topic as an interrelationship, we are rejecting theories that treat religion as *nothing but* an epiphenomenon to other social processes, though we will discuss some of these theories later in this chapter. Any effort to present this complex relationship in the brief space of a chapter necessarily entails great simplification. Suggested sources at the end of the chapter and in footnotes can direct interested readers to further examination of this important theme.

of the foremost reformist sects in Christianity, even though their initial thrust was millenarian and somewhat mystical. As the movement developed, the Quakers' values of hard work and rejection of "worldly" amusements contributed to the wealth of numerous member families. Their other values (e.g., a strong personal conscience on social and political issues) stimulated their active participation in political and social reforms. They were prominent in abolitionism and the underground railroad, established self-help projects during and after the famine in Ireland, and were active in campaigns for prison and legal reform. Thus, the ascetic, inward ideals of the original movement were transformed—first into monetary philanthropy and later into social activism (Baltzell, 1979; Isichei, 1967; Wilson, 1970:178–181). The impetus for its monetary philanthropy was the *unintentional* economic change produced by religious values, and the resulting social activism exemplified an *intentional* change orientation, or deliberate stance of opposition to the dominant society (Westhues, 1976).

RELIGION SUPPORTS THE STATUS QUO

There is an inherently conservative aspect to religion. Religion can evoke a sense of the sacred precisely because of believers' respect for tradition and continuity. Religious symbols link the believers' present experience with meanings derived from the group's tradition, and religious beliefs that are taken-for-granted truths build a strong force against new ways of thinking. Practices handed down through tradition as the god-approved ways are highly resistant to change. Although other aspects of religion promote social change, important elements in religion maintain the status quo.

A central theme in the sociology of religion is the relationship between religious ideas and the nature of the social groups that hold them. Social stratification, in particular, appears closely correlated with religious belief. **Stratification** refers to the differential distribution of prestige and privilege in a society according to criteria such as social class, age, political power, gender, or race. Religion has a different significance to various strata in a society, and different religious ideas will probably appeal to different groups in the society's stratification system. In the United States, Episcopalians and Congregationalists (i.e., United Church of Christ) draw members disproportionately from the upper classes, while Holiness and Pentecostal groups draw relatively large proportions of members from the lower classes (Kosmin and Lachman, 1993). Similarly, religious ideas of a social group often reflect social caste, as illustrated by African-American religion later in this chapter.

Interest Theories: Marx and Engels

The connection between stratification and religious ideas has often been explained by interest theories such as Marx and Engels's explanation of religion as ideology. **Ideology** is a system of ideas that explains and legitimates the actions and interests of a specific sector (i.e., class) of society. The classical

Marxian approach applies the concept of ideology only to ideas that embody the vested interests of the dominant classes; thus, the term has come to have a negative evaluative connotation. The concept of ideology applies more broadly to religion, however, if it is used neutrally. In this usage, *ideology* also refers to belief systems defending the interests of a socially subordinate group, thus justifying reform or revolution (Geertz, 1964; Lewy, 1974). From this perspective, recent sociology of religion has begun to explore some forms of popular religion as class-based production of religious meanings and practices (Aman, 1988; Aman and Parker, 1991; Gismondi, 1988; Lancaster, 1988; Maduro, 1982; Parker, 1993, 1994; Taussig, 1980).

According to the dominant strain of Marxian analysis, the fundamental basis of social action is **material interests,** referring to considerations that give economic benefits and power to a person or group. Religious and philosophical ideas are seen as mere epiphenomena, after-the-fact explanations that justify or mask real motivations for behavior.[2] Another school of Marxian thought has especially in recent years emphasized the relative autonomy of religion. This school strives to study religious belief systems in themselves, locating them in a larger social-historical context. These theorists retain the insights of Marx and especially Engels but focus on the complex *reciprocal* influences between religion and social structure. By treating the relationship as complex, this latter Marxian approach identifies both the passive and active, conservative and revolutionary elements in religion (this strain of Marxian thought developed from the interpretations of Gramsci, Lukacs, and the Frankfurt school, among others; see Beckford, 1991; Fulton, 1987; Maduro, 1977; Mansueto, 1988; Mayrl, 1976).

Another concept explaining the change-inhibiting aspects of religion is the idea of **alienation,** which is central to the Marxian definition of religion. Marx used this concept to analyze the false consciousness that he believed religion engendered. Marx (1963:122) asserted:

> All these consequences follow from the fact that the worker is related to the *product of his labor* as to an *alien* object. For it is clear on this presupposition that the more the worker expends himself in work the more powerful becomes the world of objects which he creates in face of himself, the poorer he becomes in his inner life, and the less he belongs to himself. It is just the same as in religion. The more of himself man attributes to God the less he has left in himself.

Religion, according to Marx, is the projection of human needs and desires into the realm of the fantastic. Religious alienation is thus the understandable reflection of the false consciousness inherent in this social system. Marx's insight into the change-inhibiting aspects of religious consciousness centers on

---

2. This dominant strain—"dialectical materialism"—is represented by the interpretations of Plekhanov, Bernstein, Kautsky, and later Stalin. Marx himself did not develop a careful, unified theory of religion; the writings of Engels and Kautsky were more fruitful for further development of Marxian theory of religion. For a lucid and critical examination of these early theorists, see McKown, 1975.

the sense of domination by an alien force. Religion obscures real sources of conflict of class interests. Only a society that destroys such illusions can establish itself on the "truth of this world."

In this context, Marx stated that religion is the "opium of the people." He considered the distress that people expressed in their religion to be real but religion itself as an illusion preventing people from doing anything effective to remedy their condition. Religion, like opium, soothed their distress, but any relief was illusory (Marx, 1963:43, 44). Often religion draws off dissent and zeal that might otherwise promote revolution. Thus, a Marxian reading of the history of the English working class suggests that the development of Methodism precluded revolution by harnessing nineteenth-century workers' dissent and fervor in a religious movement. This thesis was first approvingly suggested in 1906 by British historian Elie Halévy; it was reexplored as a Marxian critique of English class relations by Hobsbawm, 1959, and Thompson, 1968. Parallels have been recently drawn to explain the rise of Pentecostalism in the face of global monopoly capitalism (Csordas, 1992; see also Anderson, 1979).

Although some evidence supports Marx's contention, studies of the correlation between religion and sociopolitical action generally suggest that the relationship is more complex, varied, and unpredictable (see Bastien, 1993; Gerlach, 1974; L'Alive D'Epinay, 1969; Levine, 1990; Marx, 1967; B. Smith, 1982; C. Smith, 1994; Stoll, 1990). Religion does relieve the tension of economic deprivation by substituting the value of religious achievement for economic achievement, and this substitution may indeed have an opiate effect because pressure for change is defused. At the same time, however, religion offers greater self-esteem by informing believers that they are superior, according to these alternative values. Such sense of superiority has transformative power, as exemplified by the zeal of the Puritans (Coleman, 1956:52).

Interest Theories: Weber

Weber, too, noted the extent to which religious ideas served to legitimate existing social arrangements, especially the stratification system. Religion has historically explained and justified why the powerful and privileged should have their power and privilege. The wealthy might justify their privilege as a sign of God's approval of their hard work and moral uprightness. Weber observed that most religions provide **theodicies** both of privilege and of disprivilege.

Theodicies are religious explanations that provide meaning for problematic experiences—in this case, the discrepancies of stratification. For example, the Hindu doctrine of rebirth simultaneously justifies the privilege of the upper classes and gives meaning and some hope for the conditions of the lower classes. This belief explains that one's present condition is the result of behavior in one's former life. A favored situation (e.g., being born a man rather than a woman) is justified as the result of appropriate behavior in a former incarnation. Prescribed action for both the privileged and disprivileged is therefore to behave appropriately in one's present social situation in order to obtain a more favorable situation in the next life. This theodicy justifies and explains the social status of both privileged and disprivileged and deters the disprivileged from trying to change the existing arrangement—in this life, at least (Weber, 1946:253ff.; 1963).

Weber's interpretation of stratification was more complex than that of Marx. Weber distinguished between "class situation" (stratification based on economic factors) and "status situation" (stratification based on factors such as lifestyle, honor, and prestige). Thus, for Weber, the linkage between religion and social arrangements went beyond simple ideological justification for the material interests of a social group; it included ideas that served as a means of **status distinction,** as a focus for group cohesion, and as a basis for controlling economic opportunities (Bendix, 1960:86, 87).

Weber's conceptualization may be useful for understanding the change-promoting and change-inhibiting aspects of the "New Christian Right" in the United States. This movement (and similar movements in America and other societies) is pressing for change, albeit reactionary change. Reactionary movements typically fight the encroachments of "modernity" and urge the cultural return to values and norms of an earlier, more "pure" era. The "New Religious Right," for example, presses for traditional patterns of family life; this goal translates symbolically into crusades against the Equal Rights Amendment to the Constitution, abortion, sex education in the schools, gay rights, and other nontraditional family lifestyles. A number of studies have shown a strong correlation between religious involvement (e.g., as measured by church attendance) and antifeminist sentiments and activism (see Himmelstein, 1986).

Some observers suggest that the appeal of such movements is a form of status politics, by which people whose status (i.e., relative prestige or honor) is threatened by changing cultural norms assert their values politically in order to reestablish the ideological basis of their status (Gannon, 1981; Lipset and Raab, 1981). The differences between supporters and nonsupporters of the "New Christian Right" are less matters of economic class than of cultural lifestyle (Johnson and Tamney, 1984; Shupe and Stacey, 1982; Tamney and Johnson, 1983). One study among U.S. Protestants found that people who were dissatisfied with the amount of social respect given to groups representing traditional values (e.g., people who were involved in churches, people who worked hard and obeyed the law, people like themselves) were especially likely to support the agenda and organization of the Christian right wing (Wald et al., 1989). Another analysis suggested that religion is a major factor in these political stands *both* as a source of traditional images of women and the family *and* as a network from which political traditionalist movements can be mobilized (Himmelstein, 1986). A study of Christian Patriotism, a sometimes-violent, radical right-wing Christian identity movement, found that fundamentalist beliefs powerfully shaped the extremist actions of members through their world images: dualistic conspiratorialism, patriarchal misogyny, inner-worldly asceticism, and dispensationalist millenarianism (Aho, 1990:166–175).

Several researchers have hypothesized that the "New Religious Right" may be a political defense of a set of lifestyle criteria that forms the bases of supporters' social status (Bruce, 1987; Harper and Leicht, 1984; Lorentzen, 1980; Page and Clelland, 1978; Wallis and Bruce, 1986). This political assertiveness may be all the more plausible because of the social and political turmoil of the 1960s and 1970s; many in the New Right interpreted such events as evidence of the failure of modernity and modern values, as well as the vulnerability of the United States as a nation (Simpson, 1983; see also Beyer, 1994). As the previous description of the abortion debate shows (see Chapter 3), lifestyle aspects of status are important. Women whose identity and social status (i.e., relative honor) are bound up in traditional definitions of motherhood and women's proper roles have much to lose when the society shifts to a different set of cultural values (Luker, 1984). Thus, religion may serve to support both socioeconomic class interests and status-group interests. Whether its role was change-promoting or change-inhibiting depends on the social location of the interest group. Furthermore, beyond their purely ideological functions, religious ideas often shape or define the very criteria by which status in a given group is evaluated.

Religious Legitimation

Religious legitimation of the status quo is sometimes the result of direct collusion between the dominant classes and the dominant religious organizations. Religious organizations and their personnel frequently have vested interests, which they protect by alliance with dominant groups in political and economic spheres. At the same time, dominant groups often try to manipulate religion to serve their purposes. Machiavelli recognized the legitimating power of religion and recommended that lawgivers resort to divine authority for their laws. He proposed:

> It is therefore the duty of princes and heads of republics to uphold the foundations of the religion of their countries, for then it is easy to keep their people religious, and consequently well conducted and united (*Discourses*, Book I, Chapter 12, as quoted in Lewy, 1974).

Through religious legitimation, wars have been justified as "holy wars," obligations have been justified as "sacred duty," and domination has been explained as "divine kingship." The theory of the "divine right" of kings (a Western political theory legitimating monarchies and reaching its apex in the seventeenth century) explained that the monarch's right to rule is God-given and that the monarch's ultimate responsibility for conduct of the state is to God. This theory interpreted the process of monarchical succession and the coronation/anointment of kings as the working-out or expression of that God-given right. By implication then, a ruler who gained dominance by other means (e.g., by revolution, usurping the throne, or popular choice) was not legitimate in the eyes of God. The monarch had both a civil and sacred role. Divine right made the king priest-mediator of God's will to the state. Belief in the divine right of kings both legitimated the rule of monarchs and suppressed dissent.

In the United States, the constitutionally mandated separation of church and state reduced the direct religious legitimation of political power, but religion has been a significant source of persuasion in political and socioeconomic spheres. Religion has been used to legitimate slavery and racial segregation, industrialization and antiunionism, warfare and international policy—all of which generally served the interests of the dominant sociopolitical groups. The historical processes by which religion was drawn into legitimation of these interests is, however, complex (see sociological studies of specific historical situations such as Yinger's 1946 analysis of church responses to America's participation in World War II; Pope's classic 1942 study of religion's role in quelling a major mill strike; and Hadden's 1969 study of tensions over clergy activism, especially civil rights).

Similarly, religion has been used to legitimate changes favoring a dominant group (e.g., the wealthy or politically powerful). Imperialism and crusades have generally been supported by religious or quasi-religious belief systems. President McKinley explained the decision to wage the expansionist war against Spain by which the United States acquired Cuba and the Philippines as follows:

> I am not ashamed to tell you, gentlemen, that I went down on my knees and prayed Almighty God for light and guidance more than one night. And one night late it came to me this way. . . . There was nothing left for us to do but to take them all and to educate the Filipinos and uplift and civilize and Christianize them and by God's grace do the very best we could by them, as our fellow men for whom Christ also died (quoted in Ahlstrom, 1972:879).

Religious legitimations and institutional support have also been historically used to promote new economic arrangements such as industrialization. A South Carolina minister, supporting the industrialization of his region, said in 1927:

> To those who can read history it is unthinkable that any one fail to see in it all the hand of God bringing the many thousands from the bondage imposed upon us by social and economic forces which of ourselves we were powerless to control. . . . It is imperative that we think of Southern industry as a spiritual movement and of ourselves as instruments in a Divine plan. Southern industry is the largest single opportunity the world has ever had to build a democracy upon the ethics of Christianity. . . . Southern industry is to measure the power of Protestantism, unmolested. . . . Southern industry was pioneered by men possessing the statesmanship of the prophets of God. . . . I personally believe it was God's way for the development of a forsaken people (quoted in Pope, 1942:25).

Religious Socialization

Religion legitimates not only the social system but also specific roles and personal qualities appropriate to existing structures. By promoting certain character types (e.g., the "hard-working individualist" or the "fatalistic happy-go-lucky" type) that are appropriate to a socioeconomic system, religion further legitimates that system. In socialization the individual internalizes these roles (as discussed in Chapter 3), which then come to exert an influence often greater than external social controls. Religious socialization often indirectly supports the socioeconomic status quo by teaching attitudes and values that adapt to that system.

The norms of existing social relations may be inculcated indirectly through religious rituals and socialization. In the nineteenth century, the social relationship of the Latin American nobleman-landowner to the peasants who tilled his land was that of the *patron*. The patron granted to the peasant farmers political and economic protection with the use of a portion of land. In exchange, the peasants were honor-bound to serve the landlord and give him their loyalty. This relationship of personal submission was concretized and given moral strength by the religious practice of the "godparent." The nobleman was linked to his peasant dependents as godparent to their children at the time of their baptism. Even veneration of the saints in popular religiosity replicated this relationship of dependence relative to the patron. The attitudes and values transmitted through the institution of godparent and the popular emphasis on patron saints indirectly supported the norms of the existing socioeconomic structure (Ribeiro de Oliveira, 1979:314, 315).

By observing the moral norms encouraged by a religious group, the individual will come to "fit" better into the existing social arrangement. This effect of religion for suppressing dissent was clearly recognized by a mill official who said: "Belonging to a church, and attending it, makes a man a better worker. It makes him more complacent—no, that's not the word. It makes him more resigned—that's not the word either, but you get the general idea" (quoted in Pope, 1942:30, 31). A restudy of this community found these attitudes still prevalent. Mill managers continue to think that the most valued service of local congregations is developing traits in workers of punctuality, hard work, sobriety, and obedience (Earle et al., 1976:26).

Many sectarian religious groups opposed to "the ways of the world" also teach their members values and behaviors that make them more successful in "the world." Holiness groups, for example, resocialize their members into the values of sobriety, hard work, and forgoing of present pleasures for future rewards (Johnson, 1961). Similarly, Black Muslims internalized a disciplined and ascetic religious ethic, indirectly producing socioeconomic benefit by enabling them to obtain jobs and run small businesses more readily than other blacks (Lincoln, 1989; Mamiya, 1982; Mamiya and Lincoln, 1988). Some 1960s religious movements helped their counterculture members "get straight," according to the values of the dominant society (Robbins and Anthony, 1972). By legitimating "appropriate" roles and personal norms, religion helps provide motivation for individuals to participate in the existing socioeconomic system (Robertson, 1977).

Social Control

Religion is a potent force for social control. Although social control can be exercised for social change, it is typically change-inhibiting. External forms of religious social control are most evident. Within a religious group, sanctions for deviance from group norms are all the more potent because of their reference to the sacred. The idea of being judged not only by other humans but also by the gods is a powerful deterrent from deviance. Salvation religions such as Christianity exercise control in urging that believers conform to norms in order to assure their future salvation.

The forms of social control exercised by religious groups include informal sanctions such as ostracizing, shaming, or shunning the offender. Other measures include confession and exclusion or excommunication (Turner, 1977). Some religious groups have highly formal controls such as laws and religious courts. The social control exercised by religious organizations varies according to their power in the society. Society-wide religious controls (e.g., the Inquisition and witch-hunts) are less likely in a pluralistic society. Nevertheless, America's modern equivalent of witch-hunts, McCarthyism and related right-wing extremism, was supported largely by members of religious groups that did not (then) favor pluralism—Roman Catholics and fundamentalist Protestants (Lipset, 1963). Even in relatively pluralistic societies such as the United States and Canada, religion forms a significant basis of the laws and formal order, as well as of informal social control.

Another relatively informal aspect of religious social control is the power of group loyalty to ensure conformity. The same loyalty can also be utilized for change-promoting action. When the group supports the status quo, however, religious loyalty ensures that the individual does nothing that would upset the group or work counter to its interests. The greater the accommodation of the religious group to the larger society, the more does members' loyalty to "our kind of people" inhibit change.

Internal religious controls are especially important. A person who is socialized into a religious perspective internalizes religious controls. Although socialization is never "perfect," most people internalize much of the normative content of their upbringing. Socialization has a built-in conservative thrust. When a person does something to have a "good conscience" or refrains from certain behavior to avoid a "guilty conscience," the belief system is exercising its control. Internal social controls are likely to impede change-oriented behavior because the individual feels guilty in breaking away from the learned norms.

RELIGION PROMOTES SOCIAL CHANGE

While certain aspects of religion inhibit social change, other aspects challenge the status quo and encourage change. In some circumstances, religion is a profoundly revolutionary force, holding out a vision of how things might or ought to be. Historically, religion has been one of the most important motivations for change because of its particular effectiveness in uniting people's beliefs with their actions, their ideas with their social lives. Religious movements such as the Great Awakening (i.e., several waves of revivals in the United States beginning in the latter part of the eighteenth century) have had tremendous

impact on society, though the outcome may not have been part of the religious goals of the movement itself. These revivals were important sources of
the abolitionist movement as well as of later temperance and prohibition
movements. They also had an impact on the democratization of the American
polity, making way for popular participation in what was largely an oligarchy
of economically prosperous citizens (Ahlstrom, 1972:349, 350; Hammond,
1979; Jamison, 1961).

To understand recent developments in modern societies, sociologists have
reexamined their classical foundations.[3] Social change issues were central in
the works of Marx, Simmel, and especially Weber. These classical theorists
have provided explanations for current developments and have inspired contemporary sociologists to new understandings. One product of this new thinking has been a more profound understanding of Weber's analysis of religion
and modernization, going considerably beyond the relatively superficial
"Protestant ethic" studies of the 1950s. We will discuss this new approach to
Weber in more detail later in this chapter.

Religion and Social Dynamics: Beyond Marx

Another product of new perspectives to the classics has been a rethinking of
the Marxian approach to religion and social change. A number of neo-
Marxian sociologists insist that an appropriately complex empirical approach
to the scientific study of religion cannot be substituted for or anticipated by
theoretical constructs. Marxian theory is thus used as a perspective from which
study can be conducted rather than as a static assumption (Maduro, 1977:366).

This new Marxian approach to religion results in a more complex conclusion than classical Marxian interpretations about the relationship of religion to
social change. Starting from some of Engels's later ideas, these sociologists
view religion as being relatively autonomous from the economic substructure.
These theorists come to a more rich, complex understanding of religion by
focusing on its functions as a partially independent variable in social change.
In contrast to classical Marxism, neo-Marxian approaches suggest that:

> (1) Religion is not a mere passive effect of the social relations of produc-
> tion; it is an active element of social dynamics, both conditioning and
> conditioned by social processes. (2) Religion is not always a subordinate
> element within social processes; it may often play an important part in the
> birth and consolidation of a particular social structure. (3) Religion is not
> necessarily a functional, reproductive or conservative factor in society; it
> often is one of the main (and sometimes the only) available channel to
> bring about a social revolution (Maduro, 1977:366).

If religion were not ever autonomous from the dominant economic interests, sociology could not comprehend situations in which religious groups

---

3. See especially Beckford, 1989, 1991; Houtart, 1988; Kersevan, 1975; Maduro, 1975,
1982; Mansueto, 1988; O'Toole, 1984; Robertson, 1977, 1979; Thompson, 1986;
Turner, 1991.

worked for socioeconomic arrangements that were *against* their personal material interests. For example, in the years before World War I, the (pietist) Basel Mission actively opposed and, with the help of their supporters in Germany, effectively blocked the German plantation system in the Cameroons (Africa) on purely religious grounds, despite the fact that its leadership's social class allegiances were with the officials and investors who shaped Germany's colonial policies (Miller, 1993). Sociology of religion needs to conceptualize religion as linked with, but not reducible to, economic interests in order to analyze how religion and social change are connected. The question is no longer "Does religion promote social change?" but rather "In what ways and under what conditions does it promote rather than inhibit change?"

We will consider the change-promoting aspects of religion under three headings: religious ideas, religious leadership, and religious groups. In practice, of course, these frequently overlap (e.g., when a religious leader proclaims a new religious idea to a group of followers).

Religious Ideas and Meanings

Ideas themselves do not directly effect change. Ideas indirectly influence society through people whose interests (i.e., all those things that could benefit them) lie in pursuing those ideas and applying them to social action. Religious ideas therefore effect social action in two ways: They may form the *content* of what a group of people try to do; and they may shape people's *perception* of what their interests are. The movement to abolish slavery in the United States, for example, had important religious impetus. Religious ideas explained the evils of slavery; religious movements created a pool of adherents receptive to abolitionist ideas; and religious interests (e.g., desire for salvation) motivated them to put the ideas into action. For such reasons, it is appropriate for sociologists to examine the emergence of new religious ideas in history.

Religious Breakthrough: Weber. By *ideas,* we mean a broad concept of meanings, such as described in Chapter 2, rather than merely the formal ideas of doctrine or theology. Weber used the yet broader notion of **ethic,** referring to the total perspective and values of a religious way of thinking. Weber (1951, 1952, 1958b) examined several world religions for those aspects of their orientations to their god and their social worlds that inhibited or promoted certain socioeconomic changes, especially the process of "modernization." In these studies, Weber examined the social location of religious ideas and innovation in each society. He sought the sources of motivation for individual action, the relationship of the individual to the larger society, and the religious ideas that shaped how individual actors in that society perceived their social world. Weber was especially interested in locating historical points of **breakthrough.** These were periods in the development of a society when circumstances pushed the social group either toward a new way of action or toward reaffirmation of the old way. The movement toward innovation in the social system constituted a breakthrough. Weber noted that religion has been historically prominent in these breakthrough developments.

Like Marx, Weber held that people's vested interests, considered in the context of certain structural conditions, explained their actions. By contrast, he felt that these interests need not be purely economic ones; people could be motivated, for example, to protect their religious standing. Although religious ideas do not determine social action, according to Weber, they are significant in shaping actors' perceptions and interpretations of their material and ideal interests. Weber (1946:280) stated:

> Not ideas, but material and ideal interests, directly govern men's conduct. Yet very frequently the "world images" that have been created by "ideas" have, like switchmen, determined the tracks along which action has been pushed by the dynamic of interest. "From what" and "for what" one wished to be redeemed and, let us not forget, "could be" redeemed, depended upon one's image of the world.

Weber considered religion one of the foremost sources of these ideas that shaped the direction of social action and were "switchmen" in the course of history.

In *The Protestant Ethic and the Spirit of Capitalism* (1958a), Weber analyzed one particularly important breakthrough—the development of the capitalistic mode of socioeconomic organization. He hypothesized a link between the rise of the Protestant worldview and the subsequent emergence of capitalism in Western society. To Weber, capitalism was a significant factor in modernization, a process he viewed with almost prophetic misgivings. The key characteristic of modernization is its *means–end* (i.e., functional) *rationality*. Weber emphasized that capitalism is not mere greed or acquisitiveness but the rational, systematic investment of time and resources toward the expectation of future chances of profit. Furthermore, the expansion of this perspective to a society-wide economic system required socially available roles for performing these tasks: specifically, the role of the entrepreneur (Weber, 1958a:17–27; excellent brief treatments of Weber's thesis include Eisenstadt, 1969, and Robertson, 1977).

Ideas and Individualism: Weber. Weber argued that creation of a pool of individuals with the necessary values and characteristics to perform as entrepreneurs was essential to the emergence of capitalism. This development was made possible by a new form of individualism, an unintentional by-product of the main variant of Protestantism at that critical moment in history. Weber reasoned that the social role of the entrepreneur called for someone who valued hard work and considered deferred gratification as almost a virtue in itself. Early Protestantism, especially Lutheranism, contributed to this ethic by its interpretation of work as a vocation. The work of the layperson was thus viewed as a virtue, fulfilling a special call from God.

Capitalism further required persons willing to deny themselves rationally and systematically for the sake of achieving a future goal. The dominant strain of Protestantism (e.g., Calvinism, Puritanism, Pietism, Anabaptism) produced such qualities because of its **inner-worldly asceticism,** encouraging members to be active "in the world"—indeed, to prove their salvation by their socioeconomic actions. At the same time, these forms of Protestantism expected their members to forgo the pleasures "of the world"—not to spend their money on luxuries, drink, gambling, or entertainments. The "Protestant ethic" therefore became one of hard work, sobriety, financial care, and deferred gratification. Even though capitalistic gain was far from the goal of these Protestant values, according to Weber, the initial development of capitalism was made possible by the available pool of persons who shared these qualities.

Simultaneously, Protestantism resulted in a powerful new source of motivation for such economic action as capitalism. Protestant beliefs produced a new form of individualism—that is, a new mode of individual-to-society relationship. Under Roman Catholicism, the individual had experienced something of a "blanket" approval. The individual's standing before God and fellow believers was certified by belonging to the church and receiving its sacraments; thus, belonging to the *group* legitimated the individual member. Protestantism, especially its Calvinist varieties, offered no such security. Weber argued that the Protestant ethic made it necessary for individuals to legitimate themselves. This need for self-legitimation, combined with inner-worldly asceticism, produced a strong motivation for the kind of socioeconomic action that capitalism entailed.

Weber linked these values and motivations produced by the Protestant ethic with the emergence of capitalism, especially the critical role of the entrepreneur. Once established, however, capitalism no longer needed the Protestant ethic; mature capitalism, according to Weber, can be self-sustaining. Some sociologists, failing to recognize this point, have tried to apply Weber's terms to current attitudinal differences between Roman Catholics and Protestants in developed countries. These studies have produced conflicting results, but there is little consistent evidence of important differences between Protestants and Roman Catholics in attitudes toward work and consumption.[4] More relevant to Weber's thesis are studies of the impact of other-worldly or this-worldly religious attitudes in developing nations (see Eisenstadt, 1968).

Numerous scholars since Weber have contested his Protestant ethic hypothesis. Some have argued that the constellation of attitudes and ideas that Weber called the Protestant ethic (if such a mind-set existed) coincided with more important socioeconomic changes such as technological developments, the availability of potential laborers, and the influx of new capital resources (e.g., from colonial holdings). They suggest that these latter factors, among others, were more significant than the Protestant ethic in the development of capitalism. Probably Weber himself would have modified his thesis had he fin-

---

4. See Greeley, 1964; see Bouma, 1973, for a detailed critique of specific empirical investigations purporting to test Weber's Protestant ethic hypothesis; the best-known study exemplifying these misunderstandings is Lenski, 1963; see counterarguments in Greeley, 1963, and Schumann, 1971; a somewhat different application is made by McClelland, 1961, with counterarguments in Robertson, 1978. Systematic critiques of the Weber thesis include Green, 1959; Samuelsson, 1964; and Tawney, 1926.

ished his projected sociology of Protestantism before his death. *The Protestant Ethic* was among his earliest writings in the sociology of religion, and his later essays emphasized the change-promoting potential of religious sects as organizations more than the Protestant ethic per se (Berger, 1971; Weber, 1946:302–322).

Religious Imagery. Another aspect that contributes to social change is the capacity of religious meanings to serve as symbols for change. Religious symbols frequently present an *image* of future change. They create a vision of what could be and suggest to believers their role in bringing about that change. Change-oriented symbolism is often directed toward the social realm, as exemplified by the ideas of the "heavenly city," the "new Jerusalem," and the "chosen people." Many new religious movements, for example, have articulated visions of the "kingdom of heaven on earth" and of how they ought to proceed to realize that new order (Barker, 1988b).

Religion provides symbols of tradition and continuity, but religious idealization also gives symbols to a group's desire for change. An example of such change-promoting religious imagery is Liberation Theology's sacralization of the community of the poor and identification of this-worldly activism with Christ's struggle to liberate humanity (Lancaster, 1988). For example, in the Nicaraguan *Misa Campesina* (peasants' Mass), the "Kyrie" is sung:

> Christ, Christ Jesus
> identify yourself with us
> Lord, Lord God
> identify yourself with us
> Christ, Christ Jesus
> solidarize yourself
> not with the oppressor class
> that wrings out and devours
> the community
> but rather with the oppressed
> with my people
> thirsting for peace.

(lyrics by Carlos Mejia Godoy, 1981; quoted and translated by Roger N. Lancaster in *Thanks to God and the Revolution* © 1988; reprinted by permission of Columbia University Press)

These symbols enable people to conceptualize their situation and to manipulate change-promoting elements of their social world. Sometimes the specific symbols evoked are not very effective. The future sought by a group may be impossible to realize; but genuine social structural changes are unlikely to occur unless people have new ways of thinking about their social world—a set of symbols that evokes images of change (see Maduro, 1977:366).[5]

---

5. Maduro's theories utilize the seminal work of Pierre Bourdieu (1977), whose theories of symbolic production are useful to a sociology of religion.

Religious Leadership

Social change often requires an effective leader who can express desired change, motivate followers to action, and direct their actions into some larger movement for change. Religion has historically been a major source of such leaders largely because religious claims form a potent basis of authority. The prototype of the change-oriented religious leader is the **prophet,** whose social role is especially significant.[6]

A prophet is someone who confronts the powers that be and the established ways of doing things, claiming to be taken seriously on religious authority. There are two types of prophetic roles. One is the **exemplary prophet,** whose challenge to the status quo consists of living a kind of life that exemplifies a dramatically different set of meanings and values. The Buddha is a good example of this kind of prophet, whose very way of life is his message. The other type is the **emissary prophet,** who confronts the established powers as one who is sent by God to proclaim a message. Most of the Old Testament prophets were of this type. The emissary prophet has historically been an important source of change because the message proclaimed offered a new religious idea (i.e., ethic) and a different basis of authority. The prophetic message was often one of judgment and criticism. Whether the message called the people back to a previous way of life or directed them to some new way, it nevertheless called for *change.*

The role of the prophet is the opposite of the priestly role. A **priest** is any religious functionary whose role is to administer the established religion—to celebrate the traditional rituals, practices, and beliefs. Most clergy and church officials perform priestly roles. The basis of priestly authority is priests' location in the religious organization as representatives of that establishment, and their actions mediate between its traditions and the people. The prophet, by contrast, challenges the established way of doing things, not only by messages of criticism but also by claiming an authority outside the established authority. Thus, the role of the prophet is essentially a force for change in society.

Weber proposed that **charisma** was the authority basis of leaders such as prophets. *Charisma* refers to "a certain quality of an individual personality by virtue of which he is set apart from ordinary men and treated as endowed with . . . specifically exceptional powers or qualities . . . [which] are not accessible to the ordinary person" (Weber, 1947:358, 359). The authority of the charismatic leader rests on the acceptance of his or her claim by a group of followers and on the followers' sense of duty to carry out the normative pattern or order proclaimed by the leader. Thus, the charismatic leader is a source of both new ideas and new obligations.

The charismatic leader says, in effect, "It is written . . . , but I say unto you. . . ." By word or deed, the charismatic leader challenges the existing normative pattern, conveying to the followers a sense of crisis, then offering a solution to the crisis—a new normative order. Charismatic leaders may arise

---

6. This discussion of prophets and charismatic authority follows Weber's theories (1946; 1947:458ff.; 1963).

either outside or within the institutional framework. Charismatic authority is *extraordinary.* It breaks away from everyday bases of authority such as that of officials or hereditary rulers. Thus, in certain periods and social settings, charismatic authority is highly revolutionary.

The authority of the Islamic leader Khomeini in the 1979 revolution against the Shah of Iran illustrates this quality of charisma. Khomeini stood in opposition to the rationalization of Iranian society, the modernization of which had upset large segments of the populace. Counterposing a religious charismatic figure against the political leaders "fit" a traditional Shi'ite model of religious action (Arjomand, 1984, 1988). The combination of Khomeini's personal characteristics and his appeal to Shi'ite Islamic dogma was powerful. Although Khomeini disclaimed his revolutionary function, he was viewed as the leader who, according to prophecy, would deliver the people; the prophecy itself was, in turn, reinterpreted as support for his charismatic authority (Kimmel, 1989).

Pure charismatic authority is, however, unstable. It exists only in *originating* the normative pattern and the group that follows it. By the process of **routinization,** charismatic authority is transformed into a routine or everyday form of authority based on tradition or official capacity, and the new religious group comes to serve the "interests" of its members. Although this routinization usually compromises the ideals of the original message, it is a necessary process for the translation of the ideal into practice. Routinization frequently dilutes the "pure" ideals and sometimes results in a comparatively static organizational form; however, it is also the process by which charisma comes to have its actual impact on history (see M. Hill, 1973b; Weber, 1963:60, 61).

The history of the Christian churches is a series of charismatic innovations, routinization, and stable organization, followed by new appearances of charisma. Saint Francis of Assisi was a charismatic leader, an exemplary prophet whose message was also one of judgment—that Christian life must return to its pristine ideals of poverty, community, and simplicity. The early stages of the Franciscan movement were a serious threat to the established church and the sociopolitical order with which the church was intertwined. There was little to distinguish Franciscanism from other threats of the day—Albigenses, Waldenses, and various other pietistic movements, each with its own charismatic leaders. The other movements were eradicated, often militarily by the sheer might of the established powers. The Franciscan movement, by contrast, was encapsulated as a religious order, and its subsequent routinization made it no longer a serious threat to the church. Nevertheless, the Franciscan ideals had great impact and were periodically raised as a critical standard against the complacency of the church and the order itself.

Similar cycles of charismatic innovation and routinization into organizational stability have occurred in other religious groups (and, indeed, in nonreligious institutions, such as political groups). The Mormons moved from a highly charismatic form of authority under Joseph Smith and Brigham Young to a stable, routinized form under church officials called "bishops" (many of whom are laymen). Similarly, Black Muslims have undergone several phases of charisma and routinization, each time altering the group's ideology, tactics, and form of organization (Lincoln, 1989; Mamiya, 1982).

The development of mass communications, especially those using image technologies such as television and mass advertising, has added a new and complex dimension to the generation of charisma and the uses of public ritual (Wuthnow, 1994). Mass media possess much more flexible and powerful means to shape the presentation of a leader (or group) to generate the image of exceptional powers or qualities for which the audience might impute charismatic authority. Modern movements may use these mass media to create and sustain a sense of urgency, crisis, or imminent disaster to appeal to a mass constituency (Barkun, 1974).

Not only political figures but also religious leaders can frame their images and their messages for maximum media impact. The television coverage of the journeys of Pope John Paul II resembled a serialized narrative, evoking recurring themes. In repetitive short news clips, the television viewer "saw" the pope set in images of conquest (e.g., with the keys of the city, honor guards, red carpets, in palatial settings). Other media images included symbolic confrontations with death (e.g., located by cemeteries, concentration camps, soldiers with machine guns) and triumphal images (e.g., cathartic, expressive images of attentive, cheering people, flag-waving, applauding, attempting to touch the pope) (Guizzardi, 1989).

Charismatic figures portrayed by the media are thus transformed by the selection process inherent in those media: Images mediate the reality to the audience. Some would-be charismatic figures also deliberately manipulate mass media to enhance their position by creating a sense of awe and power unlikely to be attributed in face-to-face interaction with the leader.

Because it is the followers who impute charismatic authority to their leader and who put the charismatic ideal into practice, any analysis of the impact of religion on social change must examine the role of the religious group or community in that change.

The Religious Group

Whether the religious group is a small band following a charismatic leader, a growing religious movement, or a staid and established religious organization, it is a potential force for social change. This potential exists because religion—especially the religious community—is a source of power, which is a fundamental category in the sociology of religion. The same power can be and usually is, however, used to support the status quo. Religion is not only an experience of power but often also results in the sense of being empowered. Thus, the followers of the charismatic leader may experience a sense of power in their relationship with the leader and with fellow believers that enables them to apply the new order to their social world.

This sense of power, especially when identified with the leadership of a charismatic figure, can give a developing religious movement great dynamism. In 1525, for example, Thomas Muntzer issued this call to battle in the German Peasants' War:

> Do not despair, be not hesitant, and stop truckling to the godless villains. Begin to fight the battle for the Lord! Now is the time! Encourage your

brethren not to offend God's testimony because, if they do, they will ruin everything. Everywhere in Germany, France, and neighboring foreign lands there is a great awakening; the Lord wants to start the game, and for the godless the time is up. . . . Even if there were only three of you, you would be able to fight one hundred thousand if you seek honor in God and his name. Therefore, strike, strike, strike! This is the moment. These villains cower like dogs. . . . Have no concern for their misery; they will beg you, they will whine and cry like children. Do not show them any mercy. . . . Strike, strike, while the fire is hot. . . . Have no fear, God is on your side! (quoted in Lewy, 1974:115).

Another potential is the religious group's capacity to unite previously disparate segments of a society. Religious sentiment can bridge barriers of tribe, family, nationality, and race. Enthusiastic religious movements (e.g., Pentecostalism) often experience intense egalitarianism in their early stages. Similarly, in some developing nations, religious groups are sometimes the only forces capable of uniting people steeped in a tradition of tribalism. The followers of Simon Kimbangu in Zaire and Jehovah's Witnesses in Kenya and Zambia exemplify this unifying dynamism. Religion often provides a bridge enabling people of different social class interests to work together, as illustrated by the early stages of the Solidarity movements in Poland and Catalonia (Johnston and Figa, 1988). The religious community can produce a sense of group consciousness and solidarity that focuses a group's awareness of its needs and interests and strengthens its efforts to achieve its goals (Levine, 1990; Lewy, 1974:217–220).

Another change-related feature of religious groups is that they are microcosms of political participation. Throughout history, members of religious groups often got a "first taste" of political agency through the polity of their religious group, rather than their state. Long before they had the opportunity for democratic participation in their governments, many Presbyterians, Baptists, and Congregationalists, among others, experienced a measure of political agency in their church organization. These experiences may have shaped their eventual image of a good government and prepared them to be more active and effective than people who had no such experience. Similarly, in many countries of Latin America, small Catholic groups called *Communidades Eclesiales de Base* ("base communities") have changed members' expectations for political participation and action toward greater democratization (Cavendish, 1994; Levine and Mainwaring, 1989; see also Bruneau, 1988; Hewitt, 1990; Ireland, 1989; Levine, 1992; Mainwaring, 1986; Smith, 1994).

Certain forms of religious collectivity typically have greater change-oriented potential. Sects and cults are, by definition, in greater negative tension with society than denominations or churches (see Chapter 5). Sects are particularly likely to mobilize efforts for change because of their characteristic collective orientation and insistence that religion be applied to all spheres of members' lives (see Yinger, 1963). Churches and denominations, by contrast, generally have more substantial social bases for action but are less likely to challenge the status quo. As Chapter 5 demonstrates, different kinds of religious collectivities have different potential for change-oriented action.

Furthermore, these collectivities experience organizational transformations that reduce or promote their dissent.

FACTORS SHAPING THE INTERRELATIONSHIP OF RELIGION AND SOCIAL CHANGE

The earlier parts of this chapter explored those aspects of religion that are change-inhibiting, change-promoting, or both. A more complex question now arises: Under what conditions is religion likely to be either change-inhibiting or change-promoting? The circumstances surrounding a change-oriented effort often determine the degree to which the effort is effective. Thus, we also need to ask: Under what conditions is religion most likely to be an effective source of social change? And what are the social factors that maximize or minimize religion's influence in a society?

Qualities of Beliefs and Practices

Certain qualities of some religions' beliefs and practices make them more likely to effect change than other religions. A sociologist would seek to learn some of these qualities by asking the following questions.

Does the belief system contain a critical standard against which the established social system and existing patterns of interaction can be measured? Those religions that emphasize a critical standard (e.g., a prophetic tradition or a revolutionary myth) pose the potential of internal challenge to the existing social arrangements. The prophetic tradition of the Israelites was a basis for subsequent religious challenges to the established way of doing things (Weber, 1952, 1963). Similarly, nations such as the United States, France, or the Republic of Ireland, where the civil religion embodies a revolutionary myth, have a built-in point of reference for internal criticism (Stauffer, 1974). This prophetic aspect of American civil religion was central to the appeal of the civil rights movement. Kennedy's inaugural address was similarly built on prophetic aspects of American civil symbols.

Ethical standards also provide a basis for internal challenge to existing social arrangements. The social critic can say, in effect, "This is what we say is the right way to act, but look how far our group's actions are from these standards!" Many Americans judged their nation's conduct of the war in Vietnam as immoral. The emphasis of ethical standards within certain religions provides a regular ground for social action and change (Nelson, 1949; Weber, 1963). Furthermore, the content of norms and ethical standards influences the kinds of resulting social action. Some religious groups emphasize personal or private moral norms such as strictures against smoking, drinking, gambling, premarital and extramarital sexual activities, divorce, homosexuality, abortion, and contraception. Other religious groups put greater emphasis on public morality, focusing on issues such as social justice, poverty, corporate responsibility, ethics of public policy, and war. Both approaches may result in efforts for social change. The temperance movement of the 1920s and the antiabortion movement of recent years both exemplify movements based on personal moral norms; the civil rights movement of the 1960s and the sanctuary movement of the 1980s were efforts to apply moral judgments to the public sphere (Crittendon, 1988).

The existence of such value standards within a group's belief system sometimes provides a basis of legitimacy for leaders who try to move the collectivity to social action. For example, in the 1950s and 1960s, priests ministering to Hispanic farmworkers in California could refer to the Catholic church's social teachings (e.g., papal encyclicals) regarding social justice and workers' right to organize (Mosqueda, 1986). Despite objections from local clergy and bishops whose congregations (and financial resources) included the grower/employers, these leaders gained legitimacy from the existence of strong standards and ideals taught by their religion at an international level.

How does the belief system define the social situation? Individuals' perceptions of the social situation are shaped largely by how their belief system defines that reality. If a religion informs believers that their misfortune is part of God's plan to test their faith, they are not likely to challenge that misfortune. Believers are unlikely to try changing a situation that the belief system has defined as one that humans are powerless to change. Belief systems that embody this kind of fatalism are not conducive to social activism. Some belief systems promote fatalism as a response to the expected imminent end of the world. One Seventh Day Adventist explained why she did not vote or get involved in Hispanic political efforts in her community: ". . . this world is going to be destroyed anyway. We should be focusing our interest on heavenly things. I fear that if we become too involved in these things, we may overlook important things in life" (quoted in Hernández, 1988).

Similarly, belief systems that are voluntaristic are unlikely to result in concerted efforts for change because they define problems in the social situation as the sum of all the individual shortcomings. Accordingly, social ills can be overcome only by converting all individuals to the right way of life. Certain religious definitions of the situation (e.g., fatalism and voluntarism) are less likely than other religious definitions to result in concerted efforts for structural social change.

How does the belief system define the relationship of the individual to the social world? This question is a generalization of the Weberian theme described earlier in this chapter. Weber pointed out that certain belief systems encourage different kinds of individualism and that this individual-to-society relationship is critical to social action. Weber also distinguished between religions that promote a "this-worldly" as compared with an "other-worldly" outlook. Buddhism's interpretation of the material world and aspirations as illusion discourages this-worldly action. By contrast, many strains of Protestantism emphasize one's "working-out" of salvation in this world and one's "stewardship" (i.e., responsible use) of God-given worldly resources. While "this-worldly" religious perspectives are generally more oriented toward social action, an "other-worldly" focus can serve to withhold legitimacy from the dominant society (Fields, 1985; Goldsmith, 1989). These aspects of the belief system determine the likelihood and direction of action for social change.

The Cognitive Framework of the Culture

Relevant to any discussion of religion's role in social change is the cognitive mode of the culture. What are the main ways that people in a culture think about themselves and their actions? While these qualities of the culture may be related to specific items of religious belief, they are more general. Discovering the cultural aspects that shape the influence of religion on social change would include the following factors.

Is the religious mode of action central to the cognitive framework of the culture? And are other modes of action (e.g., political) foreign to the way people think? Religious modes of action may be the *only* channel people have for affecting their world; other modes of action may be not only foreign but literally inconceivable. In much of Latin America, most people's worldview is saturated by religion, and religion has the potential to be a viable and vital mode of social action. Economic dissatisfaction and political dissent may be expressed in religious terms and resolved in religious modes of action (Maduro, 1977; Worsley, 1968).

In cultures where the religious mode of action is pervasive, people are unlikely to conceive of other ways of organizing for change. If another mode of action is introduced, people are likely to be highly suspicious because the idea of a group with no religious identity is impossible to them; they cannot conceive of a group for whom religion is irrelevant to mutual activities. The People's Democracy, a radical independent political movement arising in Northern Ireland during the late-1960s strife, was utterly anomalous because in that social context a nonsectarian social group was inconceivable to most persons (Beach, 1977). In the United States, by contrast, nonreligious modes of action are both common and conceivable partly because of the ideological separation of church and state but also because of the high degree of institutional differentiation in modern society. Religious institutions are separated from other institutional spheres such as economic and political ones. The difference between these two cultural settings is not only in the social structure but also in the characteristic ways the people think about themselves and their actions.

Are religious roles and identification significant modes of individual action in the culture? Does it make sense in the culture for an individual to claim a religious identity or a special religious role? And are religious roles understandable forms of leadership? For Joan of Arc to make religious claims for her leadership made sense to the people of her culture, whereas similar claims in a modern society would probably result in derision or labeling as symptomatic of mental illness. In cultures where religion is important in determining how one thinks of oneself or how others identify one, religion is probably a more significant vehicle for change than in cultures where religious roles are less acceptable.

These aspects of religion suggest that not merely the belief content of religion determines its influence on social change but also the underlying cognitive framework of people in that culture. Sociological analysis of the change-promoting and change-inhibiting characteristics of religion must take into account these cognitive modes, which shape people's ways of thinking about themselves and their world.

The Social Location of Religion

Most sociological analyses of religion's impact on social change have focused on the social location of religion in various societies. They refer to the structural relationship between religion and other parts of society (i.e., where does religion "fit" into the total pattern of how things are done?). Also, the internal structure of religious groups is related to their larger social location. The following criteria suggest some of the structural aspects of religion in relation to society that are relevant for understanding the extent of its impact on social change.

Is religion relatively undifferentiated from other important elements of the society? In most simple societies, religion is diffuse. It permeates all activities, social settings, group norms, and events. Religion, not a separate institution, is undifferentiated even in people's ways of thinking. The person who plants seed with a ritual blessing does not think of the blessing as a separate religious action; it is simply part of the right way to plant seed. One of the definitive characteristics of modernization is *differentiation* of various spheres of social life. Relatively complex modern societies are characterized by structural (and often spatial) segregation of different parts of social life into identifiable separate institutions.

We can envision societies as falling somewhere on a continuum between these two polar types:

Highly undifferentiated
(religion diffuse)

Highly differentiated
(religion segregated)

In societies where religion is relatively undifferentiated, any change-oriented action is also likely to be religious action. Some of the millenarian movements described in Chapter 2 exemplify religious actions that are also sociopolitical. In relatively undifferentiated societies, religious dissent is often a way of expressing political and economic dissatisfactions. By contrast, although religious action *can* express other dissatisfactions in relatively differentiated societies such as the United States, members of such societies often think of each institutional sphere as separate and of religious action as irrelevant to the political or economic sphere.

If religion is relatively differentiated from other institutional spheres, do strong ties exist between religious institutions and other structures—especially political and economic ones? The more linkages that exist between religion and other institutional spheres, the more likely that religious movements for change are expressions of dissatisfaction with other spheres as well. Historically, dissatisfaction with political and economic spheres has been prominent, but dissatisfaction with other differentiated institutions is also possible. For example, a number of contemporary religious movements show strong currents of dissatisfaction with the dominant medical system in American society (Freund and McGuire, 1995).

Even in highly differentiated social structures, linkages often exist between institutional spheres. Such linkages range from formal ties (e.g., the state church in England) to highly informal connections (e.g., the "coincidence" of closely overlapping membership between a congregation and other social and political associations in a community). Earlier in this chapter, we saw how religion is often used to legitimate the status quo and serve the interests of the dominant group. Religious organizations themselves often have vested interests such as lands, wealth, and power to protect. These kinds of linkages with political structures and the stratification system are precisely why religion is often used as a vehicle of dissent *against* the political or stratification system.

In societies where religion is closely linked by formal or informal ties with the state, religious movements are especially likely to have political (though not *only* political) significance. The Protestant Reformation was a highly political movement, but religious ideas were also very important. In the United States, where the official linkages between religion and the state are minimized, there is still a considerable informal religiosity (e.g., Congressional prayer breakfasts) and an entire set of beliefs and practices related to civil religion (see Demerath and Williams, 1990). The strong connection between religion and America-love-it-or-leave-it in the 1960s made plausible a specifically religious mode of political reaction. Precisely because the dominant political system was supported by mainstream religiosity, religion was a significant force for political dissent.

Similarly, in situations where religion is closely linked with the stratification system, religious movements are likely to be expressions of socioeconomic dissatisfaction. As the Extended Application in Chapter 6 shows, Catholics were particularly active in the 1960s civil rights movement in Northern Ireland because persistent discrimination against Catholics was embedded in the entire stratification system.

Many theories about the development of religious movements focus on economic or status deprivation (as discussed in Chapter 5). A relationship clearly exists between religious discontent and disprivilege, but it is not so simple or direct as some of these theories imply. The utterly poor and powerless rarely form change-oriented movements because of their fatalism, lack of resources, and the sheer struggle for basic subsistence. At the other end of the socioeconomic ladder, the powerful and privileged rarely form such movements because the established ways of doing things in the society serve their interests well. The broad spectrum of persons between these two poles, however, are difficult to characterize in their likelihood of forming movements for change. Nevertheless, it is possible to generalize that where religion is structurally linked with socioeconomic privilege, religious dissent and movement for change are likely to express dissatisfaction with the existing distribution of privilege and power. Where such dissatisfactions are not adequately handled by the existing religious groups, fertile ground exists for new change-oriented religious movements.

Do other modes effectively compete with religion for the expression of human needs, the development of leadership, and organization of effort toward social changes? In some societies, religious groups may be the most effective vehicle for social change because they are better situated to mobilize change-oriented action. In others, other modes of action such as political or legal action may be better developed and more effective. In some situations, religious organizations and leaders become virtually the only available voices for change due to the co-optation or repression of other avenues. In South Africa, church leaders (except those of the Dutch Reformed churches) protested apartheid from its inception in 1948. By the 1980s, the government had imprisoned, banned, or exiled most secular antiapartheid leaders, so religious leaders (taking considerable risk) became the foremost spokespersons and activist leaders. In 1988 Anglican Bishop Desmond Tutu (who was awarded the Nobel Peace Prize in 1984), together with twenty-five other church leaders from sixteen denominations, openly defied government bans and called on all Christians to boycott the election for segregated municipal councils. They declared, "By involving themselves in the elections, Christians would be participating in their oppression and the oppression of others" (Thompson, 1990:239).

Similarly, the Catholic church became the primary voice for socioeconomic and legal-political changes in Brazil between 1964 and 1985, when a dictatorial military regime had systematically suppressed opposition from political parties and congress, the courts, the press, universities, and labor organizations (Levine, 1990; Smith, 1979). Catholic bishops denounced use of torture and absence of fair trials for political prisoners, published information demonstrating the links between government security forces and private death squads, publicly highlighted the inequitable distribution of wealth, and helped indigenous peoples protect their lands against encroachments by mining and agribusiness developments. Many of the laypeople and clergy involved in these resistance efforts were harassed, arrested, or murdered (Smith, 1979). The bishops openly denounced the military government's economic policies for the poverty they engendered and the injustice of economic development that displaced peasant farmers and indigenous peoples from their lands. The bishops declared "a preferential option for the poor," meaning that, by design, the church was identifying with the masses of poor, trying to represent their needs and interest, and structurally opening itself to their fullest participation in the church and society (Adriance, 1986). After more than a decade of Catholic church dissent against the oppressive Brazilian military regime, one observer concluded that "The church has become the primary institutional focus of dissidence in the country" (Bruneau, 1982:151).

After 1985, however, the government came under the control of a civilian government and many repressive measures were lifted. Other forms of expressing dissent became increasingly legal and viable, and the church has been less vocal on political issues (due partly to constraints from the Vatican and the Conference of Latin American Bishops). The Brazilian Catholic church has continued to speak out against elitist politics, gross inequalities of wealth and productive resources, serious and widespread poverty, and injustices to peasants and indigenous peoples (Levine and Mainwaring, 1989), but its voice is now one among many competing public claims.

What is the social location of religious leaders? The potential of charismatic or traditional religious leaders to effect change largely depends on their social location: ties with a pool of potential followers, links with resources that can be mobilized in a movement, and connections with networks of related movements or leaders. Leaders of any movement require these kinds of ties. Recognizing the importance of these linkages, social organizations threatened by dissent often isolate potential leaders in a social location where they are cut off from the people and resources necessary to establish a change-oriented movement. Church officials, for example, might assign a potential "troublemaker" to a remote and tiny parish. Although charismatic leaders are less restricted by such official pressures than other leaders, their social location also influences the likelihood of their gaining a sufficient base of support. Whether or not religious leaders or movements succeed in their efforts for social change depends on their location relative to social resources, organization resources, and a pool of potential followers.

In rural Brazil, "base communities" (*Communidades Eclesiales de Base,* or CEBs) are small Catholic groups that base their social activism on their reflections of the Bible as applied to their social situation. Most of these grassroots organizations were organized and taught this approach by pastoral agents (typically nuns or lay church workers, and sometimes priests), who live and identify with the people. Their leadership effectiveness rests upon the ability to generate local lay leadership and to open new ways of thinking—new images of social relationships in people's ways of thinking about their world. For example, CEBs in rural Brazil encouraged members to consider ways of thinking about land reform in the light of their reading of the Gospel (Adriance, 1994; see also Adriance, 1991). A similar leadership role was assumed by a handful of intellectual clerics who identified with the masses of Puerto Rican immigrants to New York (Díaz-Stevens, 1993).

A related aspect of religious leaders' location is their relative autonomy from economic interests of their followers and their opposition: How independent are they? When religious leaders must depend upon powerful interest groups for patronage, they are not likely to challenge the status quo. Oppositional movements require organizational autonomy, not only for control of material resources but especially for control of cultural resources. Local clergy were complicit with local industrialists in putting down textile workers' strikes (studied by Pope, 1942). By contrast, during coal miners' strikes of the same era religious leaders (some of whom were ministers who came to identify with the workers and some of whom were lay leaders, miner-ministers) helped legitimate miners' struggles for a union (Billings, 1990).

One important factor in the relative influence of religious leaders is the impact of pluralism and institutional differentiation on the salience of religious authority. In a pluralistic religious situation, competing worldviews each have their own spokespersons who can exercise influence that is not necessarily recognized, however, outside their immediate group. A Mormon bishop's pronouncement on a social issue may be printed in national newspapers and read with interest by non-Mormons, but it is not likely to be considered authoritative by persons outside the Mormon fold.

Similarly, institutional differentiation has often led to the idea that religious leaders should speak out only on "spiritual" matters. One survey found that 82 percent of the laity agreed that "clergymen have a responsibility to speak out as the moral conscience of this nation." Nevertheless, 49 percent felt that "clergy should stick to religion and not concern themselves with social, economic, and political questions" (reported in Hadden, 1969:148, 149). This opinion reflects the sense that religious leadership is not relevant to other institutional spheres and that "moral conscience" is generally a matter of private-sphere activities (e.g., marriage and family). Pluralism and institutional differentiation, important features of some modern societies like the United States, considerably diminish the impact of religious leadership and religious movements.

The Internal Structure of Religious Organizations and Movements

In societies where religion is relatively differentiated with distinctive religious organizations, members' access to religious power is an important variable in whether that religious group can be change promoting. A centralized priestly hierarchy that controls religious "goods" (e.g., salvation) is in a strong position of power to shape the direction and force of that group's efforts. The American Roman Catholic church was able to racially integrate its congregations and schools long before most Protestant denominations or public schools because it used an authority that superseded local authorities and opinion (see Campbell and Pettigrew, 1959, for a study of the tension between local opinion and organizational security in ministers' involvement in civil rights activism).

Centralized religious organizations typically support the societal status quo because they also have vested interests in the economic and political arrangements of that society. In some cases, however, a religious organization identifies with those out of power and privilege (e.g., when it is the religion of a colonial people). A study of turn-of-the-century labor relations in Philadelphia found that interactions between trade unionists and mainstream denominations had considerable *mutual* influence. The popular millennialism of the working-class labor movement and the Social Gospel and ideas of justice of the mainstream Protestants were mutually transformative (Fones-Wolf, 1989).

Access to religious power can be the source of social power, whether for change or stability. Similarly, religions that incorporate a more democratized sense of religious power (e.g., Pentecostalism and other forms of experiential religiosity) have the potential to convert that sense of power to social action. Typically, these groups do not organize themselves for action, however, because other elements of their belief systems do not encourage social action. While the potential for change-oriented action is present in such groups, the loose internal structure of their organizations would lead to a very different form of action than that of a centralized organization.

Another feature of a religious organization that sometimes strengthens its effectiveness in social change is the degree to which a group has larger organizational support outside its immediate situation. Roman Catholic clergy in Poland were more able to challenge those in power in the country because theirs is an organization with a supranational base (Mojzes, 1987; see also Chodak, 1987; Chrypinski, 1989). If their religious organization had only a local base of power and authority, the clergy would be more vulnerable to suppression or co-optation by the established national political-economic powers (Westhues, 1973).

In some situations, the larger organization tries to foster change in the face of local resistance. In South Africa, the National Council of Churches has been at the forefront of confrontation with the state over the issues of human rights and apartheid (i.e., between 1948 and 1994, the government policy of rigid segregation and stratification of racial groups). The Dutch Reformed Church of South Africa (the denomination of the majority of the dominant white group, the Afrikaners) generally supported apartheid and provided scriptural justification for the policy (Cowell, 1985; Jubber, 1985). When they refused to alter their proapartheid stand, two white South African Reformed churches were suspended in 1982 by the international denomination, the World Alliance of Reformed Churches (Barkat, 1983). Finally, in 1986, the Dutch Reformed Church of South Africa rescinded its support of apartheid. While not actively condemning it, as did the international denomination, the church statement did affirm that the "use of apartheid as a socio-political system . . . cannot be accepted on Christian ethical grounds . . ." (quoted in *The New York Times,* October 23, 1986).

The national centralized denomination served a similar role in supporting ministers speaking out against child labor and other working conditions in mills (Pope, 1942:195–198). A study of twenty-eight American religious groups found that the ability of church leaders to carry through controversial policies (such as racial integration) is dependent on the degree of authority given them by a central formal authority. Ministers were in especially weak positions in groups with congregational polity, where their congregations had power to dismiss them without reference to a higher authority (Wood, 1970, 1981). At the same time, however, the larger organization can stifle or suppress local initiatives for change, as illustrated by the tension between the (local) Latin American bishops and the pope at the 1978 Puebla Conference in Mexico, as well as Rome's later pressures to curb Liberation Theology in Latin America.

EXTENDED APPLICATION:
AFRICAN-AMERICAN RELIGION

As the preceding section suggests, analysis of whether or not religion promotes change in a specific situation must take into account a complex array of variables. Sociologists have analyzed numerous historical developments, such as the spread of evangelical Protestantism in Latin America and the role of religion in worker/union militancy and suppression in the United States. One recurring question is whether U.S. minority religions, such as African-American churches, serve to support the status quo or to promote social change for their people. The answer is "yes": Religion *both* supports the status quo *and* promotes active social change, often simultaneously. This section examines some developments in the religious experience of African-Americans that illustrate its change-promoting and change-inhibiting aspects.

There is no single entity that is African-American religion, but it includes a number of expressions, mostly Christian, that have a common heritage in the experiences of slavery and racism. This common background makes it possible to speak of "African-American religion" even when describing groups with very different belief systems. While the African Methodist Episcopal church is greatly different from the Spiritualist churches and from the Lost Found Nation of Islam (i.e., Black Muslims), all are outgrowths of the historical situation of African-Americans. At the same time, the enormous diversity of African-American religious expressions makes them impossible to summarize in tidy generalizations.

One source of this complexity is that religious groups themselves continually deal with built-in "dialectical tensions," and they change their resolution to these tensions over time, depending upon the sociohistorical situation. As Chapter 5 and 6 showed, these tensions exist for most Western religions, but minority religions in pluralistic societies experience them differently. The dialectic between *priestly and prophetic functions* pulls each local group between an involvement in the celebration of its own religious life on the one hand and pronouncing God's judgment (including speaking to political and social issues) to the wider community on the other hand. The dialectic between *other-worldly and this-worldly orientations* is another pair of poles simultaneously present in most African-American religious expressions. The dialectic between *universalism and particularism* is present for all Western religions, but it is particularly problematic for ethnoreligious minorities. Another pair of poles is the *communal and the privatistic;* with other Protestants, most black Christian groups share the pull of privatistic religiosity, but African-American popular religious tradition puts a unique and powerful communal "spin" on their Protestantism. Another dialectic that influences the degree of change-promoting activism is the pull between *charismatic and bureaucratic patterns of authority and organization.* Two political options, *resistance* versus *accommodation,* represent another pull experienced by African-Americans and other nondominant groups (Lincoln and Mamiya, 1990: 11–15). If we remember that these polar opposites are not one-time, either-or decisions, we can better appreciate how African-American religion can be both change-promoting and change-inhibiting.

Afro-Christian Popular Roots

African-American religion began under slavery, accounting for a considerable part of its distinctiveness but also of its diversity. The exigencies of slave life, with its omnipresent power relationships and its dialectic of domination and resistance, brought together several religious strands.

Historians disagree about the exact shape of American slave religion, but its general outline is well known. Like all American religions, it had both popular and institutional elements, but the former were stronger. As described in Chapter 4, popular religion is by definition *not* under official religious control. Thus, it has the potential for expressing a nondominant people's dissent. It is also, however, characteristically loosely organized and fragmented, each local group's beliefs and practices being poorly articulated with those of other adherents. Lacking official clergy or recognized interpreters, it is vulnerable to co-optation or suppression. Early African-American religion was highly syncretic, with different localities interweaving different elements; some of the recognizable combinations today include mainline, evangelical, and pentecostal/holiness Afro-Christian strands, Creole vodou, Spiritualist, and Afro-Caribbean practices such as Orisha vodou, Santería, and Espiritismo (Baer and Singer, 1992).[7] Afro-Christian religious expression retained elements of West African oral performance, music and rhythm, and other ritual expression (Pitts, 1993; see also Davis, 1985).[8] The tribal religions of the parts of Africa from which slaves were taken were highly diverse and local, so no single religion could have survived in North America, but African influences were kept in highly syncretized form (Raboteau, 1978).

These diverse African tribal religious elements were thoroughly blended with elements of enthusiastic evangelical Christianity spread in the early 1800s by the fervor of the Second Great Awakening—a religious revitalization movement that swept the American countryside from 1770 to 1815. Both blacks and whites were influenced by revivalistic camp meetings, days-long enthusiastic preaching, and a theology that expressed a lively sense of the Holy Spirit. Not all blacks were affected equally, however.

The conditions of Afro-Christian religious practice varied widely under slavery: Some slaves had extensive opportunities to learn Christian teachings

---

7. For sources on Afro-Atlantic and Afro-Caribbean syncretism, see Baer, 1984; Barrett, 1974; Brown, 1991; Deren, 1970; Glazier, 1983; Gonzalez-Wippler, 1975; Metraux, 1972; Murphy, 1988; Simpson, 1965, 1978; Stevens-Arroyo and Pérez y Mena, 1995.

8. Although the many studies seeking to identify African roots of American religions are a useful corrective to portrayals of slave religion as having abandoned all African influences, they distract us from appreciating the diversity, creativity, and contemporary significance of *ongoing* syncretism and *mutual* influence of religious cultures in contact. As Baer and Singer (1992:3) emphasize, "much of the current importance of these African elements derives not from their possible source but in the part they have played and continue to play in the crafting of special mechanisms for social survival, emotional comfort, and transcendent expression under the harshest of physical circumstances."

and practices, while others had only fragmented and secondhand knowledge. After a series of slave uprisings, most slave states passed laws forbidding black preaching, forcing slave religious meetings into hiding. Subsequent missions to the slaves (1820s until the Civil War) reached far more people, but also were more compromised by their need to be approved by plantation owners, who typically welcomed Christian teaching if it served to keep the slaves docile and obedient. Nevertheless, by the middle of the nineteenth century there had developed a clearly Christian religion among slaves. It afforded them a chance to gather with each other and exchange social support, to develop their own leadership and social standing, and to express emotionally and spiritually their feelings, despite a social system that treated them as nonpersons. It sometimes also provided opportunities for resistance and escape (Raboteau, 1978).

Particularly significant for our understanding of the change-promoting and change-inhibiting features of African-American religious synthesis is the centrality of its theme of liberation. The Bible—the center of Christian missionary preaching—was replete with stories of liberation, which became central in African-American Christianity. The slaves identified with the suffering Israelites in their bondage. While the Euro-American Protestants thought of their young country as "the New Israel," or even the Promised Land, the African-American experience of it was more like "the New Egypt" of bondage and oppression (Raboteau, 1995). Not only was this theme of liberation preached and sung, for example in spirituals such as "Go Down, Moses," but it was embedded in ritual practice, especially the experience of conversion (Dvorak, 1991). What slave Christians "remembered" from biblical stories, hymns, and other Christian practices, thus, was very different from what their southern white Christian counterparts "remembered," even though both practiced a very similar evangelical, intensely emotional, Bible-centered popular Christianity.

Contained within the worldview of slave Christianity were both change-promoting and change-inhibiting themes. Certainly the potential for a prophetic stance is inherent in the themes of liberation and deliverance.[9] The emphasis on communal bonds, of an identity as God's people and commensality (sharing a meal with the Lord and each other) also has potential for promoting solidarity and support—both features of successful movements for change (see Dodson and Gilkes, 1995). On the other hand, the eschatological emphases of the Afro-Christian worldview (e.g., the image of Jesus' future return setting things right) often promoted fatalism rather than activism and typically led to an other-worldly emphasis. But the same symbols have the

---

9. The theme of Moses confronting the pharoah is still prominent in the prophetic stance of African-American churches. When the National Baptist Convention, the nation's largest black religious organization, decided in 1993 to focus on social issues and self-empowerment, the convention chairman told the story of Moses leading his people to the Red Sea, asking God how to get across, and hearing that with the rod in his hand he could part the waters. The chairman concluded: "What we have in our hand is the dollar. What we have in our hand is the vote. We must use what we have in our hand" (quoted in Jones, 1993).

potential to promote change, as the example of millenarianism in Chapter 2 suggests (see Genovese, 1979; Moses, 1993).

Making Their Own Organizations

Independent denominations (e.g., African Methodist Episcopal and National Baptist) are the prominent religious organizations in the African-American community today. Seven historic denominations account for roughly 80 percent of all religiously affiliated African-Americans (Lincoln and Mamiya, 1990: 441). The historical split of these groups from their white-dominated parent denominations was neither theological nor caused by dissent over doctrinal or moral purity; it was essentially a split along lines of social caste.

The religious fervor of the Second Great Awakening (1770–1815) was intensely egalitarian. Not only did the belief systems of evangelical groups promote brotherhood, but the shared intense religious experience itself produced a strong sense of equality. In the period after the Revolutionary War, Methodists and other Protestants directly attacked slavery and welcomed African-Americans into their fellowships (Lincoln and Mamiya, 1990; Washington, 1972). In the North and Midwest there were racially mixed congregations of Presbyterians, Methodists, Quakers, Episcopalians, Baptists, and Congregationalists. This form of African-American religious affiliation was, however, typically limited to free persons (indeed, some of them freed because of the antislavery influence of the Second Great Awakening on slaveholders). It was not extended to the masses until slavery was ended (Genovese, 1974; Lincoln and Mamiya, 1990).

From these beginnings grew the major African-American denominations. After the initial egalitarian euphoria waned, African-American Christians found themselves given only limited leadership roles and often segregated participation in their churches, and they challenged these arrangements. When Richard Allen and Absalom Jones protested segregated seating by leading a band of African-American Methodists out of Philadelphia's Saint George's Methodist Episcopal Church in 1787, it was the beginning of separate U.S. African-American denominations. Allen later formed the African Methodist Episcopal Church, a prototype of African-American independent churches (Lincoln and Mamiya, 1990:49–58; Washington, 1972:36–46).

Unlike most other religious splinter groups, these churches did not differ significantly from their parent organizations in their belief systems or patterns of worship. Their dissent was based solely on the treatment of African-American members by predominantly white parent organizations. As a result, their organizations were essentially denominational from the start. They did not exhibit sectarian characteristics such as withdrawal into boundary-maintaining enclaves. These independent denominations were active in abolitionism and the underground railroad, and they organized numerous educational and social service projects to help escaped slaves arriving in northern cities. After the Civil War they were able to extend their leadership and organization to recently freed slaves; indeed, several Methodist and Baptist denominations competed vigorously to establish new congregations in the postwar South. Also, unlike the popular religious expressions of African-American slaves, denominational expressions promoted the norm of an educated clergy, networks of denominational organization, and regular regional conferences. These organizational features gave them greater coherence and structural supports for new congregations; the denominational organization also curbed idiosyncratic teachings and unorthodox interpretations of the Bible. Thus, while their denominational structure did not itself promote social activism, it increased the effectiveness of those change-promoting efforts mounted by the denominations.

A completely different pattern led to the creation of separate African-American denominational groups in the rural South. Major national denominations, including the Methodists, Baptists, and Presbyterians, had split in the early 1800s, largely over the issue of slavery. The factions of each denomination that were organized separately in the South were segregated, with the all-black congregations in an inferior and subordinate relationship. After Emancipation, the northern denominations successfully intensified their recruitment among southern blacks. In the 1860s and 1870s, partly in response to African-American members' push for greater autonomy and partly due to the white majority's wish to dispense with the black membership, southern denominational organizations spun off their African-American constituents into their own organizations, which eventually became separate denominations (Lincoln and Mamiya, 1990; Dvorak, 1991).

Unlike their northern counterparts, however, these denominations lacked experience with church organization and leadership; their members—most of them only recently slaves—had little education or decision-making experiences and virtually no economic resources. Thus, although they gained nominal local autonomy, it was a very insecure social base. Often they were completely subordinate to local (white) patrons for economic support and vulnerable to the economic injustice, intimidation, and outright violence of the caste system that developed after Emancipation. Rural African-American churches became indirectly supportive of the political and socioeconomic status quo, becoming a spiritual escape, a safe emotional outlet, a haven/home. They adopted an almost exclusively other-worldly emphasis: Rewards in the next life would make up for deprivations and abuses in this life. Although nominally denominational, these groups were thrust into sectarian withdrawal by the oppression of the larger society (see Baer and Singer, 1992, for a critical analysis of the economic structures shaping the era).

We must understand African-American denominationalism and sectarianism in the context of the caste system in which they developed. These religious organizations do not, on the surface, appear to fit the standard forms described in Chapter 5, because not even the most "established" African-American denomination is fully integrated into the dominant society. We can understand this apparent discrepancy, however, by realizing the extent to which the **caste system** imposed by the larger society positioned *all* African-American institutions in a situation of somewhat negative tension with the institutions of the larger society.

√ The caste model suggests that individuals can achieve relatively high status within the African-American community and its own institutions (Dollard, 1949). Black religious organizations are stratified in a way that parallels the social class bases of white religious organizations. There are relatively prosperous, staid, and genteel African-American denominations and congregations, supporting professional clergy and a wide variety of community programs; there are numerous middle-class and working-class congregations of established African-American denominations; and there are vast numbers of (mainly sectarian) religious groups, such as the storefront churches—appealing primarily to economically marginal persons (Wilmore, 1972:195–197). Some of the sectarian groups are often critical of both the established denominational groups and of the larger society—as are many white sectarian groups, but many groups express this tension by withdrawal into sectarian enclaves.

The black denominational organizations, however, are also able to be critical of the larger society because of their location in the subordinate caste. Indeed, continuing to the present, the more established denominational congregations are more likely than the sectarian congregations to be involved in the social, political, and economic affairs of their communities; in the 1990s, they also tend to be more socially and politically involved than their white denominational counterparts (Chaves and Higgins, 1992; Lincoln and Mamiya, 1990; Nelsen and Nelsen, 1975). Nevertheless, neither African-American denominations nor sectarian groups have historically been particularly critical of the dominant society, largely because their adherents have generally accepted the American "way of life"—the dominant cultural values and criteria for success. Although they are conscious of racial grievances and real social inequalities, they have not typically questioned the rest of the established cultural order (Wilmore, 1972:213–214).

In the last decades of the nineteenth century, the churches became the focal point of black community life, but the spiritual message was largely other-worldly and moralistic. We must keep in mind, however, the external social, legal, and economic barriers, which—even more than Afro-Christian belief systems—were responsible for this retreat to quietism.

Quietism, Oppression, and Institution-Building

Despite their increased membership, African-American denominations underwent a dramatic retrenchment from the end of the nineteenth century to the late 1950s. The specific experiences leading to this quietist phase varied from region to region but can be generalized as two major factors: *economic struggle* and *sociopolitical discrimination and repression*. In the South, newly freed blacks struggled to make a livelihood in a shattered economy. The Compromise of 1877 led to the withdrawal of federal troops from the South and abandonment of federal guarantees of African-Americans' civil and political liberties. The last two decades of the nineteenth century were marked by considerable violence toward blacks in the South, and several thousands were victims of lynch mobs (Wilmore, 1972:192). "Jim Crow" statutes created and gave legal force to segregation in every sphere of social life. These local and state laws racially segregated schools, workplaces, building entrances, waiting rooms, drinking fountains, toilets, and seating in public transportation. Statutes enforced the exclusion of African-Americans from theaters, restaurants, parks, and residential areas. Poll taxes and discriminatory "literacy" tests effectively excluded most African-Americans from voting in many southern states. In 1940, only about 2 percent of voting-age blacks in twelve southern states "qualified" to vote (Woodward, 1957). Discrimination in northern regions at the turn of the century was less overt. There was a sizable African-American middle class in many northern communities, but the vast majority were unskilled and suffered considerable discrimination in competing for jobs with waves of European immigrants.

The retrenchment of African-American churches in the face of such massive difficulties is understandable. Individual church leaders (e.g., Bishop Henry Turner) protested the situation, but most congregations retreated into the segregated African-American community where they became the dominant institution, providing alternatives to services and facilities from which their members were otherwise excluded. It was not so much a matter of social activism as of simply meeting their members' needs that inspired black churches to provide recreation and social clubs, social services, insurance, and a "decent" burial.

The massive twentieth-century rural-to-urban migrations of African-Americans created new disruptions. Before World War I, only 28 percent of the black population lived in cities; fifty years later 70 percent were urban dwellers (Report of the National Commission on Civil Disorders, 1968). Rural churches were crippled by losses of members and material support. Urban churches were overwhelmed by the needs of the huge influx of rural blacks, many of whom were uneducated, unskilled, and unprepared for the difficulties of the urban ghetto (see Nelsen and Nelsen, 1975:36–47). This period marked the beginning of the greatest increase in sectarian groups, discussed in more detail later in this chapter. The city represented a double threat to many rural immigrants: It represented their first experiences with a general nonreligious worldview and with a social situation where the norms and values of a small, tightly knit community did not hold. One commentator concluded: "With a basically rural orientation, most Black churches retreated into enclaves of moralistic, revivalistic Christianity which tried to fend off the encroaching secular gloom and the social pathology of the ghetto. . . . The socially involved, 'institutional' church was the exception rather than the rule" (Wilmore, 1972:221). From the end of reconstruction to the mid–twentieth century, African-American churches generally retreated from social and political activism.

Nevertheless, the inward turn of African-American denominations in the first half of the twentieth century may have served to build institutions for survival and resistance. In the face of discrimination and segregation, these churches developed schools, savings banks, employment services, day-care facilities, health clinics, and social clubs. They were also where members learned participation in decision making and wide-ranging discussion of social issues.

African-American women were particularly important in these institution-building efforts, and they were remarkably successful in developing alliances with white churchwomen (Higgenbotham, 1993). While not directly engaging in social activism, such institution-building and grassroots political experience within some churches may have been important platforms on which later activism built.

Religion and the Civil Rights Movement

Several features of African-American religion—especially as practiced in urban denominational congregations—account for much of the success of the 1960s civil rights movement in mobilizing resistance to the widespread segregation, discrimination, and disenfranchisement experienced by American blacks at mid-century.

Internal Organization. African-American religious groups have had particular potential for social change because of their special place in their community. Because of the strategies developed during the quietist period, the black churches are very much a symbolic expression of community (see Williams, 1974) and a potentially unifying force, though this potential has seldom been realized. Part of the problem is the internal structure; most African-American denominations (especially the Baptists) tend to be highly localized and therefore difficult to organize above the congregational level. Thus, most African-American denominational organizations at the national level were conspicuously absent from 1950s and 1960s civil rights activism. A 1962 split that created a new denomination—the Progressive Baptist Alliance—was not over doctrines or rituals but over the growing activism of progressive members.

In several localities, however, churches of various denominations began to band together to promote change. In 1958, 400 African-American ministers in Philadelphia launched a "selective patronage" campaign (i.e., boycott) against industries that depended heavily on black patronage but discriminated against them in hiring. They used the pulpit for announcing target employers and succeeded in stopping some of the blatant job discrimination. The segregation of churches may have aided this effort, since none of the "opposition" was in the audience (Lincoln, 1974). Similar organizational efforts by local churches were important in the Montgomery bus boycott, sit-ins, and freedom rides (see Fredrickson, 1995). African-American denominations, especially with their broadened national bases, have the potential for promoting nationwide programs for change. Many groups, however, have been involved with the problems of maintaining their organizations, and activism seems threatening to their stability and acceptability.

The Role of the Preacher. The example of the "Philadelphia 400" suggests another feature of African-American religion that is effective in promoting change: the role of the preacher. Religious leaders have greater influence in the black community than is characteristic in most white communities. Just as the African-American churches were often the primary institution in the black community, so too the preacher was the community leader (Hamilton, 1972). Although the preeminence of this role is declining in recent years, many African-Americans still look to their preachers for leadership on social and political issues. Studies have shown that blacks are more likely than whites to approve of political activism from their preachers and of using the pulpit to discuss political issues (Nelsen and Nelsen, 1975; see also Lincoln and Mamiya, 1990); African-Americans in the 1970s cited their church or clergyman as a major influence on their political thinking (Wald, 1987); and active clergy support for presidential candidate Jesse Jackson appears to have been influential in the voting behavior of regular churchgoers (Wilcox, 1991).

The African-American clergy activists of the 1960s civil rights movement used their influence as religious leaders, and their style of leadership and exhortation was characteristic of the preacher. Clergy such as Martin Luther King, Jr., Leon H. Sullivan, Fred Shuttlesworth, Wyatt Walker, and Ralph Abernathy did not distinguish between their roles as religious leaders and civil activists. It was not just their status as clergy, but indeed their religious experience that was recognized in the African-American community as a basis of authority (Manis, 1991). In the civil rights movement, as in some other 1960s protest movements, dissent itself had a religious quality. It was an integral part of the messianic vision implicit in the theology of the movement (see Moses, 1993). The importance of the civil rights movement was not only its specific achievements in gaining voting rights, desegregating educational opportunity, and eliminating discriminatory legislation but also in making African-Americans conscious of their power to effect change.

King's effectiveness resulted, in large part, from his ability to combine a mode of sociopolitical dissent (modeled after Gandhi—another exemplar of religious social action) with the folk religion of his people and the revival technique to which they enthusiastically responded (Franklin, 1994; Wilmore, 1972:24–52). The interweaving of religious with political images and symbols and the utilization of religious modes of expression (e.g., hymns and "shouts") were part of the appeal that enabled religious leaders to mobilize the African American community for action. King's political speeches were essentially sermons in their structures and imagery:

> We cannot walk alone . . . We cannot turn back . . . We cannot be satisfied as long as the Negro is the victim of unspeakable horrors of police brutality. We can never be satisfied as long as our bodies, heavy with the fatigue of travel, cannot gain lodging in the motels of the highways and the hotels of the cities. We cannot be satisfied as long as the Negro in Mississippi cannot vote and a Negro in New York believes that he has nothing for which to vote. We cannot be satisfied as long as the Negro's basic mobility is from a smaller ghetto to a larger one. No, no, we are not satisfied, and we will not be satisfied until "justice rolls down like water and righteousness like a mighty stream" (quoted in Spillers, 1971:24).

The leadership style of the preacher "made sense" in the African-American community, enabling clergy activists to move the people to action.

African-American Religious Diversity

Black urban communities display a fascinating variety of religious groups. In addition to representative churches of African-American denominations, there are many sectarian groups. So far we have considered only the African-American denominations, which are large and growing, but there are many vigorous sectarian and cultic groups in African-American communities. By far the most numerous of these are Pentecostal and Holiness congregations. These religious groups originated at the turn of the century in a series of southern revivals (initially among whites and later among blacks), which also spawned white religious groups such as the Nazarenes and the Assemblies of God. Both black and white pentecostal groups are characterized by their literal interpretation of New Testament practices, especially manifestations of the Holy Ghost (or Holy Spirit) such as speaking in tongues, divine healing, visions, prophecies, and testimony. Another vigorous sectarian group is Jehovah's Witnesses, whose members are expected to spend several hours each week in proselyting and witnessing to the coming millennium. This group is racially integrated, but has attracted a larger African-American following than have most other groups led by whites. Most African-American sectarian groups are loosely, if at all, connected with each other. Often the local congregations exist only because a leader has gathered a group; religious leadership is still a path to respect in the African-American community.

There are also numerous unaffiliated religious groups in African-American communities. These congregations are frequently very similar to organizationally affiliated congregations but depend solely upon local leadership to establish their beliefs and practices, often based on popular religion. Because of this independence from outside authority, some of these groups are idiosyncratic. Father Divine's Peace Mission (begun about 1932) ministered to several thousand black and white people. Members practiced communalism and extremely strict moral norms. The Mission was active in social welfare (e.g., feeding, clothing, and housing people, opening cooperatives and small business enterprises). It was also politically active, opposing Jim Crow and lynching (see Burnham, 1978; Fauset, 1944). The controversial People's Temple of Jim Jones was a sectarian group that, like Peace Mission, attracted both blacks and whites and had substantial social and political programs (see Hall, 1987).

Some sectarian groups are not specifically Christian but draw on Christian themes (at least in their initial appeal) because their potential recruits are generally from Christian backgrounds. Black Muslims and Black Jews exemplify this type. They are clearly sectarian because of their internal structure and their oppositional stance toward the dominant society and its religious organizations. One particularly good example of the deliberate appeal to Christian themes is the Shrine of the Black Madonna in Detroit. This group retained much of the faith and practice of traditional African-American denominational churches, but asserted that key biblical figures—especially the Messiah—were black. The leader stated that "the Black Muslims demand too much of a break with the past for African-Americans, most of whom have grown up in a Christian church of some sort. We don't demand a break in faith, or customs" (Cleage, 1967:208).

Afro-Christian religious traditions were, interestingly, utilized by non-Christian African-American leaders. Malcolm X, a spokesperson for Black Muslims, was very much a "prophet" in the black religious style. Since most Black Muslims came from Christian backgrounds, their relationship to their religious leaders was probably styled after Afro-Christian folk religion rather than any Islamic pattern. The message of the Nation of Islam, especially as articulated by Malcolm X, is clearly an example of the revolutionary potential of religious movements (see Lincoln, 1973). Following Malcolm's lead, the Nation of Islam (renamed the American Muslim Mission in 1980) has become more denomination-like, patterning itself on the Sunni branch of Islam. Although they have become a more staid part of the American religious scene, the Black Muslims are probably one of the most important American religious movements in the twentieth century. In reaction to the main group's move away from an extreme sectarian stance, however, Louis Farrakhan revitalized the messianic and dualistic claims of the movement's founder, W. D. Fard (Baer and Singer, 1992:110–146). Nevertheless, Farrakhan has used traditional Afro-Christian leadership style to maintain his place in the public eye, and he has drawn on themes of self-empowerment that appeal to long-standing self-help traditions nurtured in the black churches.

The religious mosaic of the African-American community includes a number of more cultic expressions, some of which are only superficially religious. As explained in Chapter 5, the cultic stance is more pluralistic and tolerant toward other religious groups than the sectarian stance, but it is nonetheless dissonant from the larger society and lacks the societal acceptance characteristic of denominational groups. Unlike sectarian allegiance, cultic adherence is more segmented; frequently members adhere to several such groups simultaneously. Such groups are diverse, ranging from a faith healer's storefront to a group gathered around a "reader." Some cultic leaders are flamboyant and manipulative, such as Prophet Jones, a healer and fortune-teller whose Detroit church "Universal Triumph, the Dominion of God, Incorporated" is remembered less for its teaching than for the wealth it generated for its leader (Robinson, 1974).

Other cultic groups, however, have fully developed religious worldviews and ritual expression based on more than a century of faith and practice. African-American Spiritual churches and vodou are examples of this cultic pattern of organization (see Chapter 4), as well as the focus on religious power for everyday concerns characteristic of the popular religions that came together in these syncretisms. The Spiritual churches amalgamated elements of (predominantly white) Spiritualism, African-American Protestantism, Roman Catholicism, and vodou or hoodoo. Because, as is typical in cultic religious groups, each congregation is autonomous, individual affiliates also mix elements from New Thought (a turn-of-the-century Metaphysical movement

also borrowed by some New Age religious groups), Islam, Judaism, Ethiopi-
anism, and astrology (Baer and Singer, 1992; Baer, 1984).

These religious groups characteristically offer adherents segmented bits of
meaning rather than a total package of belief and practice. A believer might
seek out a reader's spiritually received advice in a marital problem, go to a
healer for rheumatism, send for Reverend Ike's prayer cloth to help get a job,
and also go to a local church on Sunday. Although the cultic groups preach a
worldview, it is seldom the totalistic worldview like that of the sectarian
groups. Adherents may be fervent, but they do not typically belong exclusively to that single religious group. While the cultic groups often provide
their members with strong social support networks and a positive identity,
their emphasis is totally privatized. Therefore African-American cultic groups
rarely challenge the status quo and they typically lack the stability or influence
in the community for any activism. For this reason, the following discussion
of change-promoting and change-inhibiting influences focuses primarily upon
sectarian religious groups in the African-American community.

The Dynamism of Sectarian Religious Expressions. Sectarian congregations are among the most dynamic forces in the African-American community. Their members commit much time, energy, and money to their religious
groups. Sectarian groups are able to motivate their members to apply religious
norms to everyday life, and many of them enforce strict moral codes. Their
creativity in forms of worship and sheer enthusiasm is an important expression
of the religious group's sense of community. This dynamism has strong potential for promoting change because conversion to a sectarian worldview can *redefine* social reality for members. Converts to the Black Muslims, for example,
gain a new way of interpreting themselves and the world around them. The
belief systems of sects typically contain a negative judgment of the "ways of
the world" and a millenarian vision of a future perfect society. These beliefs
are potentially change-promoting; but only rarely, historically, have African-
American sectarian groups produced fundamental social change.

Sectarian Quietism. The reasons for this general quietism appear to lie in
the emphases of their belief systems and the social situations in which they
must operate. Many sectarian worldviews are even more other-worldly than
those of denominational religious groups. Although members may eagerly
await the millennium, few African-American sectarian groups actively promote its arrival (Jehovah's Witnesses are the major apparent exception, although their time-demanding activities may serve primarily as commitment
mechanisms; cf. Cooper, 1974). The emphasis of many sectarian groups upon
divine Providence reduces members' consciousness of human agency. Another
emphasis of much sectarian religion that inhibits members from social and political activism is their emphasis on private moral norms and individual responsibility. Many groups are strict about members' behavior. They proscribe
the use of tobacco, alcohol, and drugs; insist on hard work, honesty, sexual fi-
delity, and modesty; and forbid entertainments such as gambling, movies, or
dancing. Sometimes religious norms directly discourage activism. One woman

said, "In my religion we do not approve of anything except living like it says in the Bible; demonstrations mean calling attention to you and it's sinful" (quoted in Marx, 1967:67). These strict personal moral codes often have the indirect consequence of enabling members to fit more effectively into the dominant society.

One study of black attitudes on civil rights issues found members of sects and cults to have the least militant attitudes. Members of independent African-American denominations (e.g., Methodist and Baptist) were substantially more militant than sect and cult members; but most militant were black members of predominantly white denominations such as Episcopalian, Presbyterian, United Church of Christ, and Roman Catholic (Marx, 1967). Some exceptional sectarian groups are seriously involved in social activism. In Chicago, an African-American Pentecostal preacher organized the Woodlawn Organization, a neighborhood association that operated successful tenants' rights and consumer protection programs (Brazier, 1969). Nevertheless, the other-worldliness and emphasis on Providence and private (rather than public) morality make African-American sects generally unlikely to be involved in social and political activism.

The social location of a religious group may be an important factor in the extent to which it emphasizes activism. Direct social and political action may be a luxury that the absolutely poor have neither the material nor personal resources to consider. Many African-American sects have emphasized comforting and taking care of their members. One early Pentecostal leader said, "The members of my church are troubled and need something to make them happy. My preaching is not about sad things, but always about being saved. The singing in my church has 'swing' to it, because I want my people to swing out of themselves all the mis'ry and troubles that is heavy on their hearts" (quoted in Washington, 1972:67).

These groups provide a small, close-knit community of mutual care and concern, often in the center of an otherwise harsh and anonymous urban environment (see Kostelaros, 1995). African-American sects may have withdrawn from action in society both from opposition to the "ways of the world" and as a response to the threatening atmosphere of their social situation. A small, close-knit group of fellow believers serves as protection against the influences of "the world," but it also narrows members' lives to a "safe" enclave. These aspects of the social situation surrounding black sects and cults, as much as the content of their belief systems, help explain their general nonactivist orientation.

The Changing Effectiveness of African-American Religious Social Activism

This Extended Application has examined the factors contributing to change-promoting or change-inhibiting qualities of African-American religion in America. We have considered aspects of the belief systems, the cultural framework, the social and historical situation, and the internal structure of the

various kinds of religious groups that bear upon their orientation toward social and political change. These aspects suggest that African-American religion
is a highly appropriate vehicle for change-oriented actions, though it has not
always been used to promote change. In particular, within the cognitive framework of many (perhaps most) African-Americans, it "makes sense" to pursue
goals through religious groups and under religious leadership. Indeed, until
recently religious organizations may have been the only viable indigenous vehicle for African-American social action.

The same line of analysis, however, suggests that African-American religion may have a considerably diminished role in future social change because
of at least four factors: increased member complacency, alienated youth, a
changing U.S. social structure, and religious privatization. Many religious
leaders themselves name *members' complacency* as the most obvious factor
(Wilmore, 1972:195–197). There is a growing economic disparity between
middle-class and lower-class African-Americans. If these two communities
become disconnected from one another, analysts worry that those who are
relatively well off socioeconomically may be less likely to utilize their increased
resources for dramatic change (Lincoln and Mamiya, 1990; the nature of social class and racial ties among African-Americans is controversial; see Wilson,
1980).

Yet there is considerable evidence that many middle-class black churches
are far less complacent than their white counterparts, even though social problems are arguably more overwhelming than before the civil rights movement:
crumbling inner cities, crack and AIDS epidemics, massive unemployment
and underemployment, and vanishing social security "nets" (for health care,
childcare, food, housing, old age income, etc.) for all Americans (see Wilson,
1987). In 1986 the total revenue of all U.S. religious congregations was $49.6
billion, of which $1 billion was spent on assistance for the poor; by 1991, de-
spite the fact that their revenue had slipped to $48.4 billion, the churches in-
creased their aid to the poor to $7 billion. Overall, the black churches spent
proportionately more of their revenues. For example, Progressive Community
church, an inner-city African-American congregation in Chicago, spent about
half of its total budget on programs for the poor: an all-purpose community
center offering hot meals, day care, youth programs, clothing aid, referrals for
emergency housing, job counselling, and more (*New York Times,* 1995). In the
face of such massive societal problems, minority communities simply do not
have the resources or the social structural leverage to make meaningful changes
(Gilkes, 1995).

A second and related factor is *alienated youth*. Few of the urban youth most
affected by this serious social decline are involved in religious groups. While
both Christian and non-Christian African-American leaders have appealed for
renewal of black families and good male models and mentoring for youth,
many young people have little respect for religion of any kind. Younger Americans are generally less likely to be involved in religious groups than older persons, but in northern cities, younger blacks with lower educational levels have
particularly low rates of religious involvement (Nelsen and Kanagy, 1993; see

also Gilkes, 1995). Many religious groups have made strong appeals to involve black men, such as the 1995 "Million Man March" in Washington, D.C. Bethel African Methodist Episcopal Church in Baltimore, one of the nation's oldest denominational churches, for example, has developed active men's organizations, including men's Bible study groups, a large male choir, and programs and rituals for young men to take Christian manhood more seriously (Mamiya, 1994). Although such programs have been successful in increasing the proportion of men in middle-class and working-class religious groups, whether they can attract alienated youth of the inner city remains to be seen.

A third factor is the *changing structure of U.S. society*. The very successes of the civil rights movement meant that nonreligious modes of action (e.g., political and legal) became increasingly accessible to African-Americans. As people master these modes, religious modes of social action will have less importance and will stand to lose even more significance as people come to think in terms of alternatives. Increasingly, nonreligious modes of action are accessible and "make sense" to African-Americans.

The fourth factor in the diminished role of black religion in promoting change is that the *general forces of religious privatization* appear to be making inroads into the African-American community, much as they have among whites in the last several decades. The forced segregation of that community had protected the religious worldview of its members, even as immigrant Italian, Irish, and Jewish communities protected their religious worldview in the first half of this century. "Making it," however, means participating in a public sphere— the worlds of work, politics, education, and law—in which religion is treated as largely irrelevant. Although people can also enjoy the private sphere (e.g., community, family, religion, leisure, etc.), the meanings and values of the private sphere are increasingly voluntary and not compelling in the public sphere. The "new voluntarism" (described in Chapter 4) applies to much African-American religiosity; especially in urban and suburban contexts, religion is private and voluntary (Nelsen and Kanagy, 1993). Unlike earlier periods of quietism, however, this voluntarism is structural. In many respects, in the modern world *no* religion has sufficient structural power to undo the preeminence of public-sphere institutions, especially the economic and political ones, that both create the conditions under which people suffer and control the terms of discourse about the direction of society.

SUMMARY

Certain qualities of religion tend to support the existing socioeconomic arrangements; at the same time, however, religion also promotes social change and has considerable potential for change-promoting action in certain social situations. Interest theories explain that religion frequently supports the vested interests of the dominant social classes, legitimating their dominance, socializing believers to comply with it, and utilizing religious social controls for deviance from it. Religious elements can be potent forces for social change; religious groups, leadership, ideas, and images can promote change-oriented action.

Because the relationship between religion and social change is complex, we have focused on the conditions under which religion is likely to be change-inhibiting or change-promoting. Social structural aspects of religion are important variables that influence the impact of religion on social change in any given situation. The social location of religion, the relationship of religious institutions to other institutions in society, and the internal structure of religious organizations are all criteria in determining whether religion is change-promoting or change-inhibiting and its degree of effectiveness in social action.

These factors that shape the interrelationship of religion and social change are illustrated by the African-American religion, which both promotes and inhibits social change. At critical historical moments, the unique qualities of African-American religion and the social location of the black churches made religious leadership and organization plausible forces for human rights and socioeconomic change efforts.

RECOMMENDED READINGS

Hans A. Baer and Merrill Singer. *African-American Religion in the Twentieth Century: Varieties of Protest and Accommodation.* Knoxville: University of Tennessee Press, 1992. Traces both hegemonic and counterhegemonic elements in diverse African-American religious expressions, locating each in its historical and socioeconomic context.

Roger N. Lancaster. *Thanks to God and the Revolution: Popular Religion and Class Consciousness in the New Nicaragua.* New York: Columbia University Press, 1988. With rich ethnographic description, the author compares and contrasts the religious expressions of Nicaraguan peasants and recent rural immigrants to cities: official Catholicism, folk religion, the Popular church, and evangelical Protestantism.

Liston Pope. *Millhands and Preachers.* New Haven, Conn.: Yale University Press, 1942. A classic sociohistorical analysis of the role of religion in a famous mill strike.

Max Weber. *The Protestant Ethic and the Spirit of Capitalism.* New York: Scribner, 1958. The most readable of Weber's classical studies of religion, originally published in 1904.

Peter Worsley. *The Trumpet Shall Sound.* New York: Schocken, 1968. A Marxian analysis of Melanesian cargo cults. Worsley's critique of Weber's social change theories is the weakest part of this highly readable, well-documented study of religion's change-promoting impact in a concrete historical setting.

8

◦◊◦

Religion in the Modern World

Major changes have occurred in the social location and significance of religion in Western societies over the last several centuries. Indeed, it appears that no developed or developing country (including non-Western ones) has remained untouched by these processes. The situation of religion in the modern world is different; the linkages of religion with societal and individual life have changed. Any examination of the nature of religious change must be grounded in an understanding of other major changes in the structure of society—or indeed, the world.

The secularization thesis is an interpretive paradigm used by many sociologists to try to "make sense" of these historical processes. There is profound disagreement over the meaning of this concept, but most interpretations of *secularization* refer to a historical development by which religion has lost (or is losing) a presumed central place in society. The secularization thesis poses the issue: Does modernization inevitably reduce or eliminate the influence and importance of religion? For classical sociologists and their heirs, the secularization thesis is essentially an attempt to *explain the emergence of the modern world* because many thinkers feel that modern society differs absolutely from what came before it (Luckmann, 1977:16, 17). In their examination of the social location of religion, these theorists have highlighted several processes of social change that are useful to our understanding of the structure of modern societies: institutional differentiation, changing patterns of legitimation and authority, rationalization, privatization, and individuation.

The general concept of "secularization" is, however, imprecise and broad, lending itself to nonobjective polemics (such as arguing that the end point of the process is either desirable or undesirable). The concept of secularization itself is not very useful because it typically implies a unilinear historical development—inevitable decline of religion and/or religiosity. The nature of social change is, however, far more complex; for example, throughout this book there are many examples of historical moments when one aspect of religion is ascendant precisely as another aspect is becoming less salient. Religion is thoroughly embedded in so many facets of society that unilinear interpretations cannot possibly portray the complex ways religion reciprocally influences society. Often the empirical data used to buttress the secularization interpretation are based on narrow, substantive definitions of religion that cannot encompass new forms that religion might take. At the same time, those opposed to the secularization thesis also often define religion to fit the needs of their hypothesis, thus obscuring real changes that may be going on. In itself, it is not a particularly important *sociological* question whether religion, thus narrowly defined, does or does not decline. Far more important are questions about the social structural changes that have produced the modern religious situation and what those structural changes imply for the individual, the society, and humanity-at-large.

THE RELIGIOUS SITUATION IN MODERN SOCIETIES

Religion in the modern world *does* occupy a very different location from its place in various premodern types of societies. We should remember, too, that there have been profound changes in the structure of society even since the time of the classical sociologists, who formulated their theories during the early stages of industrialization. The pace and quality of social and technological changes in the last 100 years have resulted in new social structures and new modes of religiosity, especially in advanced industrial societies (Beckford, 1989). By trying to understand the changed situation of religion, sociologists also gain understandings of the nature of contemporary societies. For example, how has the nature of national or societal cohesion changed? How has the individual-to-society nexus changed in the face of such significant social structural changes? How does religion reflect the globalization of economic, political, and legal/moral issues?

When we think about rapid social change, especially with regard to religion, we must avoid romantic notions about the "good old days." Our image of traditional "old-time religion" is often historically very limited (e.g., referring to a period of the heyday of church religiosity that, in the United States, encompassed several decades ending in the mid-twentieth century). We conveniently forget that throughout much of America's relatively short history the majority of Americans were apparently unchurched, highly susceptible to an uncontrolled assortment of nonofficial religious teachings, superstitious, or ignorant of church doctrines and practices. Much of what we identify as "traditional" religious morality, such as norms regarding gender roles, abortion, incest, and abstinence from undesirable substances, dates primarily from Victorian times, with the domestication of religion in the face of early industrialization.

Nor were the "good old days" necessarily all that "good." Yes, the quaint little church on the village green back then had the potential to be a place of neighborly warmth and cohesion, where everyone knew everyone else, and where religious belief, practices, and shared values were more fully supported in everyday life by all in the community. That same firm sense of tradition and community, however, also gravely restricted individual freedom: Choices of marriage partners, occupations, leisure-time activities, and political options were all controlled, sometimes subtly and often overtly. The traditional societies that so firmly supported traditional religion were generally authoritarian, patriarchal, highly stratified, and nondemocratic. Indeed, the very discovery of the individual, with emotional needs and human rights and prerogatives of choice, is a peculiarly *modern* feature of a society, as discussed further below. So, in trying to understand the location of religion in modern society, let us strive for a more neutral appreciation of social changes: Some qualities we may value have been lost or transformed; other valuable features have been gained.

This chapter first examines the various processes of societal change toward modern forms of (industrial) society, and for each process we explore the implications for the individual and for the society. Subsequently, we look briefly at the effects of additional changes wrought by globalization and by certain characteristics of advanced industrial societies. These interpretations of societal changes suggest major changes in the nature and location of religion but do not necessarily imply religious decline. They do, however, emphasize significant aspects of *change* not only of the institution of religion but also of politics, work, family, and other spheres of social life.

Institutional Differentiation

Institutional differentiation refers to the process by which the various institutional spheres in society become separated from each other, with each institution performing specialized functions. For example, religious functions are focused in special "religious" institutions, separate from other institutions such as educational, political, and economic. This contrasts with an image of simple societies, in which the beliefs, values, and practices of religion directly influence behavior in all spheres of existence and religion is diffused throughout every aspect of the society. Workers might pray over their tools at the beginning of each workday, and an intragroup conflict might be expressed in a religious ritual. In complex societies, by contrast, each institutional sphere has gradually become differentiated from others. The division of labor in complex societies is similarly differentiated, with specialized roles for each different function. In a highly differentiated social system, the norms, values, and practices of the religious sphere have only *indirect* influence on other spheres such as business, politics, leisure-time activities, education, and so on (Parsons, 1971:101). This means that religion influences these other areas through the personally held and applied values and attitudes of people who are active in each sphere, rather than directly through specifically religious institutions such as the church.

Some theorists point to differentiation as evidence of the declining influence of religion. They interpret the facts that religion is not diffused throughout the society and that specifically religious institutions have limited control over other institutional spheres as evidence of religion's diminished strength and viability. Particularly important is the loss of control over the definition of deviance and the exercise of social control (Wilson, 1976:42). During the Middle Ages, the churches defined and prosecuted deviant behavior, exercising social control both through informal measures (e.g., confession or community ostracism) and formal measures (e.g., church courts). In most modern societies, however, the churches have no such direct control. Courts are more independent, and laws are the province of specialists. Medical institutions have similarly acquired much power over social control, defining deviance as "sick" and calling medical social control measures "therapy." The net effect of this differentiation is that separate institutional specialists *compete* for areas of control that previously were mainly church prerogatives. Although religious institutions still proclaim their definitions of deviance and use some measures of social control, their influence is limited in most modern societies.

Most of the evidence supporting the hypothesis of increasing institutional differentiation is from historical studies. Other evidence can be drawn from conflict between institutional spheres over control of a given function. For example, recent court conflicts over civil disobedience, euthanasia, and forced cesarean deliveries involved competing jurisdictional claims by representatives of several institutional spheres (Fenn, 1978; Irwin and Jordan, 1987; Willen, 1983). Different institutions also compete for control over the definitions of health, illness, and healing (Conrad and Schneider, 1980). These contemporary issues exemplify the process of delineating the boundaries of institutional differentiation.

Implications for the Individual. There are two important implications of extensive institutional differentiation: one for the situation of the individual and one for the larger society. For the individual, the process of differentiation involves a conflicting development. On the one hand, differentiation appears to go hand in hand with the discovery of self—the unique individual within society. Religion has been an important factor in the increased awareness of and modern emphasis on the individual (Bellah, 1964). On the other hand, differentiation results in segregation of the individual's various roles. A woman's role as mother is not considered relevant to her role as mayor; a man's role as religious believer is not considered relevant to his role as corporate manager.

Specialization in the division of labor extends not only to the separation of institutional spheres but also to specialization of roles within each sphere. One is not merely a "factory worker" but, more specifically, the one who turns the third bolt on the left side of the object coming down the assembly line. Each institutional sphere dictates its own expectations of the individual so that he or she should become an effective performer of that role. Requirements of institutional functioning, however, often conflict with the individual's personal goals, preferences, or needs. The role of advertising account executive may require the individual to wear certain kinds of clothes, to eat and drink in certain kinds of places, to have tact, to lie, to flatter, to be self-effacing, polite, or even obsequious in order to please clients and keep accounts (good examples of such role requirements are found in Berger, 1964; Hochschild, 1983; Terkel, 1975).

Role requirements of occupations vary, but all tend to exclude qualities that do not contribute to achieving the goals of the organization. Values such as moral qualms or self-realization are not necessarily negated; they are simply relegated to another institutional sphere and considered irrelevant to the job. We will discuss these developments as we find them implied in other processes. The key point is that the individual may experience a conflict between the needs and goals of the self and the demands of these social roles (see Simmel, 1959, 1971).

Implications for Society. Similarly, the processes of differentiation make it harder for society to mobilize the commitment and efforts of its members. Values from one separate sphere do not readily motivate behavior in another. Why should one vote in an election, serve in the army, or work hard on the job? In societies with relatively little institutional differentiation, behavior in work, politics, and social or military service is often motivated by values from other spheres—family, community, religion, and tradition. In contemporary society, the main motivating force in public spheres appears to be the promise of certain levels of consumption—that is, a material standard of living (Fenn, 1974:148).

The process of differentiation is not inexorable. There appear to be limits to the effectiveness of specialization. Some workplaces are experimentally reducing job specialization with hopes of increasing worker motivation. Even such experiments, however, are dictated by the criteria of the economic sphere; their goal is not to make work meaningful but to increase productivity. Nevertheless, the process of differentiation has important implications for the location of religion in contemporary society. The effective criteria of public institutional spheres—notably the economic—are separate from the values of the private sphere.

Religion is relegated to the private sphere, a development detailed further in this chapter. The individual's desire for meaning and belonging must be pursued in the private sphere. Apparently, the same differentiation that makes possible the "discovery of the self" also frees the institutions of the public sphere to ignore or counteract the autonomy of individuals under their control. The tension between the demands and needs of institutions at the societal level and the demands and needs of individuals makes this an important issue in both the sociology of religion and sociology in general.

Authority and Legitimacy

Legitimacy, as defined in Chapter 6, refers to the basis of authority of individuals, groups, or institutions by which they can expect their pronouncements to be taken seriously (Fenn, 1978:xiii). Legitimacy is not an inherent quality of individuals, groups, or institutions but is based on the *acceptance* of their claims by others. If an individual proclaims "it is absolutely imperative for Americans to reduce their consumption of gas and oil by 15 percent," on what basis does this person claim to be taken seriously? Such a statement by the president of the United States would be based on a different source of legitimacy than the same statement by a spokesperson for the Union of Concerned Scientists, the editor of *Newsweek,* the National Council of Churches, or, for that matter, a Chicago elevator operator.

The location of religion in contemporary society reflects societal changes in the bases of legitimacy. Relatively stable societies typically have stable sources of legitimacy. The key criterion in such societies is usually traditional authority such as the inherited authority of the patriarch or king. Institutional differentiation often produces a different kind of authority: the authority of the holder of a specialized role or "office." Claims to be taken seriously are based not on who one is but on what position one holds. The authority of a judge, for example, is based on the role for that office rather than the person.

Religion legitimates authority indirectly in traditional societies by its pervasive interrelationship with all aspects of society. Myth and ritual support the seriousness of all spheres of life. The chief, priest, or matriarch can speak with authority because their roles correspond to or reflect the authority of divine beings. Historic religions legitimate authority more directly, as shown in Chapter 7. Such historic religions as Christianity, Islam, and Judaism have similarly given authority to pronouncements on education, science, economic policy, law, family life, sport, art, and music. Whether directly or indirectly invoked, the images and symbols of the sacred are a source of legitimacy (Fenn, 1978:xiii). The taken-for-granted quality of religion as a source of legitimacy characterizes all of these earlier societal situations.

Conflicting Sources of Authority. The differentiation process has resulted in a rise of competition and conflict among the various sources of legitimate authority in modern society. As Bellah (1964) suggests, historic and early modern religions experienced such competition to a limited extent. Early differentiation of religious from political institutions meant that religious institutions could come into conflict with political institutions. The potential to legitimate implics the corollary ability to delegitimate power, whether in the context of government, courts, workplace, or family (see Luckmann, 1987). Historically, the churches could authoritatively evaluate whether the state was engaged in a "just" war and could authoritatively criticize business practices they judged as "usurious" (i.e., charging unfair interest rates on loans). In contemporary society, by contrast, religious institutions must actively compete with other sources of legitimacy. Personal, social, and political authority are more uncertain.

This competition for legitimacy can be seen in a number of issues in recent years. Civil disobedience over civil rights, abortion, sanctuary for political refugees, and the wars in Vietnam and Central America was often based on a different source of authority than that of the prevailing legal or political ones. Civil disobedience often claimed religion or higher human values as the basis for conflict with civil authorities. Sometimes court events involve a clash of several different sources of legitimacy. In one court case, claims were made from legal, medical, parental, and religious bases of authority. A young person had been comatose for months and "as good as dead" in the commonsense view. Her body was kept alive by technological intervention, and eventually her parents sought legal permission to terminate the "extraordinary" measures of keeping her body alive. No single authority held uncontested legitimacy. The court case was complicated. Medical experts gave their opinions on the medical definitions of death. Legal experts raised issues of the legal rights and guardianship of comatose patients. Theological experts offered briefs on the borderlines of life and death, and the girl's father made a thoughtful personal statement about his request. The relevant issue is not merely the uncertainty of the outcome but that the court is the arena in which medical, legal, religious, and parental figures vied to have their statements taken seriously (Fenn, 1982; Willen, 1983).

Pluralism. One particular source of this uncertainty of legitimacy is **pluralism,** a societal situation in which no single worldview holds a monopoly of legitimacy. This feature is an especially important characteristic of the American and Canadian religious scenes (see D. Martin, 1978:1–99, for a sophisticated description of factors related to secularization in nonpluralistic as well as pluralistic societies). Historic official religions were characteristically monolithic: They established *the* worldview of their society and had a monopoly over the ultimate legitimation of individual and collective life. Where alternative views coexisted with the dominant one, they were absorbed and co-opted or effectively suppressed and segregated. Examples of absorption include the incorporation of early monasticism into church-controlled monastic orders or the development of sects within Hinduism. The alternative worldviews of medieval Jews among Christians and Zoroastrians (i.e., Parsees) among Hindus were little threat to the monopolies of these dominant worldviews because the minority religions were effectively suppressed and segregated (Berger, 1967:134, 135).

Pluralism is sometimes used in a narrower sense to describe the political or societal tolerance of competing versions of truth. Highly pluralistic societies such as the United States, Canada, and Australia typically have several diverse groups with competing worldviews. By contrast, Spain and Russia exemplify a generally monopolistic situation. Between these two polar types, however, there is much variation. In England and Scandinavian countries, there is an official state church but relatively high tolerance of minority positions. A more rigid situation exists in Northern Ireland; although there is "official" tolerance at the societal level, there is little intergroup tolerance. In the Netherlands and Belgium, less volatile but still religiously divided societies, most aspects of society have been split into mutually exclusive "pillars" or "columns" according to people's religion. Political parties, hospitals, mass media, youth clubs, social welfare organizations, and schools were all divided according to these religious lines, largely as a protective device for the separate religious enclaves. The divided columns were given political and societal recognition, although leaders discouraged intergroup mingling (Dobbelaere, 1989; Laeyendecker, 1972).

Berger emphasizes that pluralism, in both this limited sense and in the broader sense previously described, greatly affects the situation of religion. Where worldviews coexist and compete as plausible alternatives to each other, the credibility of all is undermined. The pluralistic situation relativizes the competing worldviews and deprives them of their taken-for-granted status (Berger, 1967:151). A farm family in Spain may take the Roman Catholic worldview for granted. It is the worldview of their friends, relatives, and neighbors, and it is a part of everyday life. Even though much of their worldview may be popular (rather than "official") Catholicism, it is shared by their community to the exclusion of other views. Although members of the community may vary considerably in piety or religious activity, other worldviews are probably literally inconceivable.

In a pluralistic situation, by contrast, no single worldview is inevitable. In the United States, a committed Roman Catholic's neighbors are Baptist, Jewish, Unitarian, Lutheran, atheist, and Zen Buddhist. The government and the society do not (formally, at least) give favorite legitimacy to any of these worldviews, nor does anyone's god accommodate anybody's quandary by sending down a lightning bolt to get rid of all the "wrong" believers. If people wish to protect the belief that their own worldview is uniquely true, they must isolate themselves from alternative worldviews. In U.S. society, especially in relatively urban settings, that is not easy to do. On the job, in school, through the media, working for a political party or a social "cause," or even playing softball—and increasingly in neighborhoods, social clubs, and parties—Americans are exposed to others who hold different worldviews from their own. The impact of the pluralistic situation is thus that the various worldviews in society also compete for legitimacy. No single view has such uncontested legitimacy that a person expressing it authoritatively could be certain of being taken seriously (Berger, 1967:151).

Religious pluralism, in the broadest sense of the term, is not like having a choice between Rotary and Kiwanis. The pluralism of worldviews is qualitatively different from a pluralism of affiliations or associations because the taken-for-granted quality of any one worldview is undermined by pluralism. Differentiation made it possible to conceive of religion as an entity separate from other institutional spheres. Pluralism, furthermore, made it possible to conceive of religions; the very concept implies a stance of some distance, a meaning system that one does not personally hold (Hammond, 1974:119, 120).

Pluralism and Legitimacy in the United States. The evidence supporting this interpretation of social change is fairly strong, at least in the United States; but we need to remember that pluralization is a process that is neither inexorable nor complete. In colonial America, the principles of pluralism and religious tolerance were gradually established, but colonial society was far from pluralistic in the modern sense. It was narrowly Protestant and adamantly Christian, even though less than 10 percent of the populace actually belonged to churches (see Wood, 1988). Colonial Massachusetts, for example, was virulently anti-Catholic, with stringent laws against "popish" ideas and practices. Quakers and Baptists were also suppressed, sometimes violently. Only Pennsylvania, Rhode Island, and—for a period—Maryland and New York had some legal support for religious tolerance, but even there religious minorities were not treated equally before the law. Many of the colonies denied Catholics, Jews, and minority Protestant groups such rights as voting, holding public office, and conducting religious services in public (Ribuffo, 1988). Later, when the Constitutional Convention of 1787 prohibited any religious test for public office, it dramatically reshaped church–state relations in the new republic, eliminating preferential treatment not only for any one religion but also of religion over nonreligion (Wood, 1988). The principle of religious liberty, together with the growing multiplicity of religious views, pressed the government and especially the courts to resolve conflicts. The courts needed moral answers to resolve legal questions, yet the increasing ethnoreligious diversity of the country made it impossible for the courts to resort to the language and legitimacy of orthodox Protestantism. In 1879 the Supreme Court had to decide whether polygamy (in this case, plural marriage) would be

legally permitted if it were part of a religious group's (i.e., Mormons') practices (*Reynolds* v. *United States,* 98 U.S. 145, 1879). The court decided against
making "religious" exceptions to "the law of the land" (Hammond,
1974:129–130).

Similar conflicts arose over whether to allow exemptions on religious
grounds from national ceremonies (e.g., saluting the flag) and military service,
release time from public schools for religious instruction, prayer in public
schools, taking prohibited drugs for religious purposes, and so on (see Manwaring, 1962; see also Demerath and Williams, 1987; Hammond, 1987). It
would be a mistake to see these conflicts as merely church–state issues; the
very *process* of their resolution reflects the pressure of an increasingly pluralistic
situation in which no single religious worldview is granted legitimacy. Note
the change between the language of cases in 1892 and 1965:

> ### Church of the Holy Trinity v. United States, 143 U.S. 226, 1892:
>
> Mr. Justice Brewer wrote that events in our national life "affirm and reaf-
> firm that this is a religious nation." Deciding that a statute prohibiting im-
> porting aliens for labor was not intended to prevent a church from hiring
> a foreign Christian minister, the Court quoted with approval two earlier
> judicial opinions stating that "we are a Christian people, and the morality
> of the country is deeply ingrafted upon Christianity" and "the Christian
> religion is a part of the common law of Pennsylvania."
>
> ### United States v. Seeger, 380 U.S. 163, 1965:
>
> In a conscientious objection case, the Court determined that "belief in
> relation to a Supreme Being" (thus exemption from military service) shall
> be determined by "whether a given belief that is sincere and meaningful
> occupies a place in the life of its possessor parallel to that filled by the or-
> thodox belief in God of one who clearly qualifies for the exemption." Ex-
> emption is not limited to monotheistic beliefs. Mr. Justice Clark noted the
> prevalence of a "vast panoply of beliefs," thus Seeger's beliefs qualified as
> "religious" and he was exempted (cited in Hammond, 1974:130, 131).

One general impact of pluralism and differentiation is to create a problem
of legitimacy for both the individual and the society. The problem of legitimacy at the societal level involves the society's very basis for authoritative decision making and its grounds of moral unity or integration. Such problems
highlight the issues of church–state tensions and the nature of American civil
religion, discussed further below. At the individual level, the problem of legitimacy makes the individual's meaning system more precarious, voluntary, and
private.

Problems at the Societal Level. In a pluralistic situation, worldviews and
authoritative claims compete; this results in the diffusion of sources of legitimacy among many agents in society. These competing claims may appeal to
sacred or quasi-sacred sources of authority, even if not using explicitly reli-

gious symbols (Fenn, 1978:45). A good example is, again, the courts. Hammond suggests that the pluralistic situation creates problems of defining order (because no monolithic worldview exists to authoritatively define it), but the social situation still needs order. Legal institutions are then called on to establish and interpret a uniform order, and the result is a constellation of legal institutions with a "decided religio-moral character" (Hammond, 1974:129). Thus, the hypothesis of secularization is connected with the issue of civil religion (i.e., the national expression of unity, as discussed in Chapter 6). According to this interpretation, the courts are trying to articulate a basis for moral unity where no such foundation is given by specifically religious institutions. Some sociologists view the civil religion as attempting to overcome problems of legitimate authority and societal integration in U.S. society.

Similarly, the Indonesian national guiding principles, the *Panch Shila,* represent a government-promulgated civil religion. These principles include belief in one god, national unity, guided democracy, social justice, and humanitarianism. They are derived from Indonesia's principal religious and cultural traditions but attempt to transcend the particular religions to unify conflicting ethnoreligious communities and to link people divided by tribal, linguistic, and geographical barriers in the face of challenges to national legitimacy by militant ethnoreligious groups (see Anderson, 1991; Geertz, 1973:193–233; Demerath, 1989; Nash, 1991). Fenn (1978) argues that development of generalized religious symbols or ideology is linked with larger societal problems of unstable sources of legitimacy of social and political authority.

The problem of legitimacy results from the collapse of a societal shared conception of order (Wilson, 1976:100; Wilson specified "transcendent" order, but the collapse of any shared worldview has equivalent results). What some people decry as "secularity" is not the result of lost belief followed by immoral behavior; first lost is the overall sense of moral community and consensus (literally "thinking and feeling together"). Without agreement on the way to live together, claims of moral authority make no sense (MacIntyre, 1967:54).

This problem affects both individual and societal decision making. How is it possible for human values to determine public policy in a pluralistic society? Is the role of religion in political decision making reduced to that of one more interest group vying with opposing interest groups? Or is it even possible for a pluralistic society to agree on human values at a societal level? The obvious examples of religious groups asserting themselves on a national level have typically been issues of legislating what might be called "private morality"—homosexuality, abortion, pornography, divorce, gambling, or drinking. Less publicized but still important have been the efforts to apply religious values to decision making on "public morality"—civil rights, the Vietnam War, arms proliferation, and the human environment.

The difficulties of reaching a consensus on social values in a pluralistic society are illustrated by the issues involved in developing a national energy policy for the United States. The energy situation could be viewed as merely the conflict of competing interests: producers versus consumers, producers versus environmental protection groups, voters versus major campaign contributors, the United States versus other nations, and so on. Policy decisions could reflect a juggling of these interests according to relative power, influence, technological expertise, financial muscle, or political pressure. Nevertheless, human values have also been raised: fair distribution of scarce resources, concern for the "have–nots," human health and safety, responsibility to future generations, and responsibility in international relations. None of these interests or values specifies the exact outcome of any policy decision, but the specific factors accepted as relevant in the decision–making process do make a difference. Is consensus on values necessary or even possible? And if so, does the society consider human values relevant or important to decisions in the public sphere?

Fenn (1972:27) argues that having unstable sources of legitimacy produces two seemingly disparate social responses: the *development of minority and idiosyncratic definitions of the situation,* together with increasingly *secularized political authority.* On the one hand is pressure to desacralize the political authority— for example, removing any ideological notions of what is "good" from decision making and replacing them with criteria such as due process and technical procedures (see also Hammond, 1974). On the other hand, challenging the civil religious synthesis results in spreading access to the sacred. Thus, individuals and groups develop their own particular (i.e., "idiosyncratic") views and symbols for which they claim the same seriousness as recognized religions. Individual and group claims to social authority multiply as the uncertainty of boundaries becomes evident. Fenn (1978:55) states, "Secularization increases the likelihood that various institutions or groups will base their claims to social authority on various religious grounds, while it undermines the possibility for consensus on the meaning and location of the sacred."

Court cases present some interesting examples of the ambiguities at this stage. The use of peyote (a hallucinogen) by some Native Americans has had a long tradition. The Native American church, instituted in 1918, ties together themes from Native American religions and incorporates peyotism in its ritual. In 1962 a group of Navajos using peyote in their religious ceremony was arrested and convicted under a California law prohibiting possession of peyote. The California Supreme Court overturned the conviction (*People* v. *Woody,* 394 P 2d813, 1964), however, granting Native Americans immunity under the First Amendment of the Constitution, which guarantees free exercise of religion. Despite the promulgation of federal policies to protect Native American religious freedom (see Michaelson, 1987), few court battles have upheld those freedoms. In 1990 the U.S. Supreme Court overturned a lower court's decision and refused to grant religious exemption from Oregon's laws against possession or use of peyote (*Oregon Employment Division* v. *Smith*).[1]

---

1. The *Smith* case represented a repoliticization of the legal issues, though, in that the Court was clearly reflecting the "statist" shift that characterized the presidencies of Reagan and Bush, who had appointed most of that Court. This shift was away from tolerance of marginal religions toward greater state control of all aspects of private life. However, a coalition of religious groups, including most mainstream denominations, pressured Congress to restore the earlier criteria for legitimate "free exercise," and the 1993 Religious Freedom Restoration Act, passed overwhelmingly by both chambers of Congress, reaffirmed protection for minority religions in the United States (see Richardson, 1995a; Williams, 1995).

The courts were even more pressed on the religious freedom issue when individuals and groups in the larger culture began to use hallucinogens for religious purposes. A number of "psychedelic churches" (e.g., the Church of the Awakening and the League of Spiritual Discovery, both founded in the mid-1960s) have tried to establish their constitutional immunity for the religious use of drugs (mainly marijuana, peyote, and mescaline). The best-known case on this issue was in the mid-1960s, when Dr. Timothy Leary was convicted and sentenced to thirty years in jail and $40,000 in fines for illegal possession of marijuana. Dr. Leary, a researcher of psychedelic drugs, claimed that marijuana use was an integral part of his religious practice. On appeal, the court upheld the conviction, commenting that Leary's religious use of marijuana was occasional, private, and personal. The court declared a "compelling state interest" in drug prohibition:

> It would be difficult to imagine the harm which would result if the criminal statutes against marijuana were nullified as to those who claim the right to possess and traffic in this drug for religious purposes. For all practical purposes the anti-marijuana laws would be meaningless, and enforcement impossible. . . . We will not, therefore, subscribe to the dangerous doctrine that the free exercise of religion accords an unlimited freedom to violate the laws of the land relative to marijuana (*Leary v. United States* 383f.2d841, 1967).

This kind of conflict illustrates the lack of moral unity and problems of legitimacy in modern societies such as the United States. Similar cases involving idiosyncratic perspectives include conflict over whether Transcendental Meditation (TM) should be allowed in the schools, whether faith healers or Scientologists could be banned as fraudulent, or whether to allow movements with strong political overtones (e.g., Black Muslims) the prerogatives of religious organizations (see Alley, 1988; Burkholder, 1974; Pfeffer, 1974, 1987; Robbins, 1981, 1985; see also *The Annals of the American Academy*, Nov., 1979).

Legitimation problems are disturbing because they undermine the ability of society to maintain belief in a symbolic whole that transcends the separate identities and conflicting interests of society's component parts (Fenn, 1978:8). Pluralism and institutional differentiation are generally important factors in this process because they break down the overarching worldview—the symbolic whole. These processes make it impossible to achieve a new firm source of societal integration and legitimacy. At the same time, however, they increase the likelihood that people will need and seek this symbolic whole.

Problems at the Individual Level. Pluralism, as suggested, undermines the taken-for-granted quality of the social worldview. The individual's own meaning system then receives less social support and becomes precarious, voluntary, and private. This too can produce conflict for the individual. Pluralism increases personal ambiguity: What am I to believe? How am I to act? On what basis can I decide?

A good example of the widespread conflict and doubt created by religio-cultural pluralism is the issue of abortion. At the societal level, the main issue is whether to legally permit abortion and, if so, under which circumstances. Thus, societal resolution requires response to conflicting legal, biological, religious, and social claims. At the individual level, however, the issue includes personal decisions: Is it right for me to have an abortion? What shall I teach my children about abortion? And how shall I respond if my friend or daughter wants an abortion? Religious pluralism contributes directly to the quandary. Some religions teach that life begins at the moment of conception; others allow that intrauterine life (i.e., the developing fetus) is qualitatively different from life after birth. Accordingly, the former view would equate abortion with killing, while the latter might allow abortion but counsel that it is still not a desirable action. Many other religious groups hold still different interpretations in good conscience. Little consensus exists even within religious groups. The problem for many individuals, then, is that while the decision is often defined as a moral issue, there is profound disagreement over the "good" path. Whatever the individual chooses, there is no massive social support for that decision.

Personal value decisions are important, but a more critical issue at the individual level is the impact of the problem of legitimacy for personal *identity*, which (as developed in Chapter 3) is influenced and supported by religion. The individual's worldview is an important element of personal identity (Luckmann, 1967:70). But one need not go so far as Luckmann, who links personal identity with religiosity, to appreciate the connection between religion and identity. The individual's subjective meaning system legitimates that person's hierarchy of goals, values, and norms. What happens, then, if this key part of the individual's identity is undermined? This problem of legitimacy of the self is related to the process of "privatization" discussed later in this chapter.

Rationalization

Rationalization is the process by which certain areas of social life are organized according to the criteria of *means–ends* (or functional) rationality. This linkage was a central thread in the works of Weber (1947; 1958a), who viewed a special form of rationality as the outstanding characteristic of modern society.

The key characteristic of modernization, according to Weber, is increasing emphasis upon functional (i.e., means–ends) rationality.[2] In traditional societies, for example, a farmer prepares the soil for planting in a certain way because "that's the way it's always been done" (i.e., traditional behavior); such orientations are extremely resistant to change. By contrast, the modern farmer is more likely to consider alternative and new methods, judging each according to rational criteria such as relative productivity, costs versus benefits, and appreciation of land values. Although Weber's concept of rationality may be useful in understanding the place of religion in contemporary society, we must avoid an overrationalized conception of human action.

---

2. Note, however, that Weber used the concept of rationality in several ways, giving rise to some ambiguities in his theories; see Luckmann, 1977:16, 17.

Rationality and Modernization. According to Weber, modern Western society has a "rationalized" economy and a concomitant special "mentality." A **rational economy** is functionally organized, with decisions based on the reasoned weighing of utilities and costs. The **rational mentality** involves openness toward new ways of doing things (in contrast with traditionalism) and readiness to adapt to functionally specialized roles and universalistic criteria of performance. Although these forms of rationality originated in the economic order, they have extended into political organization and legal order—the modern state. Weber argued that religious motives and legitimations played a central role in bringing about this form of organization and mentality—for example, by the development of universalistic ethics (i.e., the norm of treating all people according to the same generalized standards) and by the development of religious drive for rational mastery over the world. Nevertheless, this rationality, once a part of societal structure, became divorced from its historical origins and acquired an impetus of its own (Weber, 1947, 1958a).

Weber (1958a) closed his early essay on the Protestant ethic with an almost prophetic evaluation of possible outcomes from the process of rationalization. He observed that the rationally organized order of modern society presents itself to the individual as an overwhelming force. All people engaged in the extensive system of market relationships are bound by the norms of functional rationality. This is exemplified by the great difficulty of establishing alternative economic arrangements, as countercultural communes have discovered. The conditions of the modern economic order have become an "iron cage" of instrumentality, which in Weber's opinion was far from desirable:

> No one knows who will live in this cage in the future, or whether at the end of this tremendous development entirely new prophets will arise, or there will be a great rebirth of old ideas and ideals, or, if neither, mechanized petrification, embellished with a sort of convulsive self-importance. For the last stage of this cultural development, it might well be truly said: "Specialists without spirit, sensualists without heart"; this nullity imagines that it has attained a level of civilization never before achieved (Weber 1958a:182).

Weber's analysis suggests that the differentiation process alone does not account for important changes in contemporary society. Rationalization specifies the direction of differentiation: Differentiation proceeds according to the criteria of means–ends rationality. Businesses can (and many do) specialize tasks according to nonrational criteria. Nevertheless, the main thrust of organization in business (and increasingly in other spheres such as schools, government, churches, and other voluntary organizations) is to extend the criteria of functional rationality: effectiveness, efficiency, cost–benefit analysis, and specialization of tasks. While differentiation produced separate norms appropriate to each institutional domain, rationalization of public-sphere institutions means that nonfunctional values (e.g., kindness, honesty, beauty, or meaningfulness) are generally irrelevant to action within these institutions. The ethical regulation of an impersonal rational organization is thus impossible (Weber, 1958a:331).

If modern society is indeed moving in the direction of increasing functional rationality, this process implies problems at two levels: the location of individual meaning and belonging and a conflict between corporate control and values versus personal autonomy and values. Personal meaning is not only relegated to the private sphere but is also undermined by the dominant rationality of other spheres. Rationalized medical practice, for example, does not deal with the "meaning" of childbirth. It is more functional to treat the woman's and child's bodies as objects to be manipulated. Personal meaning, satisfaction, and emotions in giving birth are not overtly denied by the medical process, but they are subordinated to "rational" criteria of efficiency and medical management. The individual seeking to apply meaning to personal experiences is in a weak situation relative to the powerful institutions for which individual meaning is irrelevant.

Disenchantment of the World. Another feature of rationalization undermines the individual's personal sense of meaning and belonging: Weber called it **"disenchantment"** of the world, referring to the process by which things held in awe or reverence are stripped of their special qualities and become "ordinary." Protestantism thus brought about much disenchantment of what Roman Catholicism had held in awe (Berger, 1967:111), emptying the believer's world of angels, saints, shrines, holy objects (e.g., candles, blessed water, and oil), holy days, and elaborate sacraments. Rational science also promotes disenchantment, explaining natural phenomena without reference to nonnatural categories of thought. Phenomena previously attributed to miracles are reinterpreted by rational science as natural (e.g., "spontaneous remission of disease"). The key feature of the rationalization process is not so much the particular explanations of phenomena but the *belief* that all phenomena *can* be rationally explained (Weber, 1958a:139).

Thus, rational science undermines other ways of knowing. The process of rationalization results in a dichotomy between "serious" and "nonserious" interpretations (rational explanations being serious and other forms of explanation nonserious). The cognitive style appropriate to modern bureaucratic structures in the public sphere and utilized in science, medicine, law, and so on, is not amenable to the cognitive style of religion—which, by contrast, allows reference to a transcendent, empirically nonverifiable realm and allows undifferentiated "experiencing" as a valid way of knowing. The rational cognitive style of modern public institutions has filtered down to everyday life, though it is not consistently or pervasively applied there. Consequently, not only is one's everyday knowledge more restricted to rationally knowable items, but the world itself is indeed transformed. The way in which people think of the world becomes distinct from the way in which they think of themselves and each other (Gellner, 1974:196). The process of rationalization means that the rational mode of cognition applies to those institutional spheres that "really matter" (i.e., are serious); other modes of cognition (e.g., fantasy, play, religion) are treated as frills of private life (i.e., as nonserious).

This situation on the individual level may therefore reflect societal conflict between the extension of corporate control and the degree of individual autonomy. The modern economic situation is characterized by the increasing power and scope of **corporate actors** in conflict with the power and rights of individuals. The corporate actor is one who fills a role in an institution; the role, rather than the individual, is part of the corporate world (Fenn, 1978:66), and the corporate role is functionally rational in the institutional area. There is little or no integration of the meaning of institutional roles into a subjective meaning system for the individual actor. This lack of integration does not disturb the business or political worlds; the individual is still controlled by the corporate norms. It does not matter in a teddy bear assembly line who (or why someone) attaches the eyes; all that matters is that the role is satisfactorily performed.

Increasing anonymity of functionally specialized roles makes the individual replaceable. The corporate actor is also less responsible as a person, able to claim, "I didn't do anything wrong; I was only doing my job." Individuals may consequently segregate "rational" institutional norms in their consciousness, compensating for their lack of autonomy in the public sphere by a somewhat illusory sense of autonomy in the private sphere (Luckmann, 1967:96, 97; also see Fenn, 1978:81, 82, on the legitimation of the corporate actor's nonresponsibility).

Privatization and Individuation

Privatization is the process by which certain differentiated institutional spheres (e.g., religion, family, leisure, the arts) are segregated from the dominant institutions of the public sphere (e.g., economic, political, legal) and relegated to the private sphere. This segregation means that the norms and values of the private sphere are irrelevant to the operations of public-sphere institutions (Berger, 1967:133). It also implies that the functions of providing meaning and belonging are relegated to institutions of the private sphere. The hypothesis of privatization is directly related to the previously discussed processes of differentiation, pluralization, and rationalization. Differentiation creates separate institutional spheres, in which formerly pervasive institutions such as religion and family are compartmentalized. Pluralization contributes to the complexity of society and makes it difficult, if not impossible, to speak of cultural consensus. And rationalization of the economic (and by extension, the politico-legal) institutional spheres renders the functions of meaning and belonging irrelevant in public institutions.

Whereas differentiation, pluralization, and rationalization all refer to conditions in the larger society, privatization primarily describes the residual category of self. Privatization implies that the individual finds sources of identity increasingly only in the private sphere. Concomitants of personal identity are a sense of order, an interpretation of reality, a system of meaning, and the integration of oneself into a larger community (see Mol, 1976:9–15). The individual finds fewer sources of identity in public-sphere institutions largely because the order obtaining in this sphere is functional rationality. All that contributes to making, maintaining, or changing personal identity is located in the private sphere. The individual's very self is privatized.

On the one hand, such privatization promotes some of the personal freedoms enjoyed by many in modern society. One freedom is the extent to which one's reference groups (i.e., those whose opinion of one "counts" in shaping one's behavior) are not imposed by kin or neighborhood but are more freely chosen. The very idea that one could freely choose one's religion is a peculiarly modern concept. Privatization produces greater individual autonomy because it carves out pockets of life in which individuals' motives and range of options are not fully controllable by institutions representative of the larger society. Ironically, American fundamentalism is as much a product of this feature of modernity as it is an antimodern campaign against religious pluralism and differentiation. American Christian fundamentalism's emphasis on voluntary association (e.g., joining a church as a freely chosen adult faith commitment) is a clearly modern feature, along with its emphasis on the value of the individual and on religious liberty (Lechner, 1990).

On the other hand, privatization apparently results in problems of meaning and integration for many individuals. These problems may be reflected in the relatively widespread quest for holistic worldviews, as expressed by many alternative health movements, agrarian communes, and contemporary religious movements. Holistic perspectives such as these express a desire for integrating all aspects of life, gaining a sense of wholeness in social, physical, psychic, and spiritual life.

Privatization implies problems in legitimating oneself. Identity becomes problematic. Sources of order, meaning, and community have been undermined; all have become increasingly voluntary and uncertain. Luckmann (1967:99) suggests that this voluntary quality contributes to a sense of autonomy in the private sphere, perhaps making up for the individual's lack of autonomy in institutions of the public sphere:

> Once religion is defined as a "private affair" the individual may choose from the assortment of "ultimate" meanings as he sees fit—guided only by the preferences that are determined by his social biography.
>
> An important consequence of this situation is that the individual constructs not only his personal identity but also his individual system of "ultimate" significance (used by permission).

This self-selected construction is, according to Luckmann, the contemporary social form of religion. While church-oriented religion continues to be one of the elements that some people choose for their constructions, other themes from the private sphere (e.g., autonomy, self-expression, self-realization, familism, sexuality, adjustment, and fulfillment) are also available in a supermarket of "ultimate" meanings (Luckmann, 1990, 1967:100–114; see also Fenn, 1978:xii, on identity and privatization). Luckmann identifies autonomy as one of the central themes of modern religiosity; however, Luckmann is less optimistic about its significance. He suggests that individual autonomy has been redefined to mean the absence of external restraints and traditional limitations in the private search for identity. Nevertheless, the institutions of the public sphere have real power over the individual; performance of one's roles in these spheres must conform to institutional requirements. Thus, autonomy may be limited to the private sphere (Luckmann, 1967:109–115).

Luckmann (1967:115–117) raises a critical note in a "postscript" when he observes the discrepancy between the subjective autonomy of the individual in modern society and the objective autonomy of the major institutions of the public sphere. He wonders if the removal of nonfunctionally rational considerations from these institutions has not contributed to the dehumanization of the social order. The irony, then, would be that privatized religiosity, by sacralizing the increasing subjectivity of the individual, supports this dehumanization; by withdrawing into privatized religious expressions, this new mode of religiosity fails to confront the depersonalized roles of the public sphere. If this is true, modern forms of religion do not have to legitimate society directly— they support it indirectly by motivating retreat into the private sphere.

Bellah's (1987) interpretation of privatization is similarly pessimistic. Drawing comparisons between modern Japan and the United States, he notes that both religion *and* politics appear to be increasingly relegated to the private sphere. Accordingly, the modern state could be legitimated neither by religious nor political discourse, but rather "managed" by bureaucratic and technological expertise, sometimes decorated by the (real or media-created) attractive personalities of leaders (cf. Williams, 1988).

While privatization refers to the location of religion in modern society, the process of *individuation* refers more to the development of the *form* of modern religion. **Individuation** is the process by which the individual and his or her concerns come to be seen as distinct from the social group and its concerns. Bellah (1964) views modern forms of religion as a new stage of religious evolution, in which religion as a *symbol system* evolves and develops increasing ability to adapt to its environment. Bellah describes several historical patterns of religion: primitive, archaic, historic, early modern, and modern—each characterized by a common pattern of symbol system, style of religious action, religious organization, and social implications.

Bellah's evolutionary scheme links the structural differentiation of religion with the process of individuation. Historic religions "discovered" the self; early modern religions enabled a greater acceptance of the self; and modern religion represents an even greater emphasis on the autonomy and responsibility of the self. Modern religion is characterized by an image of the dynamic multidimensional self, able (within limits) to continually change both self and the world. The mode of action implied by this image is one of continual choice, with no firm, predetermined answers, and the social implications of modern religion include the image of culture and personality as perpetually revisable. Bellah notes, however, that whether the freedom allowed by this perspective will be realized at the social rather than purely individual level remains to be seen (Bellah, 1964:374).

Robertson's synthesis of themes from Weber, Durkheim, and Simmel similarly emphasizes the twin processes of individuation (especially the quest for individual autonomy) and societalism (i.e., the growth in societal power over individual members). He suggests that the individuation process has proceeded so far that large-scale institutions of the public sphere cannot easily resist it. Some emerging religious movements of recent years may prompt a new mode of relationship between the individual and society (Robertson, 1977:305; 1978:180; 1979; cf. Dreitzel, 1981; McGuire, 1988, 1995). Indeed, the very privatized form of modern religiosity may make it more effective than traditional religiosity as a bridge between autonomous individual actors on the one hand and societal institutions and humankind on the other (Riis, 1989, 1993).

Reviewing developments of modern forms of religiosity in France, Hervieu-Léger (1990:S22) comments that these phenomena are "vehicles of an alternative rationality which is as much in harmony as in contrast with modernity." Rather than weaknesses, she observes that features of privatized religion are particularly effective strengths in modern society. For example, the purely voluntary associative basis of religious networks makes them more flexible supports for the mobile modern individual. Similarly, the contemporary religious emphasis on personal experience and the individual's "right to subjectivity" are well suited to personal motivation and decision making in a situation where values and ends are not institutionally "given." Hervieu-Léger suggests that modernity produces not the "decline" of religion but a new location—and a new work for religion to accomplish.

EXTENDED APPLICATION:
RELIGION AND HEALING

Many of the processes of change described above have affected the relationship between religion and healing.[3] In most societies before the impact of modernization, religion and healing were completely intertwined. Both are concerned with the human person, both seek wholeness, both try to restore people to a happy, fulfilling life, both stand in the face of death—to stave it off and, eventually, to give death meaning. Because we are so thoroughly socialized in modern, scientific ways of thinking, it is sometimes difficult for us to appreciate that in such a worldview, it is completely reasonable—indeed essential—to treat the sick person's body, mind, spirit, and social relations all at once. Traditional Navajos, for example, heal their sick with religious ceremonies lasting for several days. These ritual practices restore for participants the meaningful universe threatened by illness and death. In their worldview, however, this action is not "merely religious"; rather, it cures on many levels. Undifferentiated and nonrationalized, their religion and their healing are intimately bound together (see Spickard, 1991).

---

3. Portions of this discussion are adapted with permission from *Health, Illness and the Social Body* by Peter E. S. Freund and Meredith B. McGuire, 1995, Prentice-Hall, Inc., and from my essays, "Health and Spirituality as Contemporary Concerns," *Annals of the American Academy of Political and Social Science* 527 (May 1993):144–154, and "Religion and Healing," pp. 268–284 in *The Sacred in a Secular Age*, P. Hammond, ed. (Berkeley: University of California Press, 1985).

Modern biomedicine, by contrast, is utterly differentiated from religious and other healing institutions. It is highly rationalized, and based on this rationalization it has gained preeminence over religion, the family, and other private-sphere institutions in the exercise of legitimate authority. Thus, the modern treatment of illness and dying also illustrates the privatization of religion. The process by which the institution of healing has changed so thoroughly illustrates the broader concepts we have discussed above.

Differentiation

Western medicine has gradually differentiated itself from other institutions. Traditionally, healing was interwoven with numerous other institutional domains, notably the family and religion. Long before there was any distinctive occupation of medicine, healing was the function of mothers and other nurses, herbalists, folk healers (e.g., persons with a special gift for setting bones), religious persons, midwives, diviners, and so on. Even those healers who were not religious specialists took for granted the importance of combining their physical care with spiritual care, such as a blessing said over medicines or rituals to protect the healers themselves from harm while healing. Although the institutional differentiation of medicine proceeded especially rapidly in the last century, its roots can be traced at least as far back as early Greek and Persian medicine. Characteristics of this differentiation include the development of a distinctive body of knowledge, a corps of specialists with control over this body of knowledge and its application, and public acknowledgement (or legitimacy) of the specialized authority of medical specialists.

The specialized occupation of "doctor" did not develop until the Middle Ages. Healing in this period, however, was highly supernaturalized. The church attempted to control healing because the power to heal was believed to come from spiritual sources, either evil or good. Nonclerical healing was severely limited. The Lateran Council of 1215 (which represented the apex of church power in the Middle Ages) forbade physicians from undertaking medical treatment without calling in ecclesiastical advice. Healing that took place outside clerical jurisdiction was suspect of having been aided by the devil. The two main groups of healers thus suspect were Jewish doctors and "white" witches (typically women). Jewish (and, in small numbers, Moorish) physicians kept their ancient medical lore alive throughout the Middle Ages, while the dominant European medical knowledge was a stagnant and less effective form of late Greek medicine (i.e., Galenism). Therefore, Jewish physicians were sought by the wealthy, to the disgust of Christian clergy who proclaimed (for example) that "it were better to die with Christ than to be cured by a Jew doctor aided by the devil" (cited in Szasz, 1970:88).

The masses, however, were served by an assortment of folk healers such as white witches—members of the community who used herbs, potions, magic, charms, and elements of pre-Christian religions to cure disease and ward off evil influences (see Ehrenreich and English, 1973). They served as midwives and were often consulted for advice about personal problems as well as magic.

These healers were a special target of the Inquisition in several countries. A papal bull (i.e., official writ), the *Malleus Mallificarum* (1486), became a manual for witch-hunts. The *Malleus Mallificarum* singled out women healers as especially dangerous:

> All witchcraft comes from carnal lust, which in women is insatiable. . . . Wherefore for the sake of fulfilling their lusts they consort with devils . . . it is sufficiently clear that there are more women than men found infected with the heresy of witchcraft. . . . And blessed be the Highest Who has so far preserved the male sex from so great a crime. (Sprenger and Kramer, [1486] 1948).

Similarly, witch-hunts in Protestant countries (e.g., England and New England) considered the "good" witch to be especially dangerous (Szasz, 1970:89).

The specialized occupation of doctor developed largely through the establishment of university medical schools and the later state regulation of credentials for physicians and creation of medical guilds. This specialization also had the effect of legitimating the increasing monopoly of physicians against the claims of other healers. Nevertheless, the formally recognized physicians were unable to command a monopoly over healing services, largely because the state of their craft did not inspire public confidence until recent times.

Rationalization

The preeminence of the modern medical profession has its roots in the rationalization and further differentiation begun in the Renaissance and proceeding rapidly since about 1850. This rationalization occurred mainly in two spheres: the application of rational science to medical discoveries and the rational organization of the profession. Modern biomedicine is based upon a number of assumptions about the body, disease, and ways of knowing that are largely incompatible with religious belief and practice. The medical model assumes a clear dichotomy between the mind and the body; physical diseases are assumed to be located in the body and thus are the purview of the physician (Engel, 1977; Gordon, 1988). Building on the dichotomy, psychiatry claimed professional jurisdiction over illnesses of the "mind." The original Cartesian dichotomy held that the nonphysical category included both mind and spirit, but modern psychiatry and psychology generally treat the spirit part as nonempirical and unreal—often even as pathological (Gaines, 1992; Williams and Faulconer, 1994). The medical model also holds that the body can be understood and treated in isolation from other aspects of the person inhabiting it (Hahn and Kleinman, 1983).

The biomedical model not only dichotomizes body and mind, but also assumes that illness can be reduced to disordered bodily functions (e.g., biochemical or neurophysiological problems). This physical reductionism, however, excludes social, psychological, and behavioral dimensions of illness (Engel, 1977). Rationalization of medicine promoted an instrumentalist approach to the body; medical "progress" is equated with mastery and control.

The rationalization of medical knowledge meant increasing reliance on scientific methods of discovery and utilization of technology and technique. Careful observation and systematic recording, dissection and autopsies, and pharmacological experiments contributed to the formal knowledge of the field, whereas some of these methods had been previously forbidden by church authorities. Technological developments such as the invention of the microscope and stethoscope also were a part of this rationalization process. Especially important was the discovery in the latter part of the nineteenth century that specific agents (e.g., bacteria) caused specific diseases (Freidson, 1970:16). Access to specialized knowledge and specialized technologies uniquely "qualified" the physician.

The increasing rationalization of medical knowledge led to a focus on *disease,* a biophysical entity, rather than on *illness,* the complex social, psychological, and spiritual situation of the sick person (Kleinman, 1978). Rational medicine has a built-in tendency toward treating the human body as an object rather than a person. This narrow biological determinism promotes the image of the physician as active and powerful, the client as passive and objectlike. The sick person must give up control of his or her own body and depend on the benevolence and knowledgeability of the professional (Young, 1976:18, 19).

Rational organization within the medical profession went hand in hand with specialization of knowledge and technical skill. Simultaneously, physicians consolidated their control and prestige by annexing and subordinating a number of potentially competing occupations such as nursing, pharmacy, and medical technology. Other competitors such as midwives and bonesetters were effectively driven out of practice and their functions taken over by the medical profession (Freidson, 1970:52). In the United States, one approach to medicine (called *allopathy*) gained a monopoly over medical practice, education, and licensing. Important competing medical approaches such as homeopathy, osteopathy, chiropractic, and naturopathy—each positing different interpretations of causes and treatments of illness—have been effectively subordinated or suppressed by the monopoly of allopathic medicine (Wardwell, 1972).

The further differentiation of medical arenas from other aspects of people's lives resulted in isolated spaces (e.g., hospitals, asylums) and in socially separate, increasingly technological "rites of passage" for the transitions of birth, healing, and death (see Davis-Floyd, 1992; Kearl, 1993). These medicalized rites, however, serve more to affirm technological control and physician status than to provide meaning or moral significance to the patient's and family's experiences.

The dominant medical system has effectively reduced the legitimacy of actions of other "encroaching" institutional areas such as religion and the family. In 1967 the pastor of First Church of Religious Science was found guilty of illegal practice of medicine for treating members' "emotional and weight problems" with hypnosis—judged by the court to be an appropriate procedure only for medical therapy (cited in Szasz, 1970:91). A similar claim was made by a forensic psychiatrist who supported the legal claims of a man who sued a retreat center of the Roman Catholic Diocese of Brooklyn, blaming the center for his psychological trauma after a three-day religious retreat from which he emerged believing he was Jesus Christ and threw himself in front of a train (as a sacrifice for the redemption of humanity), resulting in the severing of both his legs. The psychiatrist charged that the retreat was like "practicing psychotherapy without a license" by creating an unprofessionally controlled setting of emotional intensity (Anthony and Robbins, 1995). Since the case did not get resolved in court, we will never know whether the psychiatrist's testimony would have swayed the court, but the case vividly illustrates how medical authority actively competes for dominance with other sources of authority.

Other religious groups have experienced similar problems (see Burkholder, 1974). In 1990 a Massachusetts court convicted a couple who postponed medical treatment for their sick child while using Christian Science healing approaches. Although testimony suggested that the child's problem might not have been surgically reparable even if a medical doctor had been consulted, the parents were tried for manslaughter (see Beck and Hendon, 1990, and Richardson and Dewitt, 1992, for descriptions of this and related cases).

Various courts have similarly overruled religious or moral objections to medically "required" procedures. Jehovah's Witnesses believe that blood transfusions are forbidden by Scripture, but the courts have generally upheld the medical authorization of transfusions, even for unwilling recipients (*U.S.* v. *George,* 239 F. Supp. 752, 1965). A number of court decisions have upheld medical judgments to perform cesarean sections, even over the strenuous religious objections of patients and their families, and despite the fact that subsequent medical developments showed that most of the operations were not medically necessary (Irwin and Jordan, 1987).

Medical authority supersedes parental authority in decisions presumed to determine life or death (precedent cases in 1952, 1962, and 1964 are cited in Burkholder, 1974:41; for parallel British cases, see Wilson, 1990). Even in instances where medical ability to prevent death is doubtful, greater legitimacy is given to medical rather than parental authority. In 1977, for example, a Massachusetts court ordered that parents of a child dying of leukemia submit the child to medically prescribed chemotherapy, rather than to a less unpleasant (but also less respectable) naturopathic therapy.

Legitimation and Social Control

The result of these processes of differentiation and rationalization (together with the efforts of the interested parties—the physicians themselves) is that the medical profession today has an officially approved monopoly over the definition of health and illness and the treatment of persons defined as "sick" (Freidson, 1970:5). A good case could, in fact, be made that the church authority of the Middle Ages was undermined first by legal institutions, which in turn were undermined by the medical institution. The medical profession has become a key legitimating agency in Western societies and thus has a major role in social control.

Society must deal with individuals who differ significantly from socially established norms. The concept of **social control** refers to all mechanisms a society uses to try to contain its members' behavior within those norms: deterrents, incentives, rewards, and punishments. Whether the society punishes the individual depends largely on its determination of the individual's responsibility for the deviant behavior. A person who fails to go to work for two weeks is likely to be fired, but if it is determined that the absence resulted from sickness, the person will probably be exempted from normal obligations of work. At the same time, however, a person who is permitted to take the "sick role" is obliged to try to get well, cooperate with medical help, and act appropriately "sick" (Parsons, 1972).

The key issues in understanding social control are definition of deviance, determination of responsibility, and administration of punishment. Religious, legal, and medical institutions have served as significant agents of social control in all three aspects. The definition of deviance is a particularly important aspect because the norms by which deviance is defined are themselves socially constructed. The definition essentially depends on the social group's idea of what is "normal" (Becker, 1963). If a society defines noncompetitive or unobtrusive behavior as the norm, any individual who is competitive or stands out is likely to be considered deviant. Deviance is a product of the group that labels it.

Religious, legal, and medical institutions have all contributed to the definition of deviance in society. For example, "Thou shalt not steal" defines a religious norm; stealing is a sin. Legal systems define similar norms of behavior, and violation of the norms is a crime. Medical systems also define what is normal or desirable behavior; deviance is sickness. But the relative weight of religious, legal, and medical institutions in defining deviance has shifted. As the Middle Ages waned and these three institutions became increasingly differentiated, the religious organizations still held greatest influence in defining societal deviance. This preeminence continued into the eighteenth century, but in the United States and France (and later other European countries) the legal mode of defining deviance gained ascendancy. In the United States, the increasing preeminence of the legal definitions was promoted by pluralism (as previously noted) and by increasing rationalization of the nation-state.

The significance of legal definitions of deviance has waned somewhat in the twentieth century, and medical definitions of deviance have gained preeminence. This shifting balance is clearly reflected in the 1954 precedent-setting court case *Durham* v. *United States,* 214 F 2d, 863, which decided that "an accused is not criminally responsible if his unlawful act was the product of a mental disease or mental defect." Abe Fortas, the court-appointed defense in this case and later a Supreme Court justice, stated, "Psychiatry is given a card of admission [into the courtroom] on its own merits, and because of its own competence to aid in classifying those who should be held criminally responsible and those who should be treated as psychologically or emotionally disordered" (cited in Szasz, 1970:317). The shift in balance favoring medical definitions of deviance corresponds chronologically with the period of rapid professionalization of medicine, when medical discoveries and technology proceeded rapidly and public faith in science and medicine was increasing.

The appeal of medical definitions of deviance over legal or religious definitions is understandable. Religious definitions appear too nonrational and, in a religiously pluralistic country, they lack society-wide acceptance. Legal definitions, while more rational, appear to hinge too greatly on human decisions— the judgment of twelve ordinary citizens on a jury, for example. Medical definitions of deviance, by contrast, appear more rational and scientific.

The concept of sickness, however, far from being a neutral scientific concept, is ultimately a *moral* one, establishing an evaluation of normality or desirability (Freidson, 1970:208). A wide range of disapproved behavior has been defined by the medical profession (especially its psychiatric branch) as sick: alcoholism, homosexuality, promiscuity, drug addiction, arson, suicide, child abuse, and civil disobedience (see Conrad and Schneider, 1980). The seeming rationality of medical diagnosis thus masks the evaluative process. Jurisdiction of medicine over these behaviors does not depend on medical *knowledge* of their causes or cures, for medicine is no better able to cure alcoholism or homosexuality than can religion or law. The preeminence of medical definitions is based on the popular and juridical *acceptance* of medical authority (Bittner, 1968; Freidson, 1970:251–253).

Privatization

A foremost characteristic of the institution of medicine in modern Western societies is that it has divorced the function of curing disease from the function of providing meaning and belonging to the sick person. The medical institution has limited itself to the cure of disease (a biophysical entity) and the physical tending of the diseased person. The meaning- and belonging-providing functions of healing are treated as relatively unimportant and are relegated to the private-sphere institutions of family and religion. Thus, if a person has cancer, the physician diagnoses a "malignancy" and uses surgery, radiation, or chemotherapy. The "disease" is an abnormal mass of cells in the body. The physician typically does not address, however, the person's *illness:* what cancer means to the person, how it is experienced, how it affects the person's life, feelings, and personal relationships (Kleinman, 1973, 1978). There is some recognition of problems of meaning for cases of "mental illness," but their treatment is also segregated in separate institutions with separate specialists. Most physicians, however, see the provision of meaning as totally unrelated to the cause or healing of medical problems.

Because a high percentage (roughly half) of patients' visits to physicians are for nonspecific complaints (i.e., for which no clear biophysical basis can be found), it appears that *illness* (i.e., the social and psychological problems of the person) needs as much attention as the disease or biophysical difficulties (Kleinman, 1978:63). For example, if a thirty-three-year-old mother of five children goes to a physician complaining of pains in her legs and increased difficulty with chronic asthma but no biophysical basis for the leg pain is found, what is the physician's response? The doctor may say there is nothing wrong with her legs and nothing to be done for her, or decide that her trouble is "all in her head" and prescribe placebos (i.e., pretend pills) or antidepressant drugs; or the physician may refer her to a specialist for illnesses that are "all in the head." The physician treats her physical problems as fundamentally unreal and inappropriate for medical attention.

Two recent social developments challenge the separation of healing from the provision of meaning and belonging. One is the increasing prominence of healing in new religious movements, both outside and within the churches. Many of the new religious movements emphasize physical, emotional, and spiritual healing. At the same time, quite a few established churches have renewed interest in their healing ministry. Faith healing previously carried the connotation of being invariably limited to uneducated, lower-class sectarians; now, however, it is prominent in numerous middle-class churches and movements, such as the Catholic Charismatic Renewal, Order of St. Luke (Episcopalian), and Women Aglow (an interdenominational Pentecostal association). Faith healing is typically used in addition to medical treatment. Few middle-class healing movements recommend forgoing secular medicine; rather, they generally emphasize the limitations of modern medicine—especially its inability to heal the whole person.

The second development is the parallel rise of a number of alternative therapies. Some of these alternative therapies draw heavily on the Christian tradition; others are adaptations of Eastern practices. Many more are eclectic combinations: Christian and Western occult traditions, Eastern thought and practice, Native American healing, elements of psychotherapeutic approaches, science and pseudoscience. Alternative therapeutic approaches are widespread among middle-class, well-educated persons—sometimes as the adherents' sole belief system and sometimes in combination with more traditional religious belief systems. A study in a relatively staid East Coast suburban community found alternative therapies widespread; hundreds of suburbanites in this catchment area were found to have employed nonmedical healing, and over eighty-five different healing methods were found in use (McGuire, 1988; see also English-Lueck, 1990; Foltz, 1985; Gevitz, 1988; Glik, 1988).

It appears that while members of "modern" Western societies are utilizing secularized medicine, they are also seeking something more. A number of similarities between the new Christian healing movements and the other alternative healing systems emphasize the challenge of these movements to the secularization of health and healing. Both the Christian and alternative healing groups assert the need for *wholeness*. They emphasize that true healing depends on the recognition that social, emotional, and spiritual aspects are intimately intertwined with the physical (Beckford, 1984, 1985b; McGuire, 1993a).

Another parallel between Christian healing movements and other alternative therapies is that both proffer alternative *norms for healthiness*. Their definitions of health are far broader than those used by the medical profession. These broad definitions imply that healing is an appropriate response to a wide range of physical, emotional, spiritual, social, economic, and political "problems."

The symbolic and metaphorical significance of health and healing are high-lighted. Spiritual values such as oneness with God, balance, and harmony with the universe are critical. Movements such as these are contemporary dissent against the hegemony of the dominant medical system. They argue that the provision of meaning and belonging are central to the healing of illness.

The limitations of highly rationalized medicine are evident when we ask: What are the proper ends of modern medicine? The rationalized biomedical model operates on a highly interventionist stance against disease; accordingly, the conquest of death became a major goal of medicine (Comaroff, 1984). Although this model acknowledges that ultimately all persons will die, it also makes it difficult and sometimes impossible to *decide* that any given person will be allowed to die (see Callahan, 1987; Muller and Koenig, 1988).

Due to technological developments, physicians can prolong lives, even though often they cannot actually cure the disease. Indeed, our cultural values turn technological intervention into an "imperative": If we have the technological capability to do something, then we *should* do it. In the context of medical institutions, once a technology becomes available its use becomes almost inexorably routinized and considered standard. A physician's failure to apply this standard care—no matter how inappropriate for the individual patient—would be considered reprehensible (Koenig, 1988).

The technological imperative is deeply embedded in many institutional responses to health crises, in large part because rationalized institutions give primacy to technological and expert knowledge. In recent years, serious ethical issues have developed regarding the ends of medicine (see Callahan, 1987, 1989). Should a terminally ill person be treated aggressively until the moment of death? What medical intervention should be used for an accident victim in a persistent vegetative state, for an irreparably damaged newborn, or for a medically unsalvageable patient (such as someone who has advanced untreatable cancer who subsequently has a heart attack)? Are there limits to the use of medical responses that themselves cause pain or other suffering? Under what conditions should a patient or patient's family be allowed to refuse a prescribed treatment, such as chemotherapy or surgery?

Increasingly, such issues are coming before the courts, but the courts have typically deferred to the medical profession itself as the proper authority on matters of defining life and death. On issues such as these, however, there is no technical (i.e., purely medical) answer. Callahan (1987:179) notes, "The 'sanctity of life' has to be the sanctity of personhood, not merely the possession of a body." Moral and legal rights adhere to persons, but there are no neat biomedical criteria to distinguish when or whether a body is a person. The choice of such criteria must come from outside the rationalized domain of scientific medicine. The effects of institutional differentiation, pluralism, and rationalization in modern societies, however, make it extremely difficult to engage in the necessary ethical discourse at a societal level.

Because Western medicine is focused on the curing of "disease" rather than healing of illness, the provision of meaning is privatized and undermined. Secularized medicine is generally unable to deal with illness, the complex of perceptions and experiences of the sick person. One critic, who speaks from both a medical and anthropological background, suggests that Western physicians should be trained to treat sickness in the context of the patient's culture and psychology (Kleinman, 1978:68). Although such attentiveness to the illness experience would probably produce a more humane treatment of patients, it is doubtful that it would satisfy the patient's desire for meaning. The Western medical paradigm is based on some presuppositions that cannot be readily reconciled with the belief systems of many patients, especially in a culturally and religiously pluralistic society. Although physicians could be taught to respect the beliefs and values of, for example, Jehovah's Witnesses (a worthwhile task), it is improbable that this would overcome physicians' evaluation that their own paradigm is superior.

The Western medical paradigm denies the validity of other paradigms, including the interpretive ones provided by religion, family, and the ethnic community. The Western medical system has come to conclude and defend that health, illness, or dying has no real meaning other than the biophysical realities determinable by empirical, rational means. Provision of meaning is allowed to institutions of the private sphere, as voluntary and purely subjective interpretations, only as long as that interpretive scheme does not interfere with the biophysical treatment of the disease. Thus the differentiation and rationalization of biomedicine have resulted in the loss of a publicly shared meaning system. These processes of modernization have promoted the further privatization of religion.

RELIGION, POWER, AND ORDER IN THE MODERN WORLD

World events in recent years have highlighted the extent to which religion is an appropriate focus for study of the continuing tensions between tradition and modernity. The resurgence of fundamentalist Islam in Mideastern politics, the application of Christian liberation theologies to political situations in Latin America, and the intensification of church–state issues in the United States exemplify the relationship of religion to political and economic life in modern and modernizing societies.

The first part of this chapter described some theories of social change as they apply to the situation of religion; that discussion focused on the implications of four processes for the society and for the individual member. Some social scientists suggest that it is necessary to look beyond the borders of any given society for the full impact of modernization and related social changes. One theorist of the modern situation, Roland Robertson, describes **globalization** as the "compression of the world and the intensification of consciousness of the world as a whole" (1992:8). It is the ongoing and complex process by which "the world as a whole" comes to be experienced as a "single place" (1991a:283). Globalization does not mean harmonious integration, nor that the world is becoming a better place. The *fact* of globalization can be illustrated by the 1995 incident in which a prestigious transnational university press withdrew its contract to publish a book in response to death threats toward that press's staff in an Asian country. Euro-American academics were outraged at the abrogation of one of the most cherished modern norms, freedom of expression of ideas. The militant religious group issuing the threat was, likewise, participating in the global "single place" by violent activism to prevent publication, not merely in their own country but in all countries—regardless of their particular laws and norms—of religiously objectionable material.

In the modern world, globalized compression results from systems of communication and technologies that create an immediacy impossible to imagine even three decades ago. In writing this chapter, I was able to carry on extended conversations with several European and South American colleagues by electronic "mail" in order to be sure I had their most current writing. Images on television and in print media bring to people around the world a consciousness of the larger world. Many groups that resist westernization find that the allure of modern media technologies makes isolation of their people from global influences very difficult. Urban Iranians, for example, can use their satellite dishes to receive European and North American television programming, undercutting religious authorities' efforts to limit their exposure to a "decadent" outside world.

Modern compression is particularly due to the globalization of economic activity. Whereas in the not-too-distant past most economic activity was local or regional, increasingly the scope of economic interdependence has broadened—to whole societies, whole continents, and ultimately the entire world.

The scope of a global economic situation is difficult to imagine. It means that decisions made in distant places by unknown powerful people affect most aspects of one's own everyday economic situation. In traditional societies, most economic relationships are within the community, and transactions are handled face-to-face: Producing, buying, and selling all of one's needs take place in the immediate community.

Modern Western societies have in this century been accustomed to a larger scope; societal economies are taken for granted. When a company opens a new factory in Texas and closes its older plants elsewhere, its employees in New Hampshire may be frustrated and angry about losing their jobs, but they can usually at least envision some of the societal factors influencing their plight. It takes a considerable leap of knowledge and sociological imagination for those workers to understand increasingly global operations, such as when rising global interest rates limit their company's access to capital, or when management closes their plant—even though it remains profitable—in order to keep stock prices high and prevent capital flight, or when the company closes its U.S. operations to avoid costly pollution controls by locating across the border in Mexico. Other experiences of global economic interdependence facing Americans include the impact of the federal budget deficit, trade deficits, international commodity cartels (oil, coffee, cocoa, etc.), trade embargoes, massive foreign lending and borrowing, and vast multinational corporations with assets greater than the entire gross national product of many entire countries.

People of developing countries of the Third World have a different experience of this interdependence. Their economies were initially developed as colonies of First-World nations, and the colonial legacy often makes economic, social, and political independence extremely difficult. Furthermore, developing countries are often totally dependent on the international economic powers for markets, development loans, and imported technology. Often these countries have only a few export products (typically natural resources such as minerals) and cannot even set their own prices for the sale of these in the international economy. "Development" in these contexts frequently means only that the Third-World country allows multinational corporations to build factories, extract natural resources, affect the environment, and pay reduced taxes in return for the employment of a number of that country's unskilled or semiskilled workers. The fragility and one-sidedness of this kind of "development" are linked to total dependence on a vast world economy in which the developing countries have little or no leverage.

Compression of political systems is also inevitable in the modern world. While retaining somewhat separate national cultures, the states of Europe are thrust together to work out complex political and legal issues, partly due to their economic union, but more generally due to their need for concerted action and international interdependence. The articulation and expansion of international law is, thus, closely linked with globalization (Lechner, 1991).

Furthermore, the fact of supranational economic structures makes necessary political and judicial institutions on the same scale. As noted in the first part of this chapter, modern national polities lack the moral consensus they need if they are to wrestle with their problems; this is even more true of emerging international polities. In Europe, there are new supranational political and judicial institutions that are grappling with such issues as pluralism in a multinational economic system (see Richardson, 1995b, for some of the church–state issues in a pan-European context). Religious issues are paramount for these decisions—the meaning of life, the moral legitimacy of the social order, etc. The differentiation and privatization that characterize modern life make religious institutions *less,* not more, able to address such questions. Religion is relevant to these developments because many of the changes in the modern world (described earlier in this chapter) explain the difficulty—and importance—of dealing with issues of moral order, legitimacy, and the basis of human rights in an increasingly globalized social system.

Although the fact of globalized economic and political involvement makes supranational moral discourse *necessary,* it is extremely difficult for supranational institutions to establish norms or criteria for judgment that constituent members would accept as legitimate. The problem is acute since different religious and philosophical traditions have radically different conceptions of the individual, the nature of freedom, and the nature of rights. In some religious/cultural traditions, for example, rights adhere to a group unit such as a family, rather than to individuals as in European and American conceptions of rights. By contrast, in Euro-American thought, law, nature, and religion were separated; the individual was considered to be a legal entity separate from the collectivity, conceptualized as an autonomous person and as a citizen (Lechner, 1991). There may be grounds on which a supranational, multicultural moral consensus can be established, but at present we do not have one—a fact with grave social and religious consequences (Spickard, 1994).

In general, recent history suggests that "national societies become related to religious cultures and traditions as the economic and power structure[s] of these societies become interdependent at a global level" (Kokosalakis, 1985:368). Globalization is inextricably intertwined with modernization in the sense described earlier in this chapter: institutional differentiation, shifting bases of legitimacy of authority, pluralism, rationalization, privatization, individuation (see Beyer, 1994). Robertson argues that globalization is best understood in terms of four interdependent processes, globally generalizing: the international system, a modern form of society, a modern conception of the individual/person, and a conception of humankind. The process evokes religious and political responses largely because it is profoundly relativizing in all four dimensions (Robertson, 1991a, 1992). If one were to predict important themes for future sociological research and theory, they would include (1) the religion-politics interface in each nation; (2) various specific responses to human-existential issues such as human rights, concerns for the environment, and the equitable distribution of resources for a "good life"; (3) the social significance of individuation (i.e., the social conception of individuals as persons, as relatively autonomous actors); and (4) issues of power and order at the global level, especially the problem of legitimation (Robertson, 1981, 1989).

The nature of the linkage between religion and politics depends partly on the nature of the religious belief system(s) and political system involved. Chapters 6 and 7 suggest some of the factors influencing the relative activism of a religion in sociopolitical spheres. The connection between religion and politics varies widely from one society to another. The United States is a particularly interesting arena because of its high degree of pluralization, its Constitutional specifications of a unique relationship between church and state, and its national identity and role in world politics.

Church–State Tensions in the United States

As the degree of interdependence at the global level increases, church–state tensions at the societal level are also likely to increase. This is because church–state issues are the locus of societal struggles to articulate a national identity and moral character (Robertson, 1981:193; 1989:19). Two particular areas of tension in the United States include legal conflicts over what may be properly called "religion" and political conflicts over the religious stance of the American people and its proper role in public life.

Religious Boundaries and the Law. Two historical mandates shape the church–state relationships in the United States: (1) a firm Constitutional prohibition against the state establishing any religion as *the* religion of the land (later this was further articulated as a "wall of separation" between church and state) and (2) Constitutional protection for free practice of religion (i.e., "religious liberty"). These two central norms have been the focus for various church–state tensions throughout the history of the United States (for a compendium of Supreme Court decisions regarding the Establishment Clause and the Free Exercise Clause, see Alley, 1988).

   Two societal factors made church–state relations more problematic, especially after World War II. One major change was the increased number, socioeconomic status, and influence of American Catholics; this was the first important challenge to taken-for-granted Protestant assumptions in the nation's concept of "religion." Subsequent proliferation of diverse new religious movements, both inside and outside of Christian traditions, has confronted the state with an even more varied array of beliefs and practices claimed as "religious." The second major change is the nature of the Supreme Court itself, which has been more willing to hear church–state cases in recent decades (Demerath and Williams, 1987). Furthermore, in attempts to reassert influence in the public sphere, numerous religious groups have turned to litigation and legislation, for example, over issues of public education, abortion rights, homosexual rights, and religious communication (Pfeffer, 1992). The depth of U.S. pluralism makes it unlikely that either new laws or court decisions will satisfy these groups' wish for a decisive break with state neutrality on religion (Hammond, 1987; see Hunter, 1987a, 1991). Such politicization of religion reflects changes in the nature of the modern state (discussed below) and has parallels in other countries (such as Egypt and Israel) where orthodox religions, confronting the pluralism of the modern world, are attempting to reassert control through the law.

   As shown in the first part of this chapter, numerous precedent-setting cases have arisen over whether a given practice (e.g., polygamy, taking of hallucinogenic drugs, exemption from military service, or refusal of blood transfusion) should be allowed under the protection of the principle of religious freedom. A key criterion in such decisions was whether the practice was part of what could be considered a valid religion.

   To enforce the principle of religious liberty, therefore, the courts must define religion. In this century, in several cases regarding conscientious objection (to war), the Supreme Court adopted very broad definitions. Subsequently, however, various courts have been faced with cases requiring decisions as to whether deviant religious movements (such as the Unification Church or Children of God) are indeed "religion" and thus due protection of religious liberty (see Bohn and Gutman, 1989; Kelley, 1987; Pfeffer, 1987; Richardson, 1995a; Robbins, 1985, 1987; Robbins and Beckford, 1993; Robbins et al., 1986).

   The larger issue of religious liberty for minority religions has become particularly acute and further complicated due to the diversity of many recent immigrant groups. In 1992 the Supreme Court unanimously upheld the "free exercise" case of the Church of the Lukumí Babalú Ayé against the City of Hialeah. To prevent the church's practice of Santería (an Afro-Caribbean religion), the city had legislated prohibition of animal sacrifice—a central ritual in the church's practice. Since the ordinance did not prevent other animal killing (one justice asked about boiling live lobsters to eat in restaurants), its intent to discriminate against a particular religious practice was evident (Do Campo, 1995). Further complications arise as to which aspects of a religious group's actions deserve protection. Most modern religious organizations are extensively involved in activities that are not purely spiritual—running hospitals and schools; investing in real estate and stocks; operating TV, radio, and film studios; offering life insurance; and so on. Many court cases have had to delineate which aspects of religious organization deserve protection under the "free practice" clause.

Church–state tensions have been exacerbated by the steady increase of government regulation of a wide range of activities affecting religious groups. In recent years, religious groups have been subject to many new tax laws, labor regulations, immigration investigations, and licensing and accreditation regulations for institutions and personnel. The increasing internal power of the centralized state has made the boundary between church and state yet more problematic.

In 1985 the U.S. government charged eleven clergy and laypersons with violation of federal immigration statutes; the defendants made no secret of their intent to assist Central American refugees to enter the country illegally and to provide them "sanctuary." The Sanctuary Movement held strong religious motivations for their aid to refugees, who they believed were fleeing political oppression in countries like El Salvador and Guatemala. The movement argued that the U.S. government was breaking its own immigration law, which requires asylum to victims of political oppression in other countries. Whether the activities of the Sanctuary Movement could be defined as "religious practices" became a key issue in the court case (*United States* v. *Aguilar*) because the government's primary evidence was based on testimony from paid infiltrators. Is the free practice of religion compromised by government infiltration? Spokespersons for the government argued that the "bugged" gatherings were not really church services and prayer meetings because they included discussion of nonspiritual issues; the sanctuary workers argued that these were very much religious gatherings and that their religion was not limited to a purely spiritual realm (Wogaman, 1988).[4] In the context of increased power of the centralized state, conflicts over what is properly "religion" (and thus rightfully handled by religious groups) are also conflicts over what is properly "the state" (and thus rightfully under the aegis of the government).

---

4. The court decided to allow use of the tapes obtained by government infiltrators, but ironically the government lawyers subsequently decided not to use the tapes because they were so full of church members' testimony of their moral convictions and descriptions of atrocities from which refugees sought sanctuary that the prosecution feared the tapes would actually help the defense. Subsequently, the Presbyterian Church (U.S.A.), the American Lutheran Church, and four local congregations in Arizona filed suit over damages wrought by government infiltration of churches; the suit was supported by several other religious bodies, including the National Council of Churches.

Religion and Politics. The push of the state to expand the boundaries of its influence has brought a political response from numerous religious groups of varying theological and political persuasions. Public issues such as sex education in the schools, "gay rights," or "the right to die" have become arenas for political battles in which various religious groups assert their diverse moral judgments and argue that the state is not the proper agency for deciding the matter. Religion-oriented political activism (both conservative and liberal) is one response to the state's expansion of the scope of its control (see Robbins, 1993). Direct political power is not the only impact of modern religion, however. Religion often has significant power through its cultural influence, by which it exercises symbolic roles such as shaping the terms of political discourse (Demerath and Williams, 1990). Nevertheless, in a pluralistic society such as the United States, even pervasive cultural influence cannot be monolithic as in premodern societies.

More important than the specific legislative and regulatory influences of the modern government apparatus, however, is the sense that the modern state (in all its forms and agencies) is increasingly deciding the very purpose and direction of the society (Robertson, 1981:194). Citizens perceive the state as having greater power and influence over their lives and everyday well-being. Furthermore, many citizens sense that the extent and method of potential state control and influence are unknown and unlimited. Concern over the growing power of the government appears to be linked with the political mobilization of churchgoers in the United States. One study concluded that active churchgoers were likely to become politically involved if their attitudes toward certain issues (e.g., defense, abortion, welfare services, affirmative action) were combined with the feeling that the government was generally becoming too powerful (Wuthnow, 1988:318).

A related aspect of the interface of U.S. religion and politics is the variety of civil religions and nationalisms put forth for the American people. Internal attention to civil religion and nationalism reflects concerns about national identity, especially in a world context: Who are we, as a nation?

Answers to this question have profound religio-moral implications. Are we, as a nation, the kind of people who protect the "little guy," support ruthless dictators, retaliate against terrorists, invade countries that disagree with us, stand up for our values—and, if so, what are the values that our nation collectively holds dear enough to fight for? And internally, should ours be the kind of nation whose laws permit drinking, birth control, divorce, gambling, abortion, homosexual marriages, pornography, or possession of handguns? What values should we apply in deciding priorities for spending our national monies? Are we the kind of people who care most about health care, education, nuclear arms, space exploration, scientific research, decent housing, highways, or the environment? Public issues thus reflect the working-out of national value judgments, but rarely are the specific moral and civil religious implications of such judgments made explicit.

Numerous international events such as the holding of American hostages in the Mideast, the flexing of foreign investment "muscle" in America, and the devaluation of the dollar against other foreign currencies have made Americans more acutely aware of problems in their national identity. Many Americans had taken for granted their nation's role as an economically comfortable superpower, but events of the 1970s and 1980s threatened that self-image and, perhaps for the first time since World War II, made Americans aware of their global interdependence and consequent vulnerability. This awareness is heightened due to the immediacy of people's experience of remote events through mass media, especially television. As Americans must grapple increasingly with their situation in a global setting, religious matters are highlighted: in particular church–state conflicts, in religious groups' counterassertions against the expanding power and influence of the central state, and in civil religious and nationalistic formulations.

Power and Order in the Global Context

The twin themes of power and order have been virtually ignored in much of the study of the sociology of religion. These aspects may, however, be the most important for understanding the linkage of religion to other spheres of life at all levels: individual, societal, or global (McGuire, 1983). *Power* has a range of meanings, all of which can be applied to religion in the modern world; in understanding religion, the most important advantage of an emphasis on the concept of power is that it highlights the *intentional* production of foreseen events (Beckford, 1983). Such a focus avoids overly reductionistic interpretations and recognizes the extent to which religious action often involves the deliberate assertion of power.[5]

In the global context, the concepts of power and order highlight the underlying problem of *legitimation* (comparable to societal problems of legitimation discussed on pp. 278–286). Legitimation involves the justification of power and its uses in terms other than the simple fact of holding power. Legitimation must therefore point to values, beliefs, and ideologies; the exercise of legitimate power cannot be merely a matter of technical procedures or rationalized operations of due process, but it always functions in the context of symbols held meaningful by those who accept the exercise of power as legitimate (Kokosalakis, 1985). One good example of the interface of religion and political life at the global level is the widespread attention to issues of human rights, justice, and abuses of power. Thus, problems of moral order and the wish for transnational or transcultural norms are evident. Religious groups are uniquely poised to question the legitimacy of regimes whose continued power rests on gross violations of norms of human rights.

Another issue is relative poverty and wealth among the nations. Robertson (1987) suggests that the world economy suffers *de*legitimation precisely as the

---

5. This use of the concept of power thus differs from the more reductionistic, materialistic interpretation of the global situation given by "world systems" theorists; furthermore, "world systems" approaches treat separate nation-states as the relevant units in an *inter*national rather than global order; for a critical appraisal of this approach, see Beyer, 1994; see also Robertson, 1985, 1991b.

strong national state gains legitimation because it becomes more difficult to justify the existence of enormous economic inequalities among the nations. Because it refers to values and sources of authority that transcend the economic situation, religion is one of the foremost vehicles for both expressing dissatisfaction with the global economic situation and proffering alternatives.

To understand the role of religion in the modern world, it is therefore useful to look beyond the scope of societal changes to global changes. The foremost problematic issue appears to be legitimation of the global order. Related religion–politics tensions are also promoted by global interdependence; thus, this global perspective provides another window for understanding intrasocietal religious developments, such as church–state legal conflicts, nationalisms, the New Right, and liberation theologies.

THE SOCIOLOGY OF RELIGION IN THE MODERN WORLD

Weber closed his work on the Protestant ethic with an almost prophetic statement, wondering if society were heading toward the ultimate nullity of a totally rationalized, disenchanted world. In this chapter, we explored this and related theses about the future of religion. Following Weber and Durkheim, we have used our examination of religion to provide a picture of the modern world and the individual's place in it.

Events of recent years make the sociology of religion a particularly helpful approach to understanding this relationship between individual and society. In the second half of the twentieth century, just as some observers were predicting that religion was declining and would virtually disappear, new religious movements and new patterns of spirituality arose among the very sectors of society that were supposed to be most secularized. New forms of religiosity highlight a number of questions about the future of religion: Do they presage new modes of individual-to-society relationship? Do they highlight individual autonomy, or are they new forms of authoritarianism? Do they represent religion's potential for bringing about social change, or are they so privatized that they exemplify the disempowerment of individuals in modern society? These questions remain to be addressed by further sociological research.

Similarly, the issue of religion's role in modern societal coherence is critical. Studies of civil religions, nationalism, political messianism, and church–state relations in modern and modernizing societies will be useful to our understanding of this issue. We need to know more about fundamental processes such as legitimacy and authority in the context of modern societies and global economic interdependence. Contemporary sociology of religion has the potential to contribute to our understanding of these important topics.

Although these themes suggest that our focus must be broader than the narrow, institutionally specific sociology of religion, much is yet to be learned from the study of church-oriented religion. Religious organizations continue to be important in the lives of many individuals; their membership, organiza-tion, and basis of social support make religious institutions perhaps the most important kind of voluntary association in the United States, and their direct and indirect influences in society are still worth studying. Besides official reli-gion and religiosity, however, sociologists also need to study nonofficial reli-gion and religiosity. We have been too quick to accept the official model of religion in our research (just as sociology of medicine uncritically accepted the medical profession's definition of health and illness as the basis of its approach). More important, the sociology of religious institutions needs to ask larger questions. Rather than focus on religious organizations (e.g., parishes or de-nominations) as ends in themselves, we must ask: Of what larger phenomenon is this an example? The model of religious collectivities presented in Chapter 5 illustrates such a broader perspective. Thus, the concepts of church and sect can be seen as broad developmental models of *any* group in particular kinds of tension with its social environment.

Sociology has strived for an objective approach to social reality. But sociol-ogists of religion, while guided by this ideal, sometimes find themselves un-comfortable with its implications. Specifically, we cannot assume the superiority or necessity of religious worldviews; at the same time, however, we wonder about the direction society would take if it were utterly nonreli-gious. Without higher sources of authority for appeal, is legitimacy in mod-ern society to be based on raw power and domination? What becomes of human freedom in the face of powerful public-sphere institutions, especially those of modern scope such as huge multinational corporations, which are ef-fectively free of societal normative constraints? There is no necessary religious "solution" to these issues, but the sociology of religion is uniquely poised to raise them.

SUMMARY

The social location and role of religion (including nonchurch religion) in con-temporary society suggest that important social changes have occurred. These changes have apparently resulted in qualitatively different patterns of relation-ships and meanings. Religion, as well as other institutions, has been profoundly affected by larger changes in society. Although religion appears adaptable and vital in its numerous social forms, these larger changes raise some important questions about the continued ability of religion—and other institutions of the private sphere—to influence other important aspects of society. Particu-larly apparent is that major changes in the relationship of the individual to so-ciety have occurred, and any examination of the nature of religious change must be linked with an understanding of other major changes in the structure of society.

These processes are particularly important because of their implications for the relationship between the individual and society. Several interpretations of this relationship are represented in the key themes of religious change theories: institutional differentiation, competing sources of legitimacy, rationalization, privatization, and individuation. Assessing these projections about the individual-to-society relationship is one of the most important research tasks of modern sociology of religion.

Understanding the role of religion in modern societies requires looking beyond the level of individual or society. The interface of religion with power and order in the subsocietal, societal, and international arenas constitutes one very important item on the sociological research agenda. Due to increasing transnational economic interdependence, a global context is a helpful perspective for understanding religion and modernity. From this viewpoint, certain international and intranational developments can be understood. Recent history suggests that religion is one of the foremost forces speaking to issues of legitimation of power and moral order at the global level.

RECOMMENDED READINGS

Articles

James Beckford. "The Restoration of 'Power' to the Sociology of Religion." *Sociological Analysis* 44 (1), 1983:11–32.

Robert N. Bellah. "Religious Evolution." *American Sociological Review* 29 (3), 1964:358–374.

N. J. Demerath, III, and Rhys H. Williams. "The Mythical Past and Uncertain Future." Pp. 77–90 in T. Robbins and R. Robertson (eds.), *Church–State Relations: Tensions and Transitions*. New Brunswick, N.J.: Transaction, 1987.

Frank J. Lechner. "Global Fundamentalism." Pp. 19–36 in W. Swatos (ed.),

A Future for Religion? New Paradigms for Social Analysis. Newbury Park, Calif.: Sage, 1993.

Thomas Luckmann. "Shrinking Transcendence, Expanding Religion?" *Sociological Analysis* 50 (2), 1990:127–138.

Roland Robertson. "Church–State Relations and the World System." Pp. 39–51 in T. Robbins and R. Robertson (eds.), *Church–State Relations: Tensions and Transitions*. New Brunswick, N.J.: Transaction, 1987.

Roland Robertson. "The Globalization Paradigm: Thinking Globally." *Religion and the Social Order* 1 (1991): 207–224.

Books

James Beckford. *Religion and Advanced Industrial Societies*. London: Unwin Hyman, 1989. A critical synthesis of sociological theories about religion, modernization, and the nature of advanced industrial societies.

Peter Berger. *The Sacred Canopy: Elements of a Sociological Theory of Religion*. Garden City, N.Y.: Doubleday, 1967. Berger analyzes the impact of rational-

ization, privatism, and especially pluralism on traditional religious meaning systems.

Peter Beyer. *Religion and Globalization*. Newbury Park, Calif.: Sage, 1994. This book gives a lucid thematic explanation of the contributions to globalization theory by Wallerstein, Meyer, Luhman, and Robertson, followed by succinct applications of these themes to case studies of the New Christian Right in the U.S., Liberation Theological movements in Latin America, the Islamic Revolution in Iran, new religious Zionism in Israel, and religious environmentalism.

José Casanova. *Public Religions in the Modern World*. Chicago: University of Chicago Press, 1994. Following a theoretical discussion of the impact of modernization for public and private religions, this book examines five case studies: Spain, Poland, Brazil, Evangelical Protestantism, and Catholicism in the United States.

Thomas Luckmann. *The Invisible Religion: The Problem of Religion in Modern Society*. New York: Macmillan, 1967. Luckmann's essay explores the changing social bases of religion, suggesting that modern society's chief mode of religiosity is a privatized "invisible religion."

Glossary

anomie a crisis in the moral order of a social group resulting in the inability of the group to provide order and normative regulation for individual members

authority power that is generally accepted by subjects as legitimate, not based on coercion

caste a group status position in a society stratified by ascription

charisma extraordinary personal qualities that are the basis of an audience granting the holder legitimate authority

churchly individual orientation typified by an acceptance of ordinary levels of personal religiosity and of a diffused religious role

churchly religious group type of religious group that considers itself uniquely legitimate and that maintains a relatively positive relationship with society

civil religion any set of beliefs and rituals related to the past, present, and/or future of a people ("nation") giving that people a transcendent sense of their collective destiny

cognitive minority a group of people whose worldview differs significantly from that of the dominant society

collective representation group-held meanings expressing something important about the group itself

commitment the process by which an individual increasingly identifies with the group, its meaning system, and its goals

conversion a transformation of one's self-identity concurrent with a transformation of one's basic meaning system

cultic/mystical individual orientation typified by a striving for personal spiritual virtuosity, but acceptance of a religious role segmented from other aspects of life

cultic religious group type of religious group that accepts the legitimacy claims of other groups but that maintains a relatively negative tension with the larger society

313

denominational individual orientation typified by an acceptance of ordinary levels of personal religiosity and of a religious role segmented from other aspects of life

denominational religious group type of religious group that accepts the legitimacy claims of other religious groups and in which the group maintains a positive relationship with the larger society

deviance the recognized violation of culturally defined norms

disenchantment one aspect of rationalization; process by which things held in awe or reverence are stripped of their special qualities and become "ordinary"

dualism a worldview based on a conceptualization of reality as consisting of two irreducible modes: Good versus Evil

functional definitions of religion those approaches to studying religion that delineate the social functions religion fulfills: what religion *does* for the individual and social group

globalization a process by which the world is increasingly compressed into a single social (and/or cultural) system, together with increased social consciousness of the world as a whole

heresy a religious idea that has been socially defined by an official religious group as seriously deviant and proscribed

heterodoxy a teaching that is "other" than the official teaching of a religious group; not orthodox

ideology system of ideas that explains and legitimates the actions and interests of a specific sector (e.g., class) of society

individuation the process by which cultural and social structural arrangements come to consider each individual as a separate entity with separate concerns, especially in relation to group entities such as the family, tribe, religious group, or political and judicial institutions

inner-worldly asceticism a pattern of religiosity characterized by norms of forgoing worldly pleasure but at the same time engaging in religious action in the world

institutional differentiation process by which the various institutional spheres in society become separated from each other, with each institution performing specialized functions

legitimacy social recognition of an authority's claims to be taken seriously; it implies negative social sanctions for failure to comply with authoritative commands

legitimating myth stories that individuals and groups use to justify their values, actions, and identity

legitimation a socially established explanation that is given to justify a course of action

mazeway disintegration cultural crisis in which a group's socially constructed and learned patterns and rules for interaction no longer hold

messianism a millenarian worldview that expects the new social order to be ushered in by a divinely appointed leader ("messiah")

millenarianism a worldview based on the expectation of an imminent collapse of the entire social order and its replacement with a perfect new order

mortification the process of stripping the individual of vestiges of the "old self," while the person is becoming resocialized into a new identity consistent with that group's beliefs and values

nonofficial religion set of religious and quasi-religious beliefs and practices that is not accepted, recognized, or controlled by official religious groups

occultism a worldview based on a set of claims that contradict established (i.e., official) scientific or religious knowledge and that typically emphasize "hidden" teachings

official religion set of beliefs and practices prescribed, regulated, and socialized by organized, specifically religious groups

orthodoxy   teaching that conforms to the official teaching of a religious group; not heterodox or heretical

orthopraxis   practice (such as a ritual) that conforms to the official norms for practice of a religious group

paranormal occurrences   events outside the usual range of experiences

particularism   a worldview that holds that one's own group's beliefs and practices are the only legitimate ones

plausibility structure   social processes within a network of persons sharing a meaning system that enable those persons to accept that meaning system as taken for granted and believable

pluralism   a state in which cultural subgroups (e.g., religious or ethnic groups) are given formally equal social standing; no single group holds a monopoly in the definition of beliefs, values, and practices

polity   pattern of arrangements for the exercise of legitimate authority in an organization

power   the ability to achieve desired ends despite resistance from others

priest   a religious role characterized by authority of "office," a functionary in the status quo official religion who celebrates its rituals, practices, and beliefs (in contrast to the religious role of the prophet)

privatization   process by which certain institutional spheres (including religion) become removed from effective roles in the public sphere

profane   that which is ordinary, part of everyday life [note: another usage, in common parlance, is as the opposite—even pejorative or negative-reverse—of "sacred"; i.e., that which defiles or pollutes what is defined as "sacred"]

prophet   a religious role characterized by charismatic authority and religiously based criticism of a society's established authorities; there are two kinds:
   emissary prophet   one who confronts the established powers as one who is sent by God to proclaim a message

exemplary prophet   one who challenges the status quo by living a kind of life that exemplifies a dramatically antiestablishment set of meanings and values

proselytization   the act of trying to persuade nonbelievers to become believers

rationalization   the change from tradition to rationality as the basis for thought (e.g., science), authority (e.g., rational law), and social organization (e.g., specialization and bureaucracy)

relative deprivation   a perceived disadvantage based on comparisons with other people's social resources (material or nonmaterial)

rites of passage   rituals that accompany a change of social status such as age-grade, office, marriage status, etc.

ritual   symbolic actions that represent religious meanings

routinization of charisma   the transformation of charismatic authority into some other basis of authority, such as tradition or the authority of office

sacred   that which is defined as extraordinary, awesome, and powerful—whether revered or feared, and set apart (from the profane) as holy

sectarian individual orientation   typified by a striving for personal spiritual virtuosity (together with rejection of ordinary levels of religiosity) and emphasis on a diffused religious role

sectarian religious group   type of religious group that considers itself to be uniquely legitimate and in which the group maintains a relatively negative tension with the larger society

sectarianism   a worldview that emphasizes the unique legitimacy of believers' creed and practices and that heightens negative tension with the larger society by engaging in boundary-maintaining practices

self-identity   each person's biographical arrangement of meanings and interpretations that form a somewhat coherent sense of "who am I?"

social change   the transformation over time of culture and social institutions

social control   all the ways by which members of a culture encourage and enforce conformity to cultural norms

stratification   the ranking of categories of people in a society into a hierarchy of prestige and privilege

substantive definitions of religion   those approaches to studying religion that delineate what religion *is*—what qualities and content are essentially religious

syncretism   beliefs and practices that result from the amalgamation of two or more diverse cultural elements or traditions

theodicy   religious explanation that provides meaning for meaning-threatening experiences such as death, disaster, and poverty

virtuosi, religious   those who strive for religious perfection (however defined) and are not satisfied by the normal levels of religiosity of the masses

worldview   a comprehensive meaning system, locating all experiences of the individual or social group in a single general explanatory arrangement

Appendix A

◦⟨⟩◦

Doing a Literature Search in the Sociology of Religion

1. The primary current sociology of religion journals and annuals with articles in English are:

Archives de Sciences Sociales des Religions. France, 1956 to date. Many articles in English.

Journal for the Scientific Study of Religion. United States, 1961 to date.

Journal of Contemporary Religion (formerly *Religion Today*). England, 1985 to date.

Religion and American Culture. United States, 1991 to date.

Religion and the Social Order (annual). United States, 1991 to date.

Religion, State and Society. England, 1992 to date.

Research in the Social Scientific Study of Religion (annual). United States, 1989 to date.

Review of Religious Research. United States, 1959 to date.

Social Compass. Belgium, 1953 to 1989; United States, 1990 to date. Articles in English and French.

Sociology of Religion (formerly *Sociological Analysis*). United States, 1939 to date.

Also, *Current Sociology* (International Sociological Association) has occasional entire issues with annotated bibliographies on religion.

2. These journals are indexed or abstracted in:

America: History and Life.

Expanded Academic Index. I.A.C., electronic search index.

Religion Index (previously *Index to Religious Periodical Literature*), hard copy and CD-ROM.

Social Sciences Index (previously *Social Science and Humanities Index*).

Sociological Abstracts, hard copy and CD-ROM.

3. Newspapers with frequent good coverage of religion include:

Christian Science Monitor.

Los Angeles Times.

National Catholic Reporter, indexed in *Catholic Periodical and Literature Index.*

New York Times, indexed in *New York Times Index.*

4. Newsletters with cross-references to current research:

Millennial Prophesy Report, P.O. Box 34021, Philadelphia, PA 19101-4021.

Religion Watch, P.O. Box 652, North Bellmore, NY 11710.

5. Dictionaries and encyclopedias with specialized coverage of religion:

Encyclopedia of African American Religion, 1993, L. G. Murphy, J. G. Melton, and G. L. Ward, eds. New York: Garland.

Encyclopedia of American Religions (2 volumes), 1978, J. G. Melton (ed.). Wilmington, Del.: Consortium.

Encyclopedia of Religion (16 volumes), 1987, M. Eliade (ed.). New York: Macmillan.

Encyclopedia of the American Religious Experience: Studies of Traditions and Movements (3 volumes), 1988, C. H. Lippy and P. W. Williams (eds.). New York: Scribner.

Encyclopedic Handbook of Cults in America, 1992, J. G. Melton (ed.). New York: Garland.

New Age Encyclopedia, 1990, J. G. Melton, J. Clark, and A. A. Kelly. Detroit: Gale Research.

6. Bibliographies:

Baer, Hans A. "Bibliography of Social Science Literature on Afro-American Religion in the United States." *Review of Religious Research* 29 (4), 1988:413–430.

Blasi, Anthony J., and Michael W. Cuneo, eds. *Issues in the Sociology of Religion: A Bibliography.* New York: Garland, 1986. This bibliography, while containing little annotation, is an excellent *catalogue raisonné,* well organized by sociological themes. It includes good coverage of international sources as well as the main English-language journals, but some references (such as unpublished dissertations) will be difficult to obtain.

Blasi, Anthony J., and Michael W. Cuneo, eds. *The Sociology of Religion: An Organizational Bibliography.* New York: Garland, 1990. This second volume is categorized according to religious traditions (e.g., Asian, Christian, etc.) and organizations.

Brunkow, Robert de V., ed. *Religion and Society in North America: An Annotated Bibliography.* Santa Barbara, Calif.: Clio Press, 1983. This bibliography contains useful sources mainly for historical studies of religion.

Daniels, Ted. *Millennialism: An International Bibliography.* New York: Garland, 1992. Idiosyncratic selection of articles for annotated bibliography; annotations are readable and include useful cross-references. Highly opinionated, not usually unfair, but certainly one sided. Nonannotated bibliography in second half is too long, unfocused, and nonselective.

Fraker, Anne T., ed. *Religion and American Life: Resources.* Urbana: University of Illinois Press, 1989. This bibliography focuses on the social history of American religion, especially Protestantism; useful sources on church–state issues. Because the annotations are long (and useful), the volume covers relatively few sources.

Lippy, Charles H. *Bibliography of Religion in the South.* Macon, Ga.: Mercer, 1985. Large number of references, but lacks depth and focus.

Murphy, L. G., J. G. Melton, and G. L. Ward, eds. *Encyclopedia of African American Religion.* New York: Garland, 1993. Rather narrowly limited to African-Americans in religious history; organized by the names of religious organizations and important historical figures, so it is of limited use for understanding other aspects of African-American religion. Includes a brief but useful bibliography and a directory of the primary extant religious organizations among African-Americans.

Saliba, John A. *Social Science and the Cults: An Annotated Bibliography.* New York: Garland, 1990. Fair, but very bland paragraph on each article, book, etc. Indexed by the name of author and movement, but not much else—unless a theme happens to be mentioned in the title; not useful for thematic or comparative research.

Social Compass (see journal listing above) publishes (usually in number 3 of each volume) an annual international bibliography of journal sources in sociology of religion; about two-thirds of the articles indexed are not in English. Arranged by broad subject areas, these references are useful for comparative, cross-cultural research and for current theory in the field.

Wilson, John F. *Church and State in America—A Bibliographic Guide* (2 volumes). Westport, Conn.: Greenwood, 1986 and 1987. Very useful compendium of references, topically organized, on the history of church-state relations in the United States.

Wolcott, Roger T., and Dorita Bolger. *Church and Social Action: A Critical Assessment and Bibliographical Survey.* New York: Greenwood, 1990. An excellent bibliography of comparative and historical studies on religious groups in social action; emphasis on Christian groups, but includes others (such as Southeast Asian religious activism).

Appendix B

◦◯◦

Teaching Resources

I. A GUIDE FOR BEGINNING ETHNOGRAPHIES OF LOCAL RELIGIOUS GROUPS

Visiting local religious groups provides a practical feel for religious life. It is also useful for developing a "sociological eye"—a key part of understanding the sociological approach to religion. This observation guide will help undergraduate sociology of religion students focus their first observations, and more advanced students will find it useful for beginning extensive ethnographic research in a religious context. You will, however, need to adapt this guide to your specific research objectives and the particular kind of religious group you visit.

I usually assign my classes as follows:

Attend the worship service of two religious groups other than your own. Try to observe relatively unfamiliar religious traditions to avoid too many preconceived notions. It may help to go with a friend who is more familiar with that religion and can help you know how to behave and what to expect, but remember that "insiders" have assumptions about what is going on that the observer does not want to take for granted. If you go with others, keep your groups limited to three persons to avoid being obtrusive. Each student will submit independent field notes of the event, to be written up immediately after the observation.

If the religious group is large and its services public, no permission is needed, but in smaller groups (e.g., a small Pentecostal church or a meditation circle), it's a good idea before attending to phone and ask consent to attend for course purposes. You should also ask consent of any group whose worship services may not be fully public.

Every religious group has norms about behavior before, during, and after worship services. Established religious groups have somewhat standardized expectations that any visitor ought to try to meet. A useful guide is *How to Be a Perfect Stranger: A Guide to Etiquette in Other People's Religious Ceremonies* (edited by A. J. Magida, Jewish Lights Publishing, Woodstock, Vermont, 1996). It contains a little background about the religion and its worship, recommended clothing (e.g., is head covering required, is jewelry okay, how dressed up must one be), how the sanctuary is arranged, where guests should sit, when not to enter, in which services guests may participate, and which are for members only. The guide includes Buddhist, Christian Science, Greek Orthodox, Hindu, Islam, Jewish, Mormon, Roman Catholic, and Quaker worship, as well as several mainline, pentecostal, and other Protestant forms of worship. If the group you plan to visit is not described, you might ask friends who are from that tradition to tell you what to expect. Most groups also welcome advance inquiries from prospective visitors; you should find out (minimally) when to arrive, how to dress appropriately, and where visitors should sit or stand. Usually they will anticipate any other preparations you ought to know about and welcome the chance to tell you in advance.

You will get the most out of this experience if you imagine you are a beginning professional researcher and these visits are only the first stages of a serious research project on this and similar religious groups. Before attending, think about what you intend to observe, and keep in mind your research objectives: What do you hope to learn about this group from observations? Some professors instruct their students to observe for particular aspects on each visit. For example, Jim Spickard's classes at the University of Redlands are assigned to observe specific religious groups. At one, they observe for institutional organization (specialized roles, norms for the organization of the service, etc.); at another, they focus on plausibility structures and commitment mechanisms (e.g., social features such as social class or ethnic homogeneity, ways that groups enmesh members in social activities); at another, they observe for boundary-maintaining activities and the group's images of "the World." Each chapter of this text includes several relevant themes on which to focus your observation.

Being clear about your objectives will enable you, like a professional researcher, to be alert to important occurrences that you might otherwise miss. Keeping these objectives in mind during the observation will help you stay flexible in order to adjust your plans midstream if you are surprised by unexpected events or changes or if you have unanticipated opportunities to learn more. For example, one student wanted to understand a religious group's social activism. At her first visit, not only the sermon but several contributions by members of the congregation expressed strong political and moral sentiments and urged activism. After the service, the family sitting beside her (to whom she had previously introduced herself as a student visitor) asked her how she liked the worship. Keeping her objectives in mind, she was able to ask relevant questions from these willing informants: Was today's service pretty typical? Are most folks here active in some of the church-sponsored projects, like the one on homelessness the pastor mentioned? What do you think a visitor like me ought to know to appreciate your congregation and its worship better?

Also before going, think about the ethical and methodological issues of doing participant observation. One important ethical rule of thumb is to never present yourself to others as something other than your real identity. This is not usually an issue for undergraduates who may attend one or two services as a part of a course assignment. If asked, you may simply say, "I'm a student at _____ College, and my professor assigned us to visit any religious group other than our own, and since I was most interested in learning more about your religion, I came here today." One student who attended an ethnic congregation's service, although he was clearly not of the same ethnic group, told a curious member: "My professor assigned us to visit some religious group other than my own, and I live near here and noticed what an active church you seem to have, so I wanted to visit this church"—all of which was true, and provided a perfect opening for the member to then tell the student a lot more about the congregation, its activities, and why there was a strong sense of "belonging" in that church.

If you were beginning a long-term study of a religious group, however, this same norm would require that you not disguise your research intent. Thus, fairly early in your research project you would need to negotiate with the appropriate leaders to obtain permission to do research. You don't have to explain all your research objectives, but you should be prepared to make serious assurances to the group not only of confidentiality—which applies to all social research—but especially of an earnest effort to treat their valued beliefs and practices with full respect and consideration.

This respect is, however, also an intrinsic part of doing good participant observation. The best participant-observer is one who can comprehend enough of the group's beliefs, practices, and experiences that it all becomes perfectly plausible as a way of being religious. There are thoughtful methodological points in several of the ethnographic monographs recommended as reading in Chapters 3 and 4. See also the long methodological notes in my ethnographies, *Pentecostal Catholics* (Temple University Press, 1982) and *Ritual Healing in Suburban America* (Rutgers University Press, 1988). Good ethnographic research requires considerable skill and methodological precision. For a beginning undergraduate, this can be a fun learning experience; advanced students will ideally be honing these more difficult skills.

A hint for researchers doing a full-fledged ethnography of the group: When I write up my field notes, I have found it very useful to distinguish paragraphs of "Observation Notes" (O.N., purely descriptive notes) and "Methodological Notes" (M.N., reflections about methodology linked to the adjacent observation notes) from "Theoretical Notes" (T.N., notes to myself about possible interpretations, correlations, and "big-picture" questions for further thought, observation, and analysis). Periodic review of these notes reminds me to try different ways of methodologically tapping certain features of group life and to be open to alternative interpretations of what I have been observing. The separation of other notes to myself from the more neutral descriptive observation notes enables me to be more reflexive—not only to deal with possible biases in the "eye of the beholder" but also to tease out potentially important understandings. For example, many years ago I made the methodological note that a particular part of the service of a group I was studying made me very uncomfortable, but it took me two or three more visits to identify why: Certain postures and gestures at that part of the service made me uneasy, not only because they were unfamiliar in my own religious experience, but—more importantly—because they literally embodied (i.e., expressed and impressed on each participant's body) certain religious values and attitudes that I did not personally agree with. That awareness resulted—in the short term—in my finding a way to participate that did not go against my personal convictions yet did not violate the group's norms, and—in the long term—in a stimulus for studying much more about the connection between the body and religion.

Observation Guide[1]

Guidelines for observations: Some of these must be recorded for each observation; others are just reminders of interesting features to observe in detail. *DO NOT TAKE NOTES DURING A SERVICE,* but write up your observations from memory *immediately* afterward (I usually jot rough notes as soon as I get to my car, then flesh them out as soon as I get back to my computer). Do not underestimate how much you will forget quickly, and details count in field notes.

I. Identification:

Group's name; religion or denominational affiliation (if any)

Address

Date and time attended

II. Physical features:

   a) Describe surrounding neighborhood; any indications of whether members are from that neighborhood?

---

1. This guide, developed originally for my research assistants in 1976 and 1981, and later used by my classes at Trinity University, has been amplified with material suggested by guides used by students of Jim Spickard at University of Redlands and Steve Warner at University of Illinois—Chicago Circle. Thanks to these colleagues and other participants in "Teaching Sociology of Religion" sessions of the Society for the Scientific Study of Religion.

b) Size, architectural style, and condition of buildings and their exterior (e.g., elaborate landscaping, bell-tower); Would you know the buildings were used for religious purposes just judging from the exterior? How?

c) Describe socioeconomic indicators about attenders (types of cars in parking lot, style and condition of clothing, jewelry)

d) Material culture:

—Describe the area where the worship service is held—its shape, mood-setting features (e.g., colors, lighting), seating, kneeling, or other space arrangements; is any area raised or otherwise highlighted? Describe that area's arrangement.

—What religious symbols are evident? Stained glass windows, candles, incense and other sensory mood-setting features, music (instruments, sound systems, hymnals), flags, posters, banners, vestments and robes, flowers, hymnals, etc.

III. Social features—members/attenders:

a) Approximate total attendance; estimate breakdown of total by age, gender, race, or ethnicity

b) What proportion of the congregation appears to attend in family units, couples, or singly? Do they sit together and participate in the service as a family unit or couple? If not, what seems to be the basis for where people sit and what roles they take during the service? Is the seating according to sex, age, or religious status, for example?

c) Congregation behavior during service:

—Subdued or enthusiastic? Loud or quiet?

—Lots of congregation participation or little?

—Describe how the congregation participates during the service: silently following service by reading in a prayer book, singing, "amens," shouts, responsive readings, handclapping, dancing, meditating, chanting, etc.?

—How much does the congregation participate in any of the ritual actions, if any (for example, prayers in unison)? From memory? From instructions in book? From instructions from service leader?

—Do members of the congregation use any special postures or gestures during the worship service?

—What are the norms for behavior of members (i.e., what behavior would be so out of keeping with the group's expectations that it would result in sanctions)?

—Behavior of children (if present)?

—What language(s) do participants use before, during, and after the worship?

—Describe social interaction between members before, during, and after the worship service. Do members appear to know each other personally?

IV. Social features—specialized roles:

a) Describe the specialized roles during service that some persons took: role of minister or service leader, assistants (may be several subspecialists), music roles (often several different ones), ushers and greeters, educators, etc. How professionalized are these roles?

b) Were there other specialized roles that appear to be more "spontaneous" (e.g., members of the congregation who regularly contribute hymn requests or prophesy in tongues)?

c) Do you see any patterns in the age, gender, race, or ethnicity characteristics of various leaders?

d) Describe the styles of key leader roles: subdued, flamboyant, dynamic, quiet, peppy, authoritative, stern, etc.

—Describe their clothing, demeanor, and speech styles: lots of use of street talk, formal "thees" and "thous," literate vocabulary, etc.

e) Describe leaders' speaking roles: sermon, prophesying, speaking in tongues, giving testimonies, congregational prayer requests, etc.

f) Describe interaction between the key leaders and congregation before, during, and after the worship service.

g) What indications do you observe of the exercise of authority (by leaders and/or by participants) in this group? What appear to be the bases of their authority?

V. The Worship Service:

a) Did they follow a planned order of worship? Did parts of the service follow a liturgical order (e.g., preset words and actions that are followed each time they hold that kind of worship service)?

b) How much of the leaders' actions and words appeared to be spontaneous? How much of the congregation's participation appeared to be spontaneous?

c) Describe the use of ritual actions, symbolic gestures or postures, recitations, reading of sacred texts, symbolic uses of space or time, etc.

d) How long was the service?

e) Did they take an offering? What did they say was its purpose?

f) Did they treat a certain portion of the time or certain activities as more important or more "sacred" than other parts of the service?

g) In general, what was the mood or spirit of the service? What features created that sense?

h) To what features of the service did the congregants respond most heartily?

—What seems to make for a "successful" or "good" worship service in this group?

—If the service seemed disappointing to members, what might have contributed to this lack of success?

—If you were a visiting anthropologist from a totally different culture, how would you identify this gathering as a religious one?

VI. Debriefing: How did you personally feel during this visit?

—Were there specific features or occurrences that made you feel uncomfortable?

—What limitations to the adequacy of your field notes and observation do we need to keep in mind?

VII. Notes:

—If you were beginning an extensive ethnographic study of this and similar groups, what would be some features to watch for in future observations?

—Some interpretive hypotheses worth pursuing?

II. FILM RESOURCES

Battle for the Bible. PBS ("God and Politics" series), 1987, 60 minutes, color. Illustrates a contemporary conflict between Christians who want to enforce a conservative orthodoxy in their denomination versus others who want to maintain their denomination's tradition of freedom of conscience for individual believers.

The Bible Belt: A CBS White Paper— The Politics of the Second Coming. Canadian Broadcasting Company, 1972, 90 minutes, color. Examines the rise of fundamentalist Protestant sects in western Canada during the 1920s and 1930s and their impact on Canadian politics then and now.

Born Again: Life in a Fundamentalist Baptist Church. James Ault Films, 1987, 87- and 58-minute versions, color. A sensitive documentary portraying the worldview and daily life of a fundamentalist community in Massachusetts.

Bronx Baptism. Icarus Films, 1980, 27 minutes, color. Film document, without narration, of a Pentecostal worship service in a Puerto Rican community in the South Bronx. Fine presentation of Pentecostal worship styles.

Father, Son and Holy War. Icarus, 1994, 2 parts, 60 minutes each, color. Explores the connection between religion, violence, and male identity, focusing on ethnoreligious and gender relations in India. Part I includes sati (suttee), ritual practices supporting caste distinctions, women's rights, ethnoreligious strife, and efforts for harmony; Part II emphasizes the role of religion in the cultural construction of male identity and how that is linked with war and other violence.

Flight of the Dove. University of California Media, 1989, 29 minutes, color. Religion and ethnic identity among Portuguese-Americans in rural southern California.

Holy Ghost People. McGraw-Hill, 1968, 53 minutes, black and white. Describes beliefs and practices of a snake-handling Pentecostal church and shows candid shots of congregation during service, including snake handling and glossolalia. Dramatic ending when the leader is bitten by a rattlesnake.

Holy Terror. CGuild, 1986, 58 minutes, color. Portrays the religious legitimation of New Right (especially antiabortion) activism of the 1980s.

The Hutterites. National Film Board of Canada, 1964, 28 minutes, black and white. Documentary (without shooting restrictions) of life in a Hutterite colony in western Canada (see also a color documentary, *Hutterites,* Canadian Broadcasting Company, Buller, 1984, 59 minutes).

In the Good Old Fashioned Way. Appalshop, 1993, 29 minutes, color. The Old Regular Baptist Church keeps alive, in Appalachia, its old religious practices such as foot washing, riverside baptisms, and expressive language and song.

In the Name of God. Icarus, 1992, 90 minutes, color. Illustrates how religious intolerance fanned by militant religious groups has results in violence, bloodshed, and destruction in India; filmed before the destruction of a sixteenth-century mosque by Hindu militants, it raises important questions about the coexistence of intolerant religious groups and the future of India, which has tried to maintain a neutral, secular state.

The Journey: From Faith to Action in Brazil. Icarus, 1984, 29 minutes, color. Shows the actions of Basic Ecclesial Communities in a poor neighborhood of Rio de Janeiro and includes interviews with the bishops of the area, as well as pastoral agents and residents.

Keeping the Faith. PBS ("Frontline" series), 1987, 58 minutes, color. Depicts black churches (with particular focus on one middle-class congregation) in a midwestern city as sources of vitality, activism, community, and identity.

The Long Search. BBC/Time/Life, series of 13 videotapes, 1977, 52 minutes each, color. A documentary on world religions and new religions, narrated by Donald Eyre (who is irritatingly obtrusive in several instances). Especially useful are *Orthodox Christianity—The Rumanian Solution,* which is very useful for illustrating rich symbolism and ritual, the pervasiveness of religion in people's lives, and the place of religion in one Communist country; and *African Religion: Zulu Zion,* which focuses on new religions in South Africa, emphasizing the importance of dreams, ancestors, and place. Other useful films in the series include *Protestant Spirit: U.S.A.; Catholicism;* and *Judaism.* The film on *Alternate Lifestyles in California* is disappointingly shallow.

Martin Luther King: A Man of Peace. Journal Films, 1968, 30 minutes, color. Shows portions of sermons, speeches, and interviews with Martin Luther King, Jr., linking his philosophy of nonviolent resistance with biblical themes.

My Town—Mio Paese. University of California Media, 1989, 26 minutes, color. Religion and ethnic identity among Massachusetts immigrants from Palermiti, Italy; focuses on the celebration of the patron saint's feast day.

The New Klan. Corinth Films, 1978, 58 minutes, color. Vivid images of the contemporary Ku Klux Klan. Useful for discussion of symbols and rituals in a quasi-religious secret society.

La Ofrenda: The Days of the Dead. Xochil Film Productions, 1989, 50 minutes, color, Spanish and English with English subtitles. This film illustrates the syncretism and communal values underlying this popular religious celebration in Mexico and Mexican-American communities in the United States.

Onward Christian Soldiers. Icarus, 1989, 52 minutes, color. Portrays inroads made into traditionally Catholic Latin American communities by evangelical Protestant preaching through the mass media.

The Performed Word. Center for Southern Folklore (Memphis), 1982, 58 minutes, color. African-American religious styles in their cultural context especially sermon music and preaching styles.

La Promesa. Center for Latin America, University of Wisconsin, Milwaukee, 1995, 26 minutes, English subtitles. Excellent illustration of the blending of official (Roman Catholic) and nonofficial (both popular Catholicism and Afro-Caribbean) elements that represents, all at once, religious pilgrimage, faith, celebration, and expression of social discontent in Cuba.

The Radio Priest. PBS, 1989, 58 minutes, color and black and white. An interesting documentary about Charles Coughlin, the controversial right-wing Catholic radio preacher of the 1930s.

Religion and Race in America: Martin Luther King's Lament. Films for the Humanities, 1994, 60 minutes, color. Examines the segregation of churches, as well as interracial congregations.

Romero. Cinecom Distributors, 1989, 105 minutes, color. Feature film about Archbishop Oscar Romero, who was assassinated in 1980 by agents of the military junta of El Salvador because of his stand against the social injustice and oppression.

Sanctuary. PBS ("Frontline" series), 1983, 60 minutes, color. Examines the 1980s Sanctuary Movement, which engaged in civil disobedience to Immigration Service regulations by giving assistance to refugees from political oppression in Central America.

Saturday Night, Sunday Morning. Resolution/California Newsreel, 1992, 70 minutes. With apt historical background, shows the connections between African-American religious expression and musical heritage, especially the blues; useful depiction of black churches in the South.

The Shrine. University of California Media, 1989, 46 minutes, color. Depicts the Holy Week pilgrimage to El Santuario de Chimayo in New Mexico; very useful for appreciation of popular religion, the time–space structure of pilgrimages, and the importance of sacred place both in America's Latino and Native American cultural heritages.

Shrine Under Siege. Icarus, 1987, 50 minutes, color. The Temple Mount in Jerusalem is the site of increasingly violent clashes between Moslems (whose centuries-old mosque there is one of the holiest sites in Islam) and Jews (who wish to rebuild the Jewish Temple on the same site); into the volatile mixture also go the money and energies of many American fundamentalists (who believe that Christ's second coming will follow the rebuilding of the Temple).

Solidaridad: Faith, Hope and Haven. Icarus, 1989, 57 minutes, color. A documentary about the work of the Vicaria de la Solidaridad, a Catholic organization in Chile that has provided support for victims of human rights abuses.

The Spirit of Kuna Yala. Filmmakers Library, 1990, 59 minutes, color. Although not explicitly about religion, this film's depiction of the Kuna Indians (of Panama's San Blas Islands) in relation to their rain forest homeland illustrates the importance of religion, among many cultural values, in resisting westernization and cultural extinction.

Thank God and the Revolution. Icarus, 1981, 50 minutes, color. Outstanding illustration of many theories about the role of religion in social change in Nicaragua. It includes Liberation Theology, basic Christian communities, brief historical background, and interviews with people from all strata of society—rural villagers to cabinet officials. The folk hymns in the soundtrack are moving.

Triumph of the Will. McGraw-Hill, 1936, 120 minutes, black and white. A Nazi propaganda piece documenting the Sixth Party Congress in 1934. Very useful for discussing the nature of ritual, political messianism, and whether movements like Nazism under Hitler are "religion."

Two Black Churches. Center for Southern Folklore (Memphis), 21 minutes, color. Images of black religious expression in two very different churches: one in rural Mississippi, the other in urban New Haven, Connecticut. Includes inspired preaching, faith healing, glossolalia, ecstatic dance.

References

Abercrombie, Nicholas, John Baker, Sebastian Brett, and Jane Foster
   1970   "Superstition and religion: The god of gaps." Pp. 93–129 in D. Martin and M. Hill (eds.), *Sociological Yearbook of Religion in Britain*. Volume 3. London: SCM.

Aberle, David
   1962   "A note on relative deprivation theory as applied to millenarian and other cult movements." Pp. 209–214 in S. Thrupp (ed.), *Millennial Dreams in Action*. The Hague: Mouton.

   1966   *The Peyote Religion Among the Navaho*. Chicago: Aldine.

Acquaviva, Sabino
   1993   "Some reflections on the parallel decline of religious experience and religious practice." Pp. 47–58 in E. Barker, J. A. Beckford, and K. Dobbelaere (eds.), *Secularization, Rationalism and Sectarianism*. Oxford: Clarendon.

Adriance, Madeleine
   1986   *Opting for the Poor: Brazilian Catholicism in Transition*. Kansas City, Mo.: Sheed and Ward.

   1991   "Agents of change." *Journal for the Scientific Study of Religion* 30, 3:292–305.

   1994   "Base communities and rural mobilization in northern Brazil." *Sociology of Religion* 55, 2:163–178.

Ahlstrom, Sidney
   1972   *A Religious History of the American People*. New Haven, Conn.: Yale University Press.

   1978   "From Sinai to the Golden Gate: The liberation of religion in the Occident." Pp. 3–22 in J. Needleman and G. Baker (eds.), *Understanding the New Religions*. New York: Seabury Press.

Aho, James A.
  1990   *The Politics of Righteousness: Idaho Christian Patriotism.* Seattle: University of
          Washington Press.

  1994   *This Thing of Darkness: A Sociology of the Enemy.* Seattle: University of Washing-
          ton Press.

Albanese, Catherine L.
  1990   *Nature Religion in America: From the Algonkian Indians to the New Age.* Chicago:
          University of Chicago Press.

  1991   *America: Religions and Religion.* Belmont, Calif.: Wadsworth.

  1992   "Dominant and public center: Reflections on the 'one' religion of the United
          States." Pp. 202–217 in M. Marty (ed.), *Civil Religion, Church and State.* Mu-
          nich: K. G. Saur.

Alfred, Randall
  1976   "The Church of Satan." Pp. 180–202 in C. Glock and R. Bellah (eds.), *The
          New Religious Consciousness.* Berkeley: University of California Press.

Alley, Robert S. (ed.)
  1988   *The Supreme Court on Church and State.* New York: Oxford University Press.

Allon, Natalie
  1973   "Group dieting rituals." *Society* 10, 2:36–42.

Aman, Kenneth
  1988   "Popular culture and religion in Chile." *Christian Century* 105:766–769.

Aman, Kenneth, and C. Parker (eds.)
  1991   *Popular Culture in Chile: Resistance and Survival.* Boulder, Colo.: Westview.

AmaraSingham, Lorna Rhodes
  1978   "The misery of the embodied." Pp. 101–126 in J. Hoch-Smith and A. Spring
          (eds.), *Women in Ritual and Symbolic Roles.* New York: Plenum.

Ammerman, Nancy Tatom
  1987   *Bible Believers: Fundamentalists in the Modern World.* New Brunswick, N.J.: Rut-
          gers University Press.

  1990   *Baptist Battles: Social Change and Religious Conflict in the Southern Baptist Conven-
          tion.* New Brunswick, N.J.: Rutgers University Press.

  1991   "North American Protestant Fundamentalism." Pp. 1–65 in M. E. Marty and
          R. S. Appleby (eds.), *Fundamentalisms Observed.* Chicago: University of Chicago
          Press.

Anderson, Alan, and Raymond Gordon
  1978   "Witchcraft and the status of women—the case of England." *British Journal of
          Sociology* 29:171–184.

Anderson, Benedict
  1991   *Imagined Communities: Reflections on the Origin and Spread of Nationalism.* London:
          Verso.

Anderson, Robert Mapes
  1979   *Vision of the Disinherited: The Making of American Pentecostalism.* New York: Ox-
          ford University Press.

Annals of the American Academy of Political and Social Science
  1979   Volume 446: *The Uneasy Boundary: Church and State.*

Anthony, Dick
  1990   "Religious movements and brainwashing litigation: Evaluating key testimony."
          Pp. 295–344 in T. Robbins and D. Anthony (eds.), *In Gods We Trust: New Pat-
          terns of Religious Pluralism in America.* New Brunswick, N.J.: Transaction.

Anthony, Dick, and Thomas Robbins
  1975   "From symbolic realism to structuralism." *Journal for the Scientific Study of Religion* 14:403–414.

  1978   "The effect of detente on the rise of new religions: The Unification Church of Reverend Sun Myung Moon." Pp. 80–100 in J. Needleman and G. Baker (eds.), *Understanding the New Religions*. New York: Seabury Press.

  1990   "Civil religion and recent American religious ferment." Pp. 475–502 in T. Robbins and D. Anthony (eds.), *In Gods We Trust: New Patterns of Religious Pluralism in America*. New Brunswick, N.J.: Transaction.

  1992a "Law, social science and the 'brainwashing' exception to the First Amendment." *Behavioral Sciences and the Law* 10:5–29.

  1992b "Spiritual innovation and the crisis of American civil religion." Pp. 27–49 in M. Marty (ed.), *Civil Religion, Church and State*. Munich: K. G. Saur.

  1995   "Negligence, coercion, and the protection of religious belief." *Journal of Church and State* 37, 3:509–536.

Aran, Gideon
  1991   "Jewish Zionist fundamentalism: The Bloc of the Faithful in Israel (Gush Emunim)." Pp. 265–344 in M. E. Marty and R. S. Appleby (eds.), *Fundamentalisms Observed*. Chicago: University of Chicago Press.

Arjomand, Saïd Amir
  1984   *The Shadow of God and the Hidden Imam: Religion, Political Order and Societal Change in Shi'ite Iran from the Beginning to 1890*. Chicago: University of Chicago Press.

  1988   *The Turban for the Crown: The Islamic Revolution in Iran*. New York: Oxford University Press.

  1989   "The emergence of Islamic political ideologies." Pp. 109–123 in J. A. Beckford and T. Luckmann (eds.), *The Changing Face of Religion*. London: ISA/Sage.

  1993   "Millennial beliefs, hierocratic authority, and revolution in Shi'ite Iran." Pp. 219–239 in S. A. Arjomand (ed.), *The Political Dimension of Religion*. Albany: State University of New York Press.

Atwood, Margaret
  1985   *Handmaid's Tale*. New York: Fawcett Crest.

Bacon, Margaret Hope
  1986   *Mothers of Feminism: The Story of Quaker Women in America*. San Francisco: Harper & Row.

Badone, Ellen (ed.)
  1990   *Religious Orthodoxy and Popular Faith in European Society*. Princeton, N.J.: Princeton University Press.

Baer, Hans A.
  1984   *The Black Spiritual Movement: A Religious Response to Racism*. Knoxville: University of Tennessee Press.

  1988   *Recreating Utopia in the Desert: A Sectarian Challenge to Modern Mormonism*. Albany: State University of New York Press.

  1993   "The limited empowerment of women in black spiritual churches: An alternative vehicle to religious leadership." *Sociology of Religion* 54 1:65–82.

Baer, Hans A., and Merrill Singer
  1992   *African-American Religion in the Twentieth Century: Varieties of Protest and Accommodation*. Knoxville: University of Tennessee Press.

Bahr, Howard M., and Stan L. Albrecht
1989   "Strangers once more: Patterns of disaffiliation from Mormonism." *Journal for the Scientific Study of Religion* 28, 2:180–200.

Bahr, Howard M., and Bruce A. Chadwick
1985   "Religion and Family in Middletown, USA." *Journal of Marriage and the Family* 47:407–414.

Bainbridge, William S.
1978   *Satan's Power: A Deviant Psychotherapy Cult.* Berkeley: University of California Press.

Bainbridge, William S., and Rodney Stark
1984   "Formal explanation of religion: A progress report." *Sociological Analysis* 45:145–158.

Balch, Robert W., and David Taylor
1976   "Salvation in a U.F.O." *Psychology Today* 10:58–66.

1977   "Seekers and saucers: The role of the cultic milieu in joining a U.F.O. cult." *American Behavioral Scientist* 20, 6:839–860.

Baltzell, E. Digby
1979   *Puritan Boston and Quaker Philadelphia: Two Protestant Ethics and the Spirit of Class Authority and Leadership.* New York: Free Press.

Barkat, Anwar
1983   "Churches combatting racism in South Africa." *Journal of International Affairs* 36:297–306.

Barker, Eileen
1982   (ed.), *New Religious Movements: A Perspective for Understanding Society.* New York: Edwin Mellen.

1983   "New religious movements in Britain: The context and membership." *Social Compass* 30, 1:33–48.

1984   *The Making of a Moonie: Choice or Brainwashing.* Oxford, England: Basil Blackwell.

1988a   "Defection from the Unification Church: Some statistics and distinctions." Pp. 166–184 in D. Bromley (ed.), *Falling from the Faith: Causes and Consequences of Religious Apostasy.* Newbury Park, Calif.: Sage.

1988b   "Kingdoms of heaven on earth: New religious movements and political orders." Pp. 17–39 in A. Shupe and J. Hadden (eds.), *The Politics of Religion and Social Change.* New York: Paragon.

Barkun, Michael
1974   *Disaster and the Millennium.* New Haven, Conn.: Yale University Press.

1985   "The awakening-cycle controversy." *Sociological Analysis* 46, 4:425–443.

1986   *Crucible of the Millennium: The Burned-Over District of New York in the 1840s.* Syracuse, N.Y.: Syracuse University Press.

1994   *Religion and the Racist Right: The Origins of the Christian Identity Movement.* Chapel Hill: University of North Carolina Press.

Barnes, Nancy Schuster
1987   "Buddhism." Pp. 105–135 in A. Sharma (ed.), *Women in World Religions.* Albany: State University of New York Press.

Barnett, Homer G.
1957   *Indian Shakers: A Messianic Cult of the Pacific Northwest.* Carbondale: Southern Illinois University Press.

Barrett, Leonard
1974 *Soul Force: African Heritage in Afro-American Religion.* Garden City, N. Y.: Doubleday.

Barrett, Stanley R.
1987 *Is God a Racist? The Right Wing in Canada.* Toronto: University of Toronto Press.

Bastian, Jean-Pierre
1993 "The metamorphosis of Latin American Protestant groups." *Latin American Religious Review* 2:33–61.

Bauman, Zygmunt
1992 "Soil, blood and identity." *Sociological Review* 40:675–701.

Beach, Stephen
1977 "Religion and political change in Northern Ireland." *Sociological Analysis* 38, 1:37–48.

Beck, Rosalie, and David W. Hendon
1990 "Notes on church state affairs." *Journal of Church and State* 32, 3:684–685.

Becker, Howard
1932 *Systematic Sociology on the Basis of the Beziehungslehre and Gebildelehre of Leopold Van Wiese.* New York: Wiley.

1963 *Outsiders: Studies in the Sociology of Deviance.* New York: Free Press.

Beckett, J. C.
1966 *A Short History of Ireland.* London: Hutchinson.

Beckford, James A.
1975a "Religious organization." *Current Sociology* 21, 2. The Hague: Mouton.

1975b *The Trumpet of Prophecy: A Sociological Study of the Jehovah's Witnesses.* New York: Halsted-Wiley.

1978a "Accounting for conversion." *British Journal of Sociology* 29, 2:249–262.

1978b "Cults and cures." Unpublished paper presented to Ninth World Congress for Sociology, Research Committee for the Sociology of Religion.

1983 "The restoration of 'power' to the sociology of religion." *Sociological Analysis* 44, 1:11–32.

1984 "Holistic imagery and ethics in new religious and healing movements." *Social Compass* 31, 2–3:259–272.

1985a *Cult Controversies: The Societal Response to the New Religious Movements.* London: Tavistock.

1985b "The world images of new religious and healing movements." Pp. 72–93 in R. K. Jones (ed.), *Sickness and Sectarianism.* London: Gower.

1989 *Religion and Advanced Industrial Societies.* London: Unwin Hyman.

1990 "The sociology of religion, 1945–1989." *Social Compass* 37, 1:45–64.

1991 "Quasi-Marxisms and the sociology of religion." *Religion and the Social Order* 1:17–35.

1994 "Religion and solidarity." Unpublished paper presented to the World Congress of Sociology, International Sociological Association.

Bednarowski, Mary Farrell
1983 "Women in occult America." Pp. 177–195 in H. Kerr and C. L. Crow (eds.), *The Occult in America: New Historical Perspectives.* Urbana: University of Illinois Press.

1992   "The New Age movement and feminist spirituality: Overlapping conversations at the end of the century." Pp. 167–178 in J. R. Lewis and J. G. Melton (eds.), *Perspectives on the New Age*. Albany: State University of New York Press.

Bell, David A.
1995   "Lingua populi, lingua Dei: Language, religion, and the origins of French Revolutional nationalism." *American Historical Review* 100, 5:1403–1437.

Bell, Desmond
1990   *Acts of Union: Youth Culture and Sectarianism in Northern Ireland*. London: Macmillan.

Bellah, Robert N.
1964   "Religious evolution." *American Sociological Review* 29, 3:358–374.

1967   "Civil religion in America." *Daedalus* 96:1–21.

1970   "Christianity and symbolic realism." *Journal for the Scientific Study of Religion* 9, 2:89–96.

1973   *Emile Durkheim on Morality and Society*. Chicago: University of Chicago Press.

1978   "Religion and legitimation in the American Republic." *Society* 15:16–23.

1987   "Legitimation processes in politics and religion." *Current Sociology* 35, 2:89–99.

Benavides, Gustavo
1994   "Resistance and accommodation in Latin American popular religiosity." Pp. 37–61 in A. M. Stevens-Arroyo and A. M. Díaz-Stevens (eds.), *An Enduring Flame: Studies on Latino Popular Religiosity*. New York: PARAL/City University of New York.

Bendix, Reinhard
1960   *Max Weber: An Intellectual Portrait*. Garden City, N.Y.: Doubleday.

Bennett, David
1969   *Demagogues in the Depression*. New Brunswick, N.J.: Rutgers University Press.

Benson, Peter L., Michael J. Donahue, and Joseph A. Erickson
1989   "Adolescence and religion: A review of the literature from 1970 to 1986." *Research in the Social Scientific Study of Religion* 1:153–181.

Ben-Yehuda, Nachman
1985   *Deviance and Moral Boundaries: Witchcraft, the Occult, Science Fiction, Deviant Sciences and Scientists*. Chicago: University of Chicago Press.

Berger, Bennett M.
1969   "The new stage of American man—almost endless adolescence." *New York Times Magazine,* November 2, 1969:32, 33.

Berger, Joseph
1985   "Catholic dissent on church rules found." *New York Times,* November 25, 1985: A7.

Berger, Peter
1958   "Sectarianism and religious sociation." *American Journal of Sociology* 64:41–44.

1961   *The Noise of Solemn Assemblies: Christian Commitment and the Religious Establishment in America*. Garden City, N.Y.: Doubleday.

1964   *The Human Shape of Work*. New York: Macmillan.

1967   *The Sacred Canopy: Elements of a Sociological Theory of Religion*. Garden City, N.Y.: Doubleday.

Berger, Peter, Brigitte Berger, and Hansfried Kellner
1973   *The Homeless Mind: Modernization and Consciousness*. New York: Random House.

Berger, Peter, and Thomas Luckmann
  1963   "Sociology of religion and sociology of knowledge." *Sociology and Social Research* 47:417–427.

  1966   *The Social Construction of Reality: A Treatise in the Sociology of Knowledge.* Garden City, N.Y.: Doubleday.

Berger, Stephen
  1971   "The sects and the breakthrough into the modern world: On the centrality of the sects in Weber's Protestant ethic thesis." *Sociological Quarterly* 12:486–499.

Bettis, Joseph, and S. K. Johannesen (eds.)
  1984   *The Return of the Millennium.* Barrytown, N.Y.: International Religious Foundation.

Bew, Paul, and Henry Patterson
  1983   "Protestant and Catholic conflict in Ulster." *Journal of International Affairs* 36:223–234.

Beyer, Peter
  1994   *Religion and Globalization.* Newbury Park, Calif.: Sage.

Bibby, Reginald W.
  1983   "Searching for the invisible thread: Meaning systems in contemporary Canada." *Journal for the Scientific Study of Religion* 22, 2:101–119.

  1993   *Unknown Gods: The Ongoing Story of Religion in Canada.* Toronto: Stoddart.

Bibby, Reginald, and Harold R. Weaver
  1985   "Cult consumption in Canada: A further critique of Stark and Bainbridge." *Sociological Analysis* 46, 4:445–460.

Billings, Dwight B.
  1990   "Religion as opposition: A Gramscian analysis." *American Journal of Sociology* 96, 1:1–31.

Bird, Frederick
  1978   "Charisma and ritual in new religious movements." Pp. 173–189 in J. Needleman and G. Baker (eds.), *Understanding the New Religions.* New York: Seabury Press.

  1979   "The pursuit of innocence: New religious movements and moral accountability." *Sociological Analysis* 40, 4:335–346.

  1993   "Charisma and leadership in new religious movements." *Religion and the Social Order* 3A:75–92.

Bird, Frederick, and Bill Reimer
  1982   "Participation rates in new religious and para-religious movements." *Journal for the Scientific Study of Religion* 21, 1:1–14.

Bittner, Egon
  1968   "The structure of psychiatric influence." *Mental Hygiene* 52:423–430.

Blasi, Anthony J.
  1989   "Sociological implications of the great Western schism." *Social Compass* 36, 3:311–325.

Bohn, Ted, and Jeremiah S. Gutman
  1989   "The civil liberties of religious minorities." Pp. 257–289 in M. Galanter (ed.), *Cults and New Religious Movements.* Washington, D.C.: American Psychiatric Association.

Borker, Ruth
  1978   "To honor her head: Hats as a symbol of women's position in three evangelical churches in Edinburgh, Scotland." Pp. 55–73 in J. Hoch-Smith and A. Spring (eds.), *Women in Ritual and Symbolic Roles.* New York: Plenum.

Botev, Nikolai
  1994   "Where East meets West: Ethnic intermarriage in the former Yugoslavia,
         1962–1989." *American Sociological Review* 59, 3:461–480.

Bouma, Gary D.
  1973   "Beyond Lenski: A critical review of recent 'Protestant ethic' research." *Journal
         for the Scientific Study of Religion* 12, 2:141–155.

Bourdieu, Pierre
  1977   *Outline of a Theory of Practice*. New York: Cambridge University Press.

Braude, Ann D.
  1985   "Spirits defend the rights of women: Spiritualism and changing sex roles in
         nineteenth-century America." Pp. 419–431 in Y. Y. Haddad and E. B. Findly
         (eds.), *Women, Religion and Social Change*. Albany: State University of New York
         Press.

  1989   *Spiritualism and Women's Rights in Nineteenth Century America*. Boston: Beacon
         Press.

Brazier, Arthur
  1969   *Black Self-Determination: The Story of Woodlawn Organization*. Grand Rapids:
         Eerdmans.

Brittan, Arthur
  1977   *The Privatised World*. London: Routledge & Kegan Paul.

Bromley, David, and Anson D. Shupe
  1979   "The Tnevnoc cult." *Sociological Analysis* 40, 4:361–366.

Brown, Karen McCarthy
  1991   *Mama Lola: A Vodou Priestess in Brooklyn*. Berkeley: University of California
         Press.

Brown, Peter
  1981   *The Cult of the Saints: Its Rise and Function in Latin Christianity*. Chicago: Univer-
         sity of Chicago Press.

Brozan, Nadine
  1990   "Telling the story of the Seder in a woman's voice." *New York Times,* April 9.

Bruce, Steve
  1987   "The Moral Majority: The politics of fundamentalism in secular society."
         Pp. 177–194 in L. Caplan (ed.), *Fundamentalism*. London: Macmillan.

Brundage, James A.
  1982   "Concubinage and marriage in medieval Canon Law." Pp. 118–128 in V. L.
         Bullough and J. Brundage (eds.), *Sexual Practices and the Medieval Church*. Buffalo:
         Prometheus.

Bruneau, Thomas C.
  1982   *The Church in Brazil: The Politics of Religion*. Austin: University of Texas Press.
  1988   "The role and response of the Catholic Church in the redemocratization of
         Brazil." Pp. 87–109 in A. Shupe and J. Hadden (eds.), *The Politics of Religion and
         Social Change*. New York: Paragon.

Bubis, Gerald B., Harry Wasserman with Alan Lert
  1983   *Synagogue Havurot: A Comparative Study*. Washington, D.C.: University of
         America Press.

Buckley, Anthony D., and Mary Catherine Kenney
  1995   *Negotiating Identity: Rhetoric, Metaphor, and Social Drama in Northern Ireland*.
         Washington, D.C.: Smithsonian Institution Press.

Bullough, Vern L.
  1982   "Introduction: The Christian inheritance." Pp. 1–12 in V. L. Bullough and
         J. Brundage (eds.), *Sexual Practices and the Medieval Church*. Buffalo: Prometheus.

Burke, Kenneth
  1935   *Permanence and Change.* New York: New Republic.

  1953   *A Rhetoric of Motives.* Englewood Cliffs, N.J.: Prentice-Hall.

Burkholder, John Richard
  1974   "The law knows no heresy: Marginal religious movements and the courts."
          Pp. 27–52 in I. Zaretsky and M. Leone (eds.), *Religious Movements in Contempo-
          rary America.* Princeton, N.J.: Princeton University Press.

Burnham, Kenneth
  1978   *God Comes to America: Father Divine and the Peace Mission Movement.* Boston:
          Lambeth.

Bynum, Caroline Walker
  1986   "'. . . And woman his humanity': Female imagery in the religious writing of the
          later Middle Ages." Pp. 257–288 in C. W. Bynum, S. Harrell, and P. Richman
          (eds.), *Gender and Religion: On the Complexity of Symbols.* Boston: Beacon Press.

  1991   *Fragmentation and Redemption.* New York: Zone Books.

Cadena, Gilbert R.
  1995   "Religious ethnic identity: A socio-religious portrait of Latinas and Latinos in
          the Catholic Church." Pp. 33–59 in A. M. Stevens-Arroyo and G. R. Cadena
          (eds.), *Old Masks, New Faces: Religion and Latino Identities.* New York:
          PARAL/Bildner Center for Western Hemisphere Studies.

Calhoun, Craig
  1993   "Nationalism and ethnicity." *Annual Review of Sociology* 19:2111–2239.

Callahan, Daniel
  1987   *Setting Limits: Medical Goals in an Aging Society.* New York: Simon & Schuster.

  1989   *What Kind of Life: The Limits of Medical Progress.* New York: Simon & Schuster.

Camara, Evandro M.
  1988   "Afro-American religious syncretism in Brazil and the United States." *Sociological
          Analysis* 48, 4:299–318.

Campbell, Colin
  1971   *Toward a Sociology of Irreligion.* London: Macmillan.

  1977   "Clarifying the cult." *British Journal of Sociology* 28, 3:375–388.

  1978   "The secret religion of the educated classes." *Sociological Analysis* 39, 2:146–156.

Campbell, Colin, and Shirley McIver
  1987   "Cultural sources of support for contemporary occultism." *Social Compass* 34,
          1:41–60.

Campbell, Ernest Q., and Thomas F. Pettigrew
  1959   *Christians in Racial Crisis: A Study of Little Rock's Ministry.* Washington, D.C.:
          Public Affairs Press.

Capps, Donald
  1992   "Religion and child abuse: Perfect together." *Journal for the Scientific Study of Re-
          ligion* 31, 1:1–14.

Carnes, Mark C.
  1989   *Secret Ritual and Manhood in Victorian America.* New Haven, Conn.: Yale Univer-
          sity Press.

Carrasco, David
  1995   "Jaguar Christians in the contact zone." Pp. 69–79 in A. M. Stevens-Arroyo and
          A. I. Pérez y Mena (eds.), *Enigmatic Powers: Syncretism with African and Indigenous
          Peoples' Religions Among Latinos.* New York: PARAL/City University of New
          York.

Carroll, Jackson W., Barbara Hargrove, and Adair Lummis
1983 *Women of the Cloth: A New Opportunity for the Churches.* New York: Harper &
Row.

Casanova, José
1993 "Church, state, nation, and civil society in Spain and Poland." Pp. 101–153 in
S. A. Arjomand (ed.), *The Political Dimensions of Religion.* Albany: State University of New York.

1994 *Public Religions in the Modern World.* Chicago: University of Chicago Press.

Castelli, Jim, and Joseph Gremillion
1987 *The Emerging Parish: The Notre Dame Study of Catholic Life Since Vatican II.* San
Francisco: Harper & Row.

Cavendish, James C.
1994 "Christian base communities and the building of democracy: Brazil and Chile."
Sociology of Religion 55, 2:179–195.

Chaves, Mark, and Lynn M. Higgins
1992 "Comparing the community involvement of black and white congregations."
Journal for the Scientific Study of Religions 31, 4:425–440.

Cherry, Conrad
1969 "Two American sacred ceremonies." *American Quarterly* 21:739–754.

1992 "Two American sacred ceremonies: Their implications for the study of religion
in America." Pp. 218–233 in M. Marty (ed.), *Civil Religion, Church and State.*
Munich: K. G. Saur.

Chidester, David
1988 *Salvation and Suicide: An Interpretation of Jim Jones, the Peoples Temple, and
Jonestown.* Bloomington: Indiana University Press.

1991 *Shots in the Streets: Violence and Religion in South Africa.* Boston: Beacon.

Chodak, Szymon
1987 "People and the church versus the state: The case of the Roman Catholic
Church in Poland." Pp. 280–307 in R. L. Rubenstein (ed.), *Spirit Matters: The
Worldwide Impact of Religion on Contemporary Politics.* New York: Paragon.

Christ, Carol P.
1987 *The Laughter of Aphrodite.* San Francisco: Harper & Row.

Chrypinski, Vincent C.
1989 "Church and nationality in postwar Poland." Pp. 241–263 in P. Ramet (ed.),
Religion and Nationalism in Soviet and East European Politics. Durham, N.C.: Duke
University Press.

Clarity, James F.
1994 "The Irish Church: Under strain but still strong." *New York Times,* Decem-
ber 11.

Clarke, Aidan
1967 "The colonisation of Ulster and the rebellion of 1641." Pp. 189–203 in T. W.
Moody and F. X. Martin (eds.), *The Course of Irish History.* Cork, Ireland:
Mercier.

Cleage, Albert G., Jr.
1967 *The Black Messiah.* New York: Sheed and Ward.

Cohen, Erik
1990 "The missionary as stranger: A phenomenological analysis of Christian missionaries' encounter with the folk religions of Thailand." *Review of Religious Research*
31, 4:337–350.

Cohn, Norman
  1970   *The Pursuit of the Millennium: Revolutionary Millenarians and Mystical Anarchists of the Middle Ages.* New York: Oxford University Press.

  1993   *Cosmos, Chaos and the World to Come: The Ancient Roots of Apocalyptic Faith.* New Haven: Yale University Press.

Coleman, James S.
  1956   "Social cleavage and religious conflict." *The Journal of Social Issues* 12:44–56.

Comaroff, Jean
  1984   "Medicine, time and the perception of death." *Listening* 19, 2:155–169.

  1985   *Body of Power, Spirit of Resistance: The Culture and History of a South African People.* Chicago: University of Chicago Press.

Conrad, Peter, and Joseph W. Schneider
  1980   *Deviance and Medicalization: From Badness to Sickness.* St. Louis: Mosby.

Conzen, Kathleen N., David A. Gerber, Eva Morawska, G. E. Possetta, and Rudolph J. Vecoli
  1990   "The invention of ethnicity." *Altreitalie* 3:37–62.

Cooper, Lee R.
  1974   "'Publish' or perish: Negro Jehovah's Witnesses' adaptation in the ghetto." Pp. 700–721 in I. Zaretsky and M. Leone (eds.), *Religious Movements in Contemporary America.* Princeton, N.J.: Princeton University Press.

Cornwall, Marie
  1987   "The social bases of religion: A study of factors influencing religious belief and commitment." *Review of Religious Research* 29, 1:44–68.

  1994   "The institutional role of Mormon women." Pp. 239–264 in M. Cornwall, T. B. Heaton, and L. A. Young (eds.), *Contemporary Mormonism: Social Science Perspectives.* Urbana: University of Illinois Press.

Cornwall, Marie, and Perry H. Cunningham
  1989   "Surveying Latter-Day Saints: A review of methodological issues." *Review of Religious Research* 31, 2:162–174.

Coser, Lewis
  1974   *Greedy Institutions: Patterns of Undivided Commitment.* New York: Free Press.

Cowell, Alan
  1985   "Churches on cutting edge of Apartheid battle." *New York Times,* March 15, 1985:A2.

Crittendon, Ann
  1988   *Sanctuary: A Story of American Conscience and the Law in Collision.* New York: Weidenfeld and Nicolson.

Cross, Robert D.
  1958   *The Emergence of Liberal Catholicism in America.* Cambridge: Cambridge University Press.

Csordas, Thomas J.
  1992   "Religion and the world system: The Pentecostal ethic and the spirit of monopoly capital." *Dialectical Anthropology* 17:3–24.

  1994   *The Sacred Self: A Cultural Phenomenology of Charismatic Healing.* Berkeley: University of California Press.

Daly, Mary
  1973   *Beyond God the Father.* Boston: Beacon Press.

  1978   *Gyn/Ecology: The Meta-Ethics of Radical Feminism.* Boston: Beacon Press.

D'Antonio, William V., and Joan Aldous (eds.)
  1983   *Families and Religions: Conflict and Changes in Modern Society*. Beverly Hills, Calif.:
         Sage.

D'Antonio, William V., James D. Davidson, Dean R. Hoge, and Ruth A. Wallace
  1989   *American Catholic Laity in a Changing Church*. New York: Sheed and Ward.

Danzger, M. Herbert
  1989   *Returning to Tradition: The Contemporary Revival of Orthodox Judaism*. New
         Haven, Conn.: Yale University Press.

Darnton, John
  1995   "Protestant and paranoid: In Northern Ireland." *New York Times Magazine,* Jan-
         uary 15:32–35.

Davidman, Lynn
  1990a "Accommodation and resistance to modernity: A comparison of two contempo-
         rary Orthodox Jewish groups." *Sociological Analysis* 51, 1:35–51.

  1990b "Women's search for family and roots: A Jewish religious solution to a modern
         dilemma." Pp. 385–407 in T. Robbins and D. Anthony (eds.), *In Gods We
         Trust: New Patterns of Religious Pluralism in America*. New Brunswick, N.J.:
         Transaction.

  1991   *Tradition in a Rootless World: Women Turn to Orthodox Judaism*. Berkeley: Uni-
         versity of California Press.

Davie, Jody Shapiro
  1995a "Speaking without tongues: Women's communicative styles in contemporary
         mainline Protestantism." Unpublished paper presented to the Society for the
         Scientific Study of Religion.

  1995b *Women in the Presence: Constructing Community and Seeking Spirituality in Mainline
         Protestantism*. Philadelphia: University of Pennsylvania Press.

Davis, Gerald L.
  1985   *I Got the Word in Me and I Can Sing It, You Know: A Study of the Performed
         African-American Sermon*. Philadelphia: University of Pennsylvania Press.

Davis, Richard
  1994   *Mirror Hate: The Convergent Ideology of Northern Ireland Paramilitaries, 1966–1992*.
         Aldershot (Hants), England: Dartmouth.

Davis-Floyd, Robbie
  1992   *Birth as an American Rite of Passage*. Berkeley: University of California Press.

Demerath, N. J., III
  1965   *Social Class in American Protestantism*. Chicago: Rand McNally.

  1989   "Religious capital and capital religions: Cross-cultural and non-legal factors in
         the separation of church and state." Unpublished paper presented to the Sociète
         Internationale de Sociologie des Religions.

Demerath, N. J., III, and Rhys H. Williams
  1987   "The mythical past and uncertain future." Pp. 77–90 in T. Robbins and
         R. Robertson (eds.), *Church–State Relations: Tensions and Transitions*. New
         Brunswick, N. J.: Transaction.

  1990   "Religion and power in the American experience." Pp. 427–448 in T. Robbins
         and D. Anthony (eds.), *In Gods We Trust: New Patterns of Religious Pluralism in
         America*. New Brunswick, N.J.: Transaction.

Deren, Maya
  1970   *Divine Horsemen: The Voodoo Gods of Haiti*. New York: Chelsea House.

Diaz-Albertini, Javier
  1993   "Note on the Shining Path and modernized millennialism in Peru."

Pp. 169–174 in S. A. Arjomand (ed.), *The Political Dimensions of Religion.*
Albany, N.Y.: State University of New York Press.

Díaz-Stevens, Ana María
  1993   *Oxcart Catholicism on Fifth Avenue: The Impact of the Puerto Rican Migration upon
           the Archdiocese of New York.* Notre Dame: University of Notre Dame Press.

  1994   "Latinas and the church." Pp. 240–277 in J. P. Dolan and A. F. Deck (eds.),
           Hispanic Catholic Culture in the U.S.: Issues and Concerns. Notre Dame: University
           of Notre Dame Press.

Dobbelaere, Karel
  1989   "The secularization of society? Some methodological suggestions." Pp. 27–44 in
           J. K. Hadden and A. Shupe (eds.), *Religion and the Political Order: Secularization
           and Fundamentalism Reconsidered.* New York: Paragon.

Do Campo, Orlando
  1995   "The Free Exercise clause and ritual bloodletting: Church of the Lukumí Babalú
           Ayé vs. City of Hialeah." Pp. 159–179 in A. M. Stevens-Arroyo and A. I. Pérez
           y Mena (eds.), *Enigmatic Powers: Syncretism with African and Indigenous Peoples'
           Religions Among Latinos.* New York: PARAL/Bildner Center/City University of
           New York.

Dodson, Jualynne E., and Cheryl Townsend Gilkes
  1995   "'There's nothing like church food': Food and the U.S. Afro-Christian tradi-
           tion: re-membering community and feeding the embodied S/spirit." *Journal of
           the American Academy of Religion* 63, 3:519–538.

Dollard, John
  [1937] *Caste and Class in a Southern Town.* Garden City, N.Y.: Doubleday.
  1949

Dougherty, Molly C.
  1978   "Southern lay midwives as ritual specialists." Pp. 151–164 in J. Hoch-Smith and
           A. Spring (eds.), *Women in Ritual and Symbolic Roles.* New York: Plenum.

Douglas, Mary
  1966   *Purity and Danger: An Analysis of Concepts of Pollution and Taboo.* London: Rout
           ledge & Kegan Paul.

  1970   *Natural Symbols: Explorations in Cosmology.* New York: Random House.

  1986   *How Institutions Think.* Syracuse, N.Y.: Syracuse University Press.

Dreitzel, Hans P.
  1981   "The socialization of nature: Western attitudes towards body and emotions."
           Pp. 205–223 in P. Heelas and A. Lock (eds.), *Indigenous Psychologies: The Anthro-
           pology of the Self.* New York: Academic Press.

Droogers, André, and Hans Siebers
  1991   "Popular religion and power in Latin America." Pp. 1–25 in A. Droogers,
           G. Huizer, and H. Siebers (eds.), *Popular Power in Latin American Religion.* Saar-
           brücken: Verlag Breitenbuch.

Dudley, Donald M., and Margaret G. Dudley
  1986   "Transmission of religious values from parents to adolescents." *Review of Reli-
           gious Research* 28, 1:3–15.

Durkheim, Emile
  1938   *The Rules of the Sociological Method.* Chicago: University of Chicago Press.

  [1897] *Suicide: A Study in Sociology.* Tr. J. A. Spaulding and G. Simpson. Glencoe, Ill.:
  1951   Free Press.

  [1915] *Elementary Forms of the Religious Life.* Tr. J. W. Swain. New York: Free Press.
  1965

[1898] "Individualism and the intellectuals." Tr. S. Lukes. *Political Studies* 17:14–30.
1969

[1915] *The Elementary Forms of Religious Life.* Tr. K. Fields. New York: Free Press.
1995

Dvorak, Katharine L.
1991 *An African-American Exodus: The Segregation of the Southern Churches.* Brooklyn: Carlson.

Earle, John, Dean Knudsen, and Donald Shriver
1976 *Spindles and Spires: A Re-Study of Religion and Social Change in Gastonia.* Atlanta: John Knox.

Ebaugh, Helen Rose
1988 "Leaving Catholic convents: Toward a theory of disengagement." Pp. 100–121 in D. Bromley (ed.), *Falling from the Faith: Causes and Consequences of Religious Apostasy.* Newbury Park, Calif.: Sage.

Eco, Umberto
1980 *The Name of the Rose.* New York: Harcourt Brace Jovanovich.

Ehrenreich, Barbara, and Dierdre English
1973 *Witches, Midwives and Nurses: A History of Women Healers.* New York: Feminist Press.

Eisenstadt, S. N.
1968 (ed.), *The Protestant Ethic and Modernization: A Comparative View.* New York: Basic Books.

1969 "The Protestant ethic thesis." Pp. 297–317 in R. Robertson (ed.), *Sociology of Religion.* Baltimore: Penguin Books.

Eisenstadt, S. N., and Stein Rokkan (eds.)
1973 *Building States and Nations.* Beverly Hills, Calif.: Sage.

Eliade, Mircea
1958 *Rites and Symbols of Initiation: The Mysteries of Birth and Rebirth.* New York: Harper & Row.

1964 *Shamanism: Archaic Techniques of Ecstasy.* Tr. W. R. Trask. New York: Pantheon Books.

Elkind, David
1964 "Age changes in the meaning of religious identity." *Review of Religious Research* 6, 1:36–40.

Elliott, Dyan
1993 *Spiritual Marriage: Sexual Abstinence in Medieval Wedlock.* Princeton: Princeton University Press.

Ellison, Christopher G.
1991 "Religious involvement and subjective well-being." *Journal of Health and Social Behavior* 32:80–99.

forthcoming "Religious involvement and the subjective quality of family life among African Americans." In R. J. Taylor, L. M. Chatters, and J. S. Jackson (eds.), *Family Life in Black America.* Newbury Park, Calif.: Sage.

Ellwood, Robert S., Jr.
1978 "Emergent religion in America: An historical perspective." Pp. 267–284 in J. Needleman and G. Baker (eds.), *Understanding the New Religions.* New York: Seabury Press.

Elzey, Wayne
1975 "Liminality and symbiosis in popular American Protestantism." *Journal of the American Academy of Religion* 43:741–756.

1976 "The most unforgettable magazine I've ever read: Religion and social hygiene in *The Reader's Digest." Journal of Popular Culture*. 10:181–190.

1988 "Popular culture." Pp. 1727–1741 in C. H. Lippy and P. W. Williams (eds.), *The Encyclopedia of American Religious Experience,* Vol. III. New York: Scribner.

Engel, George L.
1977 "The need for a new medical model: A challenge for biomedicine." *Science* 196:129–136.

English-Lueck, June Anne
1990 *Health in the New Age: A Study in California Holistic Practices.* Albuquerque: University of New Mexico Press.

Espín, Orlando O.
1994 "Popular Catholicism among Latinos." Pp. 308–359 in J. P. Dolan and A. F. Deck (eds.), *Hispanic Catholic Culture in the U.S.: Issues and Concerns.* Notre Dame: University of Notre Dame Press.

Eversley, David
1989 *Religion and Employment in Northern Ireland.* London: Sage.

Fabian, Johannes
1974 "Genres in an emerging tradition: An anthropological approach to religious communication." Pp. 249–272 in A. Eister (ed.), *Changing Perspectives in the Scientific Study of Religion.* New York: Wiley.

Farnsley, Arthur Emery
1994 *Southern Baptist Politics: Authority and Power in the Restructuring of an American Denomination.* University Park: Pennsylvania State University Press.

Fauset, Arthur H.
1944 *Black Gods of the Metropolis.* Philadelphia: University of Pennsylvania Press.

Feher, Shoshanah
1992 "Who looks to the Stars? Astrology and its constituency." *Journal for the Scientific Study of Religion* 31, 1:88–93.

1994 "The hidden truth: Astrology as worldview." *Religion and the Social Order* 4:165–177.

Feltey, Kathryn M., and Margaret M. Poloma
1990 "From sex differences to gender role beliefs: Exploring effects on six dimensions of religiosity." Unpublished paper presented to the Association for the Sociology of Religion.

Fenn, Richard
1972 "Toward a new sociology of religion." *Journal for the Scientific Study of Religion* 11, 1:16–32.

1974 "Religion and the legitimation of social systems." Pp. 143–161 in A. Eister (ed.), *Changing Perspectives in the Scientific Study of Religion.* New York: Wiley.

1978 *Toward a Theory of Secularization.* Monograph Series 1. Storrs, Conn.: Society for the Scientific Study of Religion.

1982 *Liturgies and Trials: The Secularization of Religious Language.* Oxford, England: Basil Blackwell.

Festinger, Leon
1957 *A Theory of Cognitive Dissonance.* Palo Alto, Calif.: Stanford University Press.

Festinger, Leon, Henry W. Riecken, and Stanley Schachter
1956 *When Prophecy Fails.* New York: Harper & Row.

Fichter, Joseph
1951 *Dynamics of a City Church.* Chicago: University of Chicago Press.

1954 *Social Relations in the Urban Parish.* Chicago: University of Chicago Press.

Fields, Karen E.
   1985   *Revival and Rebellion in Colonial Central Africa.* Princeton, N.J.: Princeton University Press.

   1993   "Antinomian conduct at the Millennium: Metaphorical conceptions of time in social science and social life." Pp. 175–203 in S. A. Arjomand (ed.), *The Political Dimensions of Religion.* Albany, N.Y.: State University of New York Press.

Finke, Roger, and Rodney Stark
   1992   *The Churching of America: Winners and Losers in the Economic Struggle, 1776–1990.* New Brunswick, N.J.: Rutgers University Press.

Finley, Nancy J.
   1991   "Political activism and feminist spirituality." *Sociological Analysis* 52, 4:349–362.

Fischler, Claude
   1974   "Astrology and French society: The dialectic of archaism and modernity."
          Pp. 281–293 in E. Tiryakian (ed.), *On the Margin of the Visible.* New York: Wiley.

Flores, Richard R.
   1994   "Para el Niño Dios: Sociability and commemorative sentiment in popular religious practice." Pp. 171–190 in A. M. Stevens-Arroyo and A. M. Díaz-Stevens (eds.), *An Enduring Flame: Studies on Latino Popular Religiosity.* New York: PARAL/City University of New York.

Foltz, Tanice
   1985   "An alternative healing group as a new religious form: The use of ritual in becoming a healing practitioner." Pp. 144–157 in R. K. Jones (ed.), *Sickness and Sectarianism.* London: Gower.

Fones-Wolf, Ken
   1989   *Trade Union Gospel: Christianity and Labor in Industrial Philadelphia, 1865–1915.* Philadelphia: Temple University Press.

Frankenberg, Ronald
   1986   "Sickness as cultural performance: Drama, trajectory and pilgrimage, root metaphors and the making social of disease." *International Journal of Health Services* 16, 4:603–626.

Franklin, Robert Michael
   1994   "The safest place on earth: The culture of black congregations." Pp. 257–284 in J. P. Wind and J. W. Lewis (eds.), *American Congregations,* Vol. 2. Chicago: University of Chicago Press.

Frazier, E. Franklin
   1974   *The Negro Church in America.* New York: Schocken.

Fredrickson, George M.
   1995   *Black Liberation: A Comparative History of Black Ideologies in the United States and South Africa.* New York: Oxford University Press.

Freeman, James M.
   1980   "The ladies of Lord Krishna: Rituals of middle-aged women in Eastern India." Pp. 110–126 in N. A. Falk and R. M. Gross (eds.), *Unspoken Worlds: Women's Religious Lives in Non-Western Cultures.* San Francisco: Harper & Row.

Freidson, Eliot
   1970   *Profession of Medicine.* New York: Dodd, Mead.

Freund, Peter E. S., and Meredith B. McGuire
   1995   *Health, Illness and the Social Body: A Critical Sociology.* Englewood Cliffs, N.J.: Prentice-Hall.

Friedl, Ernestine
   1975   *Women and Men: An Anthropologist's View.* New York: Holt, Rinehart & Winston.

Friedman, Thomas L.
1986   "No illusions: Israel reassesses its chances for peace." *New York Times Magazine,* Jan. 26, 1986:13–18ff.

1987   "Fight builds over the shape of religious future in Israel." *New York Times,* June 29.

Fulton, John
1987   "Religion and politics in Gramsci: An introduction." *Sociological Analysis* 48, 3:197–216.

1991   *The Tragedy of Belief: Division, Politics, and Religion in Ireland.* Oxford: Clarendon.

Gaines, Atwood D.
1992   "From DSM-I to III-R; Voices of self, mastery, and the other: A cultural constructivist reading of U.S. psychiatric classification." *Social Science and Medicine* 35, 1:3–24.

Gallup, George, Jr., and Jim Castelli
1989   *The People's Religion: American Faith in the 90s.* New York: Macmillan.

Gallup Poll
1976   "Americans taking up religious, spiritual experimentation." Princeton, N.J.: Gallup International. See also *New York Times,* November 18, 1976.

1984   *Religion in America: 1984.* Report 222. Princeton, N.J.: The Princeton Religion Research Center.

1990   *Gallup Poll Monthly,* September.

1993   *Gallup Poll Monthly,* August.

Gallup Poll and Princeton Religion Research Center
1988   *The Unchurched American—Ten Years Later.* Princeton, N.J.: Religious Research Center.

Gallup Poll/Social Surveys, Limited
1984   *Human Values and Beliefs.* London: European Value Systems Study Group; Gallup International.

Gannon, Thomas M.
1981   "The New Christian Right in America." *Archives de Sciences des Religions* 52, 1:69–83.

Gans, Herbert J.
1994   "Symbolic ethnicity and symbolic religiosity: Towards a comparison of ethnic and religious acculturation." *Ethnic and Racial Studies* 17, 4:577–592.

Gardner, Hugh
1978   *The Children of Prosperity.* New York: St. Martin's Press.

Garrett, Clarke
1987   *Spirit Possession and Popular Religion: From the Camisards to the Shakers.* Baltimore: The Johns Hopkins University Press.

Garrett, William R.
1975   "Maligned mysticism: The maledicted career of Troeltsch's third type." *Sociological Analysis* 36, 3:205–223.

Geertz, Clifford
1957   "Ritual and social change: A Javanese example." *American Anthropologist* 59:23–54.

1964   "Ideology as a cultural system." Pp. 47–76 in D. Apter (ed.), *Ideology and Discontent.* New York: Free Press.

1966   "Religion as a cultural system." Pp. 1–46 in M. Banton (ed.), *Anthropological Approaches to the Study of Religion.* London: Tavistock.

1973   *The Interpretation of Cultures.* New York: Basic Books.

Gellner, Ernest
  1974   *Legitimation of Belief.* Cambridge, England: Cambridge University Press.

Genovese, Eugene D.
  1974   *Roll, Jordan, Roll: The World the Slaves Made.* New York: Vintage.
  1979   *From Rebellion to Revolution: Afro-American Slave Revolts in the Making of the Modern World.* Baton Rouge: Louisiana State University Press.

Gerlach, Luther
  1974   "Pentecostalism: Revolution or counter-revolution." Pp. 669–699 in I. Zaretsky and M. Leone (eds.), *Religious Movements in Contemporary America.* Princeton, N.J.: Princeton University Press.

Gerlach, Luther, and Virginia Hine
  1968   "Five factors crucial to the growth and spread of a modern religious movement." *Journal for the Scientific Study of Religion* 7, 1:23–40.

  1970   *People, Power, and Change: Movements of Social Transformation.* Indianapolis: Bobbs-Merrill.

Gevitz, Norman
  1988   *Other Healers: Unorthodox Medicine in America.* Baltimore: The Johns Hopkins University Press.

Giddens, Anthony
  1991   *Modernity and Self-Identity: Self and Society in the Late Modern Age.* Stanford: Stanford University Press.

Gilkes, Cheryl Townsend
  1987   "'Together and in harness': Women's traditions in the Sanctified Church." *Signs: Journal of Women in Culture and Society* 10, 4:678–699.

  1995   "The storm and the light: Church, family, work, and social crisis in the African-American experience." Pp. 177–198 in N. T. Ammerman and W. C. Roof (eds.), *Work, Family, and Religion in Contemporary Society.* New York: Routledge.

Gismondi, Michael
  1988   "Conceptualizing religion from below: The Central American experience." *Social Compass* 35, 2–3:343–370.

Glanz, David, and Michael Harrison
  1977   "Varieties of identity transformation: The case of newly Orthodox Jews." Unpublished paper presented to Israeli Sociological Association.

Glazer, Barney G., and Anselm L. Strauss
  1971   *Status Passage.* London: Routledge & Kegan Paul.

Glazier, Stephen
  1983   *Marchin' the Pilgrims Home: Leadership and Decision-Making in an Afro-Caribbean Faith.* Westport, Conn.: Greenwood Press.

Glenn, Norval D.
  1982   "Interreligious marriage in the United States: Patterns and recent trends." *Journal of Marriage and the Family* 44, 3:555–566.

Glik, Deborah C.
  1988   "Symbolic, ritual and social dynamics of spiritual healing." *Social Science and Medicine* 27:1197–1206.

Glock, Charles Y.
  [1962] "On the study of religious commitment." Research Supplement, *Religious*
  1965   *Education* 57, 4. Reprinted in C. Glock and R. Stark, *Religion and Society in Tension.* Chicago: Rand McNally.

1973   (ed.), *Religion in Sociological Perspective*. Belmont, Calif.: Wadsworth.

Glock, Charles Y., and Robert Bellah (eds.)
1976   *The New Religious Consciousness*. Berkeley: University of California Press.

Glock, Charles Y., and Rodney Stark
1966   *Christian Beliefs and Anti-Semitism*. New York: Harper & Row.

Glock, Charles Y., and Robert Wuthnow
1979   "Departures from conventional religion: The nominally religious, the nonreligious, and the alternatively religious." Pp. 47–68 in R. Wuthnow (ed.), *The Religious Dimension: New Directions in Quantitative Research*. New York: Academic Press.

Goffman, Erving
1961   *Asylums*. Garden City, N.Y.: Doubleday.

Gold, Penny Schine
1985   *The Lady and the Virgin: Image, Attitude and Experience in Twelfth-Century France*. Chicago: University of Chicago Press.

Goldenberg, Judith
1974   "Epilogue: The coming of Lilith." Pp. 341–343 in R. Ruether (ed.), *Religion and Sexism*. New York: Simon & Schuster.

Goldsmith, Peter
1989   *When I Rise Cryin' Holy: African-American Denominationalism on the Georgia Coast*. New York: AMS Press.

Gonzalez-Wippler, Migene
1975   *Santería: African Magic in Latin America*. Garden City, N.Y.: Doubleday.

Gordon, Deborah
1988   "Tenacious assumptions in Western medicine." Pp. 19–56 in M. Lock and D. R. Gordon (eds.), *Biomedicine Examined*. Dordrecht, Netherlands: Kluwer.

Gorsuch, Richard, and D. Aleshire
1974   "Christian faith and ethnic prejudice." *Journal for the Scientific Study of Religion* 13:281–307.

Gravely, Will B.
1989   "The rise of African churches in America (1786–1822): Re-examining the contexts." Pp. 301–317 in G. S. Wilmore (ed.), *African American Religious Studies*. Durham, N.C.: Duke University Press.

Greeley, Andrew
1963   *Religion and Career: A Study of College Graduates*. New York: Sheed and Ward.

1964   "The Protestant ethic: Time for a moratorium." *Sociological Analysis* 25:20–33.

1972a   *The Denominational Society: A Sociological Approach to Religion in America*. Glenview, Ill.: Scott, Foresman.

1972b   *Unsecular Man: The Persistence of Religion*. New York: Schocken.

1975   *The Sociology of the Paranormal: A Reconnaissance*. Beverly Hills, Calif.: Sage.

1977   *The American Catholic: A Social Portrait*. New York: Basic Books.

1979   "Towards a secular theory of religious behavior." Unpublished paper presented to American Sociological Association.

1987   "Mysticism goes mainstream." *American Health*, Jan./Feb.:47–49.

1990   *The Catholic Myth: The Behavior and Beliefs of American Catholics*. New York: Charles Scribners Sons.

1991   "The demography of American Catholics: 1965–1990." *Religion and the Social Order* 2:37–56.

Green, Robert W. (ed.)
  1959   *Protestantism and Capitalism: The Weber Thesis and Its Critics.* Boston: Heath.

Greenfeld, Liah
  1993   *Nationalism: Five Roads to Modernity.* Cambridge, Mass.: Harvard University Press.

Greil, Arthur
  1977   "Previous disposition and conversion to perspectives of social and religious movements." *Sociological Analysis* 38, 2:115–125.

  1993   "Explorations along the sacred frontier: Notes on para-religions, quasi-religions, and other boundary phenomena." *Religion and the Social Order* 3A:153–172.

Greil, Arthur, and David Rudy
  1984   "What have we learned from process models of conversion? An examination of ten case studies." *Sociological Focus* 17, 4:305–323.

Greven, Philip
  1991   *Spare the Child: The Religious Roots of Punishment and the Psychological Impact of Physical Abuse.* New York: Alfred A. Knopf.

Griffin, Wendy
  1995   "The embodied goddess: Feminist witchcraft and female divinity." *Sociology of Religion* 56, 1:35–48.

Guizzardi, Gustavo
  1989   "Religion in the television era." *Social Compass* 36, 3:337–353.

Hadaway, C. Kirk
  1989   "Will the real Southern Baptist please stand up: Methodological problems in surveying Southern Baptist congregations and members." *Review of Religious Research* 31, 2:149.

  1993   "Church growth in North America: The character of a religious marketplace." Pp. 346–357 in D. Roozen and C. K. Hadaway (eds.), *Church and Denominational Growth.* Nashville: Abingdon.

Haddad, Yvonne Yazbeck, and Adair Lummis
  1987   *Islamic Values in the United States.* New York: Oxford University Press.

Hadden, Jeffrey K.
  1969   *The Gathering Storm in the Churches: The Widening Gap Between Clergy and Laymen.* Garden City, N.Y.: Doubleday.

  1993   "The rise and fall of American televangelism." *Annals of the American Academy of Political and Social Science* 527 (May):113–143.

Hadden, Jeffrey K., and Charles E. Swann
  1981   *Prime-Time Preachers: The Rising Power of Televangelism.* Reading, Mass.: Addison-Wesley.

Hahn, Robert A., and Arthur Kleinman
  1983   "Biomedical practice and anthropological theory: Frameworks and directions." *American Review of Anthropology* 12:305–333.

Hall, John R.
  1987   *Gone from the Promised Land: Jonestown in American Cultural History.* New Brunswick, N.J.: Transaction.

Hamilton, Charles
  1972   *The Black Preacher in America.* New York: Morrow.

Hammond, John L.
  1979   *The Politics of Benevolence: Revival Religion and American Voting Behavior.* Norwood, N.J.: Ablex.

  1983   "The reality of revivals." *Sociological Analysis* 44:111–116.

Hammond, Phillip E.
  1974 "Religious pluralism and Durkheim's integration thesis." Pp. 115–142 in
       A. Eister (ed.), *Changing Perspectives in the Scientific Study of Religion*. New York:
       Wiley.

  1976 "The sociology of American civil religion: A bibliographical essay." *Sociological
       Analysis* 37, 2:169–182.

  1987 "The courts and secular humanism." Pp. 91–101 in T. Robbins and R. Robert-
       son (eds.), *Church–State Relations: Tensions and Transitions*. New Brunswick, N.J.:
       Transaction.

  1992 *Religion and Personal Autonomy: The Third Disestablishment in America*. Columbia:
       University of South Carolina Press.

  1994 "American civil religion revisited." *Religion and American Culture* 4 (Winter):
       1–23.

Hammond, Phillip E., and Kee Warner
  1993 "Religion and ethnicity in late-twentieth-century America." *Annals of the Amer-
       ican Academy of Political and Social Science* 527 (May):55–66.

Hanf, Theodor
  1994 "The sacred marker: Religion, communalism, and nationalism." *Social Compass*
       41, 1:9–20.

Happold, F. C.
  1970 *Mysticism*. Baltimore: Penguin Books.

Hargrove, Barbara, Jean Miller Schmidt, and Sheila Greeve Davaney
  1985 "Religion and the changing role of women." *Annals of the American Academy of
       Political and Social Science* 480:117–131.

Harper, Charles L., and Bryan F. Le Beau
  1993 "The social adaptation of marginal religious movements in America." *Sociology of
       Religion* 54, 2:171–192.

Harper, Charles L., and Kevin Leicht
  1984 "Explaining the New Religious Right: Status politics and beyond." Pp. 101–110
       in D. Bromley and A. Shupe (eds.), *New Christian Politics*. Macon, Ga.: Mercer
       University Press.

Harrell, David Edwin
  1984 "Dispensational premillennialism and the religious right." Pp. 9–34 in J. Bettis
       and S. K. Johannesen (eds.), *The Return of the Millennium*. Barrytown, N.Y.:
       New ERA.

Harrison, Michael
  1974 "Sources of recruitment to Catholic Pentecostalism." *Journal for the Scientific
       Study of Religion* 13:49–64.

  1975 "The maintenance of enthusiasm: Involvement in a new religious movement."
       Sociological Analysis 36, 2:150–160.

Hart, Laurie Kain
  1992 *Time, Religion and Social Experience in Rural Greece*. Lanham, Md.: Rowman &
       Littlefield.

Hay, David, and Ann Morisy
  1985 "Secular society, religious meanings: A contemporary paradox." *Review of Reli-
       gious Research* 26, 3:213–227.

Haywood, Carol Lois
  1983 "The authority and empowerment of women among spiritualist groups." *Journal
       for the Scientific Study of Religion* 22, 2:157–166.

Heelas, Paul
  1991   "Cults for capitalism: Self religions, magic, and the empowerment of business."
          Pp. 27–41 in P. Gee and J. Fulton (eds.), *Religion and Power Decline and Growth:
          Sociological Analyses of Religion in Britain, Poland, and the Americas*. London:
          B.S.A., Sociology of Religion Study Group.
  1992   "The sacralization of the self and New Age capitalism." Pp. 139–166 in N.
          Abercrombie and A. Warde (eds.), *Social Change in Contemporary Britain*. Cam-
          bridge: Polity Press.

Heilman, Samuel C.
  1990   "The Jews: Schism or division?" Pp. 185–198 in T. Robbins and D. Anthony
          (eds.), *In Gods We Trust: New Patterns of Religious Pluralism in America*. New
          Brunswick, N.J.: Transaction.

Heilman, Samuel C., and Steven M. Cohen
  1989   *Cosmopolitans and Parochials: Modern Orthodox Jews in America*. Chicago: Univer-
          sity of Chicago Press.

Heirich, Max
  1977   "Change of heart: A test of some widely held theories about religious conver-
          sion." *American Journal of Sociology* 83, 3:653–680.

Heller, David
  1986   *The Children's God*. Chicago: University of Chicago Press.

Herberg, Will
  1960   *Protestant-Catholic-Jew: An Essay in American Religious Sociology*. Garden City,
          N.Y.: Doubleday.

Herbert, Frank
  1965   *Dune*. New York: Berkley Books.
  1969   *Dune Messiah*. New York: Berkley Books.
  1976   *Children of Dune*. New York: Berkley Books.
  1981   *God Emperor of Dune*. New York: Putnam.
  1984   *Heretics of Dune*. New York: Putnam.
  1985   *Chapterhouse Dune*. New York: Putnam.

Heriot, M. Jean
  1994   *Blessed Assurance: Beliefs, Actions, and the Experience of Salvation in a Carolina Bap-
          tist Church*. Knoxville: University of Tennessee Press.

Hernández, Edwin I.
  1988   "Eschatological hope and the Hispanic evangelical experience." Unpublished
          paper presented to the Society for the Scientific Study of Religion.

Hertel, Bradley R.
  1980   "Inconsistency of beliefs in the existence of heaven and afterlife." *Review of Reli-
          gious Research* 21, 2:171–183.

Hervieu-Léger, Danièle
  1990   "Religion and modernity in the French context: For a new approach to secular-
          ization." *Sociological Analysis* 51, S:S15–25.
  1993   "Present-day emotional renewals: The end of secularization or the end of reli-
          gion?" Pp. 129–148 in W. Swatos (ed.), *A Future for Religion?* Newbury Park,
          Calif.: Sage.

Hewitt, W. E.
  1990   "Religion and the consolidation of democracy in Brazil: The role of the Com-
          munidades Eclesiais de Base (CEBs)." *Sociological Analysis* 50, 2:139–152.

Hewlett, Sylvia
  1991   *When the Bough Breaks*. New York: Basic Books.

Higgenbotham, Evelyn Brooks
   1993   *Righteous Discontent: The Women's Movement in the Black Baptist Church,*
       1880–1920. Cambridge: Harvard University Press.

Hill, Herbert
   1973   "Anti-Oriental agitation and the rise of working-class racism." *Society* 10,
       2:43–54.

Hill, Michael
   1973a *The Religious Order: A Study of Virtuoso Religion and Its Legitimation in the Nine-*
       teenth Century Church of England. London: Heinemann.

   1973b *A Sociology of Religion.* New York: Basic Books.

Himmelstein, Jerome L.
   1983   "The New Right." Pp. 13–30 in R. C. Liebman and R. Wuthnow (eds.), *The*
       New Christian Right. New York: Aldine.

   1986   "The social basis of antifeminism: Religious networks and culture." *Journal for*
       the Scientific Study of Religion 25, 1:1–15.

Hobsbawm, Eric J.
   1959   *Primitive Rebels.* Manchester, England: Manchester University Press.

   1992   "Ethnicity and nationalism in Europe today." *Anthropology Today* 8, 1:3–13.

Hochschild, Arlie Russell
   1983   *The Managed Heart: Commercialization of Human Feeling.* Berkeley: University of
       California Press.

Hoch-Smith, Judith, and Anita Spring (eds.)
   1978   *Women in Ritual and Symbolic Roles.* New York: Plenum.

Hoge, Dean R.
   1981   *Converts, Dropouts, Returnees: A Study of Religious Change Among Catholics.* New
       York: Pilgrim.

   1987   *The Future of Catholic Leadership: Responses to the Priest Shortage.* Kansas City,
       Mo.: Sheed and Ward.

Hoge, Dean R., Benton Johnson, and Donald A Luidens
   1994   *Vanishing Boundaries: The Religion of Mainline Protestant Baby Boomers.* Louisville,
       Ky.: Westminster/John Knox Press.

Holt, Michael
   1973   "The Antimasonic and Know-Nothing Parties." Pp. 575–620 in A. M.
       Schlesinger, Jr. (ed.), *History of U.S. Political Parties,* Vol. I. New York: Chelsea
       House.

Hood, Ralph W., Jr.
   1985   "Mysticism." Pp. 285–297 in P. Hammond (ed.), *The Sacred in a Secular Age.*
       Berkeley: University of California Press.

Hood, Ralph W., Jr., Ronald J. Morris, and Paul J. Watson
   1989   "Prayer experience and religious orientation." *Review of Religious Research* 31,
       1:39–45.

Hout, Michael, and Andrew Greeley
   1982   "The center doesn't hold." *American Sociological Review* 52:325–347.

Houtart, François
   1977   "Theravada Buddhism and political power—construction and destruction of its
       ideological function." *Social Compass* 24, 2–3:207–246.

   1988   "Towards a sociology of Marxist atheism." *Social Compass* 35, 2–3:161–173.

Hunter, James Davison
   1983   *American Evangelicalism: Conservative Religion and the Quandary of Modernity.* New
       Brunswick, N.J.: Rutgers University Press.

1987a "American Protestantism: Sorting out the present, looking toward the future." *This World* 17:53–76.

1987b *Evangelicalism: The Coming Generation*. Chicago: University of Chicago Press.

1991 *Culture Wars: The Struggle to Define America*. New York: Basic Books.

Hurh, Won Moo, and Kwang Chung Kim
1990 "Religious participation of Korean immigrants in the United States." *Journal for the Scientific Study of Religion* 29, 1:19–34.

Hyde, Kenneth E.
1990 *Religion in Childhood and Adolescence: A Comprehensive Review of the Research.* Birmingham, Ala.: Religious Education Press.

Hyman, Paula
1973 "The other half: Women in the Jewish tradition." *Response: A Contemporary Jewish Review* 8, 18:67–76.

Iadarola, Antoinette
1985 "The American Catholic bishops and woman: From the Nineteenth Amendment to ERA." Pp. 457–476 in Y. Y. Haddad and E. B. Findly (eds.), *Women, Religion and Social Change*. Albany: State University of New York Press.

Idler, Ellen L.
1994 *Cohesiveness and Coherence: Religion and the Health of the Elderly*. New York: Garland.

Idler, Ellen L., and Stanislav V. Kasl
1992 "Religion, disability, depression, and the timing of death." *American Journal of Sociology* 97, 4:1052–1079.

Ignatieff, Michael
1993 *Blood and Belonging: Journeys into the New Nationalism*. New York: Farrar, Strauss, and Giroux.

Imbens, Annie, and I. Jorker
1992 *Christianity and Incest*. Minneapolis: Fortress Press.

Ireland, Rowan
1989 "Catholic Base Communities, spiritist groups and the deepening of democracy in Brazil." Pp. 224–250 in S. Mainwaring and A. Wilde (eds.), *The Progressive Church in Latin America*. Notre Dame, Ind.: University of Notre Dame Press.

Irwin, Susan, and Brigitte Jordan
1987 "Knowledge, practice and power: Court-ordered cesarean section." *Medical Anthropology Quarterly* 1, 3:319–334.

Isichei, Elizabeth
1967 "From sect to denomination among English Quakers." Pp. 161–181 in B. Wilson (ed.), *Patterns of Sectarianism*. London: Heinemann. Also, "Organization and power in the Society of Friends, 1852–59," pp. 182–212.

Jacobs, Janet L.
1989a *Divine Disenchantment: Deconverting from New Religions*. Bloomington: Indiana University Press.

1989b "The effects of ritual healing on female victims of abuse: A study of empowerment and transformation." *Sociological Analysis* 50, 3:265–279.

1990 "Women-centered healing rites: A study of alienation and reintegration." Pp. 373–383 in T. Robbins and D. Anthony (eds.), *In Gods We Trust: New Patterns of Religious Pluralism in America*. New Brunswick, N.J.: Transaction.

1995 "The violated self and the search for religious meaning." *Religion and the Social Order* 5:237–250.

Jacobson, Cardell K., Tim B. Heaton, and Rutledge M. Dennis
  1990   "Black-white differences in religiosity: Item analysis and a formal structural
          test." *Sociological Analysis* 51, 3:257–270.

Jahoda, Gustav
  1969   *The Psychology of Superstition.* London: Allen Lane.

James, Janet Wilson
  1980   *Women in American Religion.* Philadelphia: University of Pennsylvania Press.

James, William
  [1902] *The Varieties of Religious Experience.* New York: New American Library.
  1958

Jamison, A. Leland
  1961   "Religions on the Christian perimeter." Pp. 162–231 in J. W. Smith and A. L.
          Jamison (eds.), *The Shaping of American Religion.* Princeton, N.J.: Princeton Uni-
          versity Press.

Johnson, Benton
  1961   "Do holiness sects socialize in dominant values?" *Social Forces* 39:309–316.
  1963   "On church and sect." *American Sociological Review* 28:539–549.
  1971   "Church–sect revisited." *Journal for the Scientific Study of Religion* 10:124–137.
  1977   "Sociological theory and religious truth." *Sociological Analysis* 38, 4:368–388.

Johnson, Doyle Paul
  1979   "Dilemmas of charismatic leadership: The case of the People's Temple." *Socio-
          logical Analysis* 40, 4:315–323.

Johnson, Stephen D.
  1987   "Factors related to intolerance of AIDS victims." *Journal for the Scientific Study of
          Religion* 26, 1:105–110.

Johnson, Stephen D., and Joseph B. Tamney
  1984   "Support for the Moral Majority: A test of a model." *Journal for the Scientific
          Study of Religion* 23, 2:183–196.

Johnson, Weldon T.
  1971   "The religious crusade: Revival or ritual?" *American Journal of Sociology* 76,
          5:873–880.

Johnston, Hank, and Josef Figa
  1988   "The church and political opposition: Comparative perspectives on mobilization
          against authoritarian regimes." *Journal for the Scientific Study of Religion* 27,
          1:32–47.

Jones, Charisse
  1993   "Black Baptists focusing on social ills." *New York Times,* September 7.

Jones, R. Kenneth
  1975   "Some sectarian characteristics of therapeutic groups." Pp. 190–210 in R. Wallis
          (ed.), *Sectarianism: Analyses of Religious and Non-Religious Sects.* London: Peter
          Owen.
  1978   "Paradigm shifts and identity theory: Alternation as a form of identity manage-
          ment." Pp. 59–82 in H. Mol (ed.), *Identity and Religion.* London: Sage.

Jubber, Ken
  1985   "The prodigal church: South Africa's Dutch Reformed Church and the
          apartheid policy." *Social Compass* 32, 2–3:273–285.

Kanter, Rosabeth Moss
  1972   *Commitment and Community: Communes and Utopias in Sociological Perspective.*
          Cambridge, Mass.: Harvard University Press.

Kantrowitz, Barbara
   1994   "In search of the Sacred." *Newsweek,* Nov. 28:53–55.

Kasulis, Thomas P.
   1993   "The body—Japanese Style." Pp. 299–320 in Kasulis et al. (eds.), *Self as Body in Asian Theory and Practice.* Albany: State University of New York Press.

Kearl, Michael
   1980   "Time, identity, and the spiritual needs of the elderly." *Sociological Analysis* 41, 2:172–180.

   1993   "Dying American style: From moral to technological rite of passage." *American Journal of Ethics & Medicine* (Fall):12–18.

Kelley, Dean M.
   1987   "The Supreme Court redefines tax exemption." Pp. 115–124 in T. Robbins and R. Robertson (eds.), *Church–State Relations: Tensions and Transitions.* New Brunswick, N.J.: Transaction.

Kemp, Alice Abel
   1994   *Women's Work: Degraded and Devalued.* Englewood Cliffs, N.J.: Prentice Hall.

Kephart, William M., and William W. Zellner
   1991   *Extraordinary Groups: An Examination of Unconventional Life-Styles.* New York: St. Martin's Press.

Kerr, Howard, and Charles L. Crow (eds.)
   1983   *The Occult in America: New Historical Perspectives.* Urbana-Champagne: University of Illinois Press.

Kersevan, Marko
   1975   "Religion and the Marxist concept of social formation." *Social Compass* 22, 3–4:323–352.

Kertzer, David
   1988   *Ritual, Politics and Power.* New Haven, Conn.: Yale University Press.

Khalsa, Kirpal Singh
   1986   "New religious movements turn to worldly success." *Journal for the Scientific Study of Religion* 25, 2:233–247.

Kifner, John
   1994   "Through the Serbian mind's eye." *New York Times,* April 10.

Kilbourne, Brock, and James T. Richardson
   1989   "Paradigm conflict, types of conversion, and conversion theories." *Sociological Analysis* 50, 1:1–21.

Kim, Andrew E.
   1993   "The absence of pan-Canadian civil religion: Plurality, duality, and conflict in symbols of Canadian culture." *Sociology of Religion* 54, 3:257–275.

Kimmel, Michael
   1989   "'New prophets' and 'old ideals': Charisma and tradition in the Iranian revolution." *Social Compass* 36, 4:493–510.

Kipp, Rita Smith
   1995   "Conversion by affiliation: The history of the Karo Batak Protestant Church." *American Ethnologist* 22, 4:868–882.

Kivisto, Peter
   1993   "Religion and the new immigrants." Pp. 92–108 in W. H. Swatos (ed.), *A Future for Religion? New Paradigms for Social Analysis.* Newbury Park, Calif.: Sage.

Klaniczay, Gábor
   1990   *The Uses of Supernatural Power: The Transformation of Popular Religion in Medieval and Early-Modern Europe.* Princeton: Princeton University Press.

Klapp, Orrin
  1969   *Collective Search for Identity*. New York: Holt, Rinehart & Winston.

Kleinman, Arthur
  1973   "Some issues for a comparative study of medical healing." *International Journal of Social Psychiatry* 19, 3–4:159–165.

  1978   "The failure of Western medicine." *Human Nature,* Nov.:63–68.

  1988   *The Illness Narratives: Suffering, Healing and the Human Condition*. New York: Basic Books.

Knudsen, Dean, John Earle, and D. W. Shriver
  1978   "The conception of sectarian religion: An effort at clarification." *Review of Religious Research* 20, 1:44–60.

Koenig, Barbara
  1988   "The technological imperative in medical practice: The social creation of 'routine' treatment." Pp. 465–496 in M. Lock and D. R. Gordon (eds.), *Biomedicine Examined*. Dordrecht, Netherlands: Kluwer

Kokosalakis, Nikos
  1985   "Legitimation power and religion in modern society." *Sociological Analysis* 46, 4:367 376.

  1987   "The political significance of popular religion in Greece." *Archives des Sciences Sociales des Religions* 64:37–52.

Kosmin, Barry A., and Seymour P. Lachman
  1993   *One Nation Under God*. New York: Harmony Books.

Kostelaros, Frances
  1995   *Feeling the Spirit: Faith and Hope in an Evangelical Black Storefront Church*. Columbia: University of South Carolina Press.

Kraybill, Donald B.
  1989   *The Riddle of Amish Culture*. Baltimore: Johns Hopkins University Press.

Kroll-Smith, Stephen
  1980   "Testimony as performance." *Journal for the Scientific Study of Religion* 19, 1:16–25.

Kuhn, Thomas S.
  1970   *The Structure of Scientific Revolutions*. Chicago: University of Chicago Press.

Kurtz, Lester
  1986   *The Politics of Heresy: The Modernist Crisis in Roman Catholicism*. Berkeley: University of California Press.

Labaki, Boutros
  1988   "Confessional communities, social stratification and wars in Lebanon." *Social Compass* 35, 4:533–561.

Laeyendecker, Leo
  1972   "The Netherlands." Pp. 325–363 in H. Mol (ed.), *Western Religion: A Country by Country Sociological Inquiry*. The Hague: Mouton.

Lakoff, Robin
  1975   *Language and Women's Place*. New York: Harper & Row.

L'Alive D'Epinay, Christian
  1969   *Haven of the Masses: A Study of the Pentecostal Movement in Chile*. London: Lutterworth.

Lancaster, Roger N.
  1988   *Thanks to God and the Revolution: Popular Religion and Class Consciousness in the New Nicaragua*. New York: Columbia University Press.

Lanternari, Vittorio
  1963   *The Religions of the Oppressed: A Study of Modern Messianic Cults.* Tr. L. Sergio.
          New York: Knopf.

Larson, Lyle E., and J. Walter Goltz
  1989   "Religious participation and marital commitment." *Review of Religious Research*
          30, 4:387–400.

Larson, Lyle E., and Brenda Munro
  1985   "Religious intermarriage in Canada, 1974–1982." *International Journal of the Soci-
          ology of the Family* 15:31–49.

Latkin, Carl A., Norman D. Swindberg, Richard A. Littman, Melissa G. Katsikis, and
Richard A. Hagan
  1994   "Feelings after the fall: Former Rajneeshpuram commune members' perceptions
          of and affiliation with the Rajneeshee Movement." *Sociology of Religion* 55,
          1:65–73.

Lebra, Sugiyama
  1972   "Millenarian movements and resocialization." *American Behavioral Scientist* 16,
          2:195–217.

Lechner, Frank J.
  1990   "Fundamentalism revisited." Pp. 77–97 in T. Robbins and D. Anthony (eds.),
          In Gods We Trust: New Patterns of Religious Pluralism in America. New Brunswick,
          N.J.: Transaction.

  1991   "Religion, law and global order." Pp. 263–281 in R. Robertson and W. R.
          Garrett (eds.), *Religion and Global Order.* New York: Paragon House.

  1993   "Global Fundamentalism." Pp. 19–36 in W. Swatos (ed.), *A Future for Religion?
          New Paradigms for Social Analysis.* Newbury Park, Calif.: Sage.

Lee, Richard Wayne
  1995   "Strained bedfellows: Pagans, New Agers, and 'starchy humanists' in Unitarian
          Universalism." *Sociology of Religion* 56, 4:379–396.

Lehman, Edward C., Jr.
  1985   *Women Clergy: Breaking Through Gender Barriers.* New Brunswick, N.J.:
          Transaction.

  1987   *Women Clergy in England: Sexism, Modern Consciousness, and Church Viability.*
          Lewiston, N.Y.: Edwin Mellen.

  1993   *Gender and Work: The Case of the Clergy.* Albany: State University of New York
          Press.

Lemercinier, Geneviève
  1981   "Relationship between means of production, caste, and religion." *Social Compass*
          28, 2–3:163–199.

Lemert, Charles
  1975   "Defining non-church religion." *Review of Religious Research* 16, 3:186–197.

Lenski, Gerhard
  1963   *The Religious Factor: A Sociological Study of Religion's Impact on Politics, Economics
          and Family Life.* Garden City, N.Y.: Doubleday.

Lerner, Max
  1937   "The Constitution and the Court as symbols." *Yale Law Journal* 46:1290–1319.

Levine, Daniel H.
  1990   "From church and state to religion and politics and back again." *Social Compass*
          37, 3:331–351.

  1992   *Popular Voices in Latin American Catholicism.* Princeton: Princeton University
          Press.

Levine, Daniel H., and Scott Mainwaring
  1989 "Religion and popular protest in Latin America: Contrasting experiences."
        Pp. 203–240 in S. Eckstein (ed.), *Power and Popular Protest: Latin American Social
        Movements*. Berkeley: University of California Press.

Lewis, James R.
  1989 "Apostates and the legitimation of repression: Some historical and empirical per-
        spectives on the cult controversy." *Sociological Analysis* 49, 4:386–396.

Lewis, James R., and J. Gordon Melton, eds.
  1992 *Perspectives on the New Age*. Albany: State University of New York Press.

Lewis, Robert E., Mark W. Fraser, and Peter J. Pecora
  1988 "Religiosity among Indochinese refugees in Utah." *Journal for the Scientific Study
        of Religion* 27, 2:272–283.

Lewy, Guenther
  1974 *Religion and Revolution*. New York: Oxford University Press.

Liebman, Charles S.
  1988 *Deceptive Images: Toward a Redefinition of American Judaism*. New Brunswick, N.J.:
        Transaction.

  1993 "Jewish Fundamentalism and the Israeli Polity." Pp. 68–87 in M. E. Marty and
        R. S. Appleby (eds.), *Fundamentalisms and the State*. Chicago: University of
        Chicago Press.

Liebman, Charles S., and Eliezar Don-Yehiya
  1983 *Civil Religion in Israel: Traditional Judaism and Political Culture in the Jewish State*.
        Berkeley: University of California Press.

Lifton, Robert J.
  1963 *Thought Reform and the Psychology of Totalism*. New York: Norton.

Lincoln, C. Eric
  1974 "The power in the black church." *Cross Currents* 14, 1:3–21.

  1989 "The Muslim Mission in the context of American social history." Pp. 340–356
        in G. S. Wilmore (ed.), *African American Religious Studies*. Durham, N.C.: Duke
        University Press.

Lincoln, C. Eric, and Lawrence H. Mamiya
  1990 *The Black Church in the African American Experience*. Durham, N.C.: Duke Uni-
        versity Press.

Lippy, Charles H.
  1994 *Being Religious, American Style: A History of Popular Religiosity in the United States*.
        Westport, Conn.: Greenwood.

Lipset, Seymour M.
  1963 "Three decades of the radical right: Coughlinites, McCarthyites, and Birchers."
        Pp. 373–446 in D. Bell (ed.), *The Radical Right*. Garden City, N.Y.: Doubleday.

Lipset, Seymour M., and Earl Raab
  1981 "The election and the Evangelicals." *Commentary* 71 (March):25–32.

Lofland, John
  1966 *Doomsday Cult: A Study of Conversion, Proselytization and Maintenance of Faith*.
        Englewood Cliffs, N.J.: Prentice-Hall.

  1977 "Becoming a world-saver revisited." *American Behavioral Scientist* 20, 6:805–818.

Lofland, John, and James T. Richardson
  1984 "Religious movement organizations: Elemental forms and dynamics." *Research
        in Social Movements, Conflict and Change* 7:29–51.

London Sunday Times Insight Team
  1972 *Northern Ireland: A Report on the Conflict*. New York: Vintage.

Long, Theodore E., and Jeffrey K. Hadden
   1983   "Religious conversion and the concept of socialization: Integrating the brain-washing and drift models." *Journal for the Scientific Study of Religion* 22, 1:1–14.

Lorentzen, Louise J.
   1980   "Evangelical life-style concerns expressed in political action." *Sociological Analysis* 41, 2:144–154.

Lucas, Phillip Charles
   1995   *The Odyssey of a New Religion: The Holy Order of MANS from New Age to Orthodoxy*. Bloomington, Ind.: Indiana University Press.

Luckmann, Thomas
   1967   *The Invisible Religion: The Problem of Religion in Modern Society*. New York: Macmillan.

   1973   "Comments on the Laeyendecker et al. research proposal." Pp. 55–68 in *The Contemporary Metamorphosis of Religion: Acts of the 12th Conference Internationale de Sociologie Religieuse*. Lille, France: CISR.

   1977   "Theories of religion and social change." *Annual Review of the Social Sciences of Religion* 1:1–28.

   1987   "Comments on legitimation." *Current Sociology* 35, 2:109–117.

   1990   "Shrinking transcendence, expanding religion?" *Sociological Analysis* 50, 2:127–138.

Luker, Kristen
   1984   *Abortion and the Politics of Motherhood*. Berkeley: University of California Press.

Lummis, Adair
   1994   "Feminist values and other influences on pastoral leadership styles: Does gender matter?" Unpublished paper presented to the Society for the Scientific Study of Religion.

Luria, Keith P.
   1991   *Territories of Grace: Cultural Change in the Seventeenth-Century Diocese of Grenoble*. Berkeley: University of California Press.

Lustick, Ian S.
   1988   *The Land and the Lord: Jewish Fundamentalism in Israel*. New York: Council on Foreign Relations.

McAllister, Ian
   1982   "The devil, miracles, and the afterlife: The political sociology of religion in Northern Ireland." *British Journal of Sociology* 33, 3:330–347.

McAvoy, Thomas
   1957   *The Great Crisis in American Catholic History, 1895–1900*. Chicago: Henry Regnery.

McClelland, David C.
   1961   *The Achieving Society*. Princeton, N.J.: Princeton University Press.

McCracken, J. L.
   1967   "Northern Ireland, 1921–1966." Pp. 313–323 in T. W. Moody and F. X. Martin (eds.), *The Course of Irish History*. Cork, Ireland: Mercier.

McCready, William, and Andrew Greeley
   1976   *The Ultimate Values of the American Population*. Beverly Hills, Calif.: Sage.

McCutcheon, Alan L.
   1988   "Denominations and religious intermarriage: Trends among white Americans in the twentieth century." *Review of Religious Research* 29, 3:213–227.

McDannell, Colleen
1986   *The Christian Home in Victorian America, 1840–1900.* Bloomington: Indiana University Press.
1995   *Material Christianity: Religion and Popular Culture in America.* New Haven: Yale University Press.

MacDonald, Jeffery L.
1995   "Inventing traditions for the New Age: A case study of the Earth Energy tradition." *Anthropology of Consciousness* 6, 4:31–45.

MacEoin, Gary
1974a   "Irish Catholicism: What Protestant Christians fear." *Cross Currents* 23:397–417.

1974b   *Northern Ireland.* New York: Holt, Rinehart & Winston.

1987   "Catholics, Catholicism, and the Northern Ireland crisis." Pp. 124–140 in A. J. Ward (ed.), *Northern Ireland: Living with the Crisis.* New York: Praeger.

McFarland, Sam G.
1989   "Religious orientations and the targets of discrimination." *Journal for the Scientific Study of Religion* 28, 3:324–336.

McGaw, Douglas B.
1980   "Meaning and belonging in a Charismatic congregation: An investigation into sources of neo-Pentecostal success." *Review of Religious Research* 21:284–301.

McGuire, Meredith B.
1972   "Toward a sociological interpretation of the Underground Church movement." *Review of Religious Research* 14, 1:41–47.

1974   "An interpretive comparison of elements of the Pentecostal and Underground Church movements in American Catholicism." *Sociological Analysis* 35, 1:57–65.

1977   "Testimony as a commitment mechanism in Catholic Pentecostal prayer groups." *Journal for the Scientific Study of Religion* 16, 2:165–168.

1982   *Pentecostal Catholics: Power, Charisma and Order in a Religious Movement.* Philadelphia: Temple University Press.

1983   "Discovering religious power." *Sociological Analysis* 44, 1:1–10.

1985   "Religion and healing." Pp. 268–284 in P. Hammond (ed.), *The Sacred in a Secular Age.* Berkeley: University of California Press.

1988   *Ritual Healing in Suburban America.* New Brunswick, N.J.: Rutgers University Press.

1993a   "Health and healing in new religious movements." *Religion and the Social Order* 3B:139–155.

1993b   "Health and spirituality as contemporary concerns." *Annals of the American Academy of Political and Social Science* 527 (May):144–154.

1994   "Gendered spirituality and quasi-religious ritual." *Religion and the Social Order* 4:273–287.

1996   "Religion and healing the Mind/Body/Self." *Social Compass* 43, 1:101–116.

MacIntyre, Alisdair
1967   *Secularisation and Moral Change.* London: Oxford University Press.

MacIver, Martha Abele
1989   "A clash of symbols in Northern Ireland: Division between extremist and modern Protestant elites." *Review of Religious Research* 30, 4:360–374.

McKinney, William, and Wade Clark Roof
1990   "Liberal Protestantism's struggle to recapture the heartland." Pp. 167–176 in T. Robbins and D. Anthony (eds.), *In Gods We Trust: New Patterns of Religious Pluralism in America.* New Brunswick, N.J.: Transaction.

McKown, Delos B.
  1975   *The Classical Marxist Critiques of Religion: Marx, Engels, Lenin, Kautsky.* The Hague: Martinus Nijhoff.

McLaughlin, Eleanor C.
  1974   "Equality of souls, inequality of sexes: Women in medieval theology." Pp. 213–266 in R. Ruether (ed.), *Religion and Sexism: Images of Women in the Jewish and Christian Traditions.* New York: Simon & Schuster.

McLoughlin, William G.
  1983   "Timepieces and butterflies: A note on the Great-Awakening-Construct and its critics." *Sociological Analysis* 44:103–110.

McNamara, Patrick H.
  1985   "Conservative Christian families and their moral world view." *Sociological Analysis* 46, 2:93–99.

  1992   *Conscience First, Tradition Second: A Study of Young American Catholics.* Albany: State University of New York Press.

Machalek, Richard, and David A. Snow
  1993   "Conversion to new religious movements." *Religion and the Social Order* 3B:53–74.

Mack, Phyllis
  1992   *Visionary Women: Ecstatic Prophecy in Seventeenth-Century England.* Berkeley: University of California Press.

Maduro, Otto
  1975   "Marxist analysis and the sociology of religion." *Social Compass* 22, 3–4: 305–322.

  1977   "New Marxist approaches to the relative autonomy of religion." *Sociological Analysis* 38, 4:359–367.

  1982   *Religion and Social Conflicts.* Tr. Robt. R. Barr. Maryknoll, N.Y.: Orbis.

Magida, Arthur J., ed.
  1995   *How To Be a Perfect Stranger: A Guide to Etiquette in Other People's Religious Ceremonies.* Woodstock, Vt.: Jewish Lights Publishing.

Mainwaring, Scott
  1986   *The Catholic Church and Politics in Brazil, 1916–1985.* Stanford, Calif.: Stanford University Press.

Malinowski, Bronislaw
  [1925] "Magic, science and religion." Pp. 17–92 in *Magic, Science and Religion and Other*
  1948   *Essays.* New York: Free Press.

Mamiya, Lawrence H.
  1982   "From Black Muslim to Bilalian: The evolution of a movement." *Journal for the Scientific Study of Religion* 21, 2:138–152.

  1994   "A social history of the Bethel African Methodist Episcopal Church in Baltimore: The house of God and the struggle for freedom." Pp. 221–292 in J. P. Wind and J. W. Lewis (eds.), *American Congregations,* Vol. 1. Chicago: University of Chicago Press.

Mamiya, Lawrence H., and C. Eric Lincoln
  1988   "Black militant and separatist movements." Pp. 755–771 in C. H. Lippy and P. W. Williams (eds.), *Encyclopedia of the American Religious Experience,* Vol. II. New York: Scribner.

Manis, Andrew Michael
  1991   "Religious experience, religious authority, and civil rights leadership: The case of Birmingham's Reverend Fred Shuttlesworth." Pp. 143–154 in C. R. Wilson

(ed.), *Cultural Perspectives on the American South,* Vol. 5: *Religion.* New York: Gordon and Breach.

Mansueto, Anthony
1988   "Religion, solidarity and class struggle: Marx, Durkheim, and Gramsci on the religion question." *Social Compass* 35, 2–3:261–277.

Manwaring, David
1962   *Render unto Caesar: The Flag Salute Controversy.* Chicago: University of Chicago Press.

Marglin, Frédérique Apffel
1985   "Female sexuality in the Hindu world." Pp. 39–60 in C. W. Atkinson, C. H. Buchanan, and M. R. Miles (eds.), *Immaculate and Powerful: The Female in Sacred Image and Social Reality.* Boston: Beacon Press.

Markoff, John, and Daniel Regan
1981   "The rise and fall of civil religion: Comparative perspectives." *Sociological Analysis* 42, 4:333–352.

Marler, Penny Long
1995   "Lost in the Fifties: The changing family and the nostalgic church." Pp. 23–60 in N. T. Ammerman and W. C. Roof (eds.), *Work, Family, and Religion in Contemporary Society.* New York: Routledge.

Marler, Penny Long, and David Roozen
1993   "From church tradition to consumer choice." Pp. 253–277 in D. Roozen and C. K. Hadaway (eds.), *Church and Denominational Growth.* Nashville: Abingdon.

Martin, Bernard (ed.)
1978   *Movement and Issues in American Judaism: An Analysis and Sourcebook of Developments since 1945.* Westport, Conn.: Greenwood.

Martin, David A.
1962   "The denomination." *British Journal of Sociology* 13, 2:1–14.

1965   *Pacificism: An Historical and Sociological Study.* New York: Schocken.

1978   *A General Theory of Secularization.* New York: Harper & Row.

Martin, John
1994   "Theories of Practice: Inquisition and the Discovery of Religion." Unpublished paper presented to Trinity University Humanities Symposium.

Marty, Martin
1970   "The occult establishment." *Social Research* 37:212–230.

1974   "Two kinds of two kinds of civil religion." Pp. 139–160 in R. Richey and D. Jones (eds.), *American Civil Religion.* New York: Harper & Row.

Marx, Gary T.
1967   "Religion: Opiate or inspiration of civil rights militancy among Negroes?" *American Sociological Review* 32, 1:64–72.

Marx, Karl
[1844] "Contribution to the critique of Hegel's philosophy of right." Pp. 43–59 in
1963   T. B. Bottomore (ed.), *Early Writings.* New York: McGraw-Hill.

Maryknoll Fathers (eds.)
1957   *Daily Missal of the Mystical Body.* New York: Kennedy.

Maslow, Abraham
1964   *Religions, Values and Peak-Experiences.* New York: Viking.

Mauss, Armand L.
1994   *The Angel and the Beehive: The Mormon Struggle with Assimilation.* Urbana: University of Illinois Press.

Mayrl, William
   1976   "Marx' theory of social movements and the church–sect typology." *Sociological Analysis* 37, 1:19–31.

Mead, George Herbert
   1918   "The psychology of punitive justice." *American Journal of Sociology* 23:577–602.

Melaugh, Martin
   1995   "Majority-minority differentials: Unemployment, housing and health." Pp. 131–148 in S. Dunn (ed.), *Facets of the Conflict in Northern Ireland*. New York: St. Martin's Press.

Melton, J. Gordon, Jerome Clark, and Aidan A. Kelly
   1990   *New Age Encyclopedia*. Detroit: Gale Research.

Menendez, A. J.
   1973   *The Bitter Harvest: Church and State in Northern Ireland*. Washington, D.C.: Luce.

Mess, Zulkarnaina M., and W. Barnett Pearce
   1986   "Dakwah Islamiah: Islamic revivalism in the politics of race and religion in Malaysia." Pp. 196–220 in J. K. Hadden and A. Shupe (eds.), *Prophetic Religions and Politics: Religion and the Political Order*. New York: Paragon.

Metraux, Alfred
   1972   *Voodoo in Haiti*. New York: Schocken.

Meyer, Donald
   1965   *The Positive Thinkers*. Garden City, N.Y.: Doubleday.

Michaelson, Robert S.
   1987   "Civil rights, Indian rights." Pp. 125–133 in T. Robbins and R. Robertson (eds.), *Church–State Relations: Tensions and Transitions*. New Brunswick, N.J.: Transaction.

Milhaven, John G.
   1993   *Hadewijch and Her Sisters*. Albany: State University of New York Press.

Millennial Prophesy Report
   1994   (November) Philadelphia, Pa.: Ted Daniels, Publisher.

Miller, Jon
   1993   "Missions, social change, and resistance to authority: Notes toward an understanding of the relative autonomy of religion." *Journal for the Scientific Study of Religion* 32, 1:29–50.

Miller, Walter M., Jr.
   1959   *A Canticle for Liebowitz*. New York: Harold Matson.

Mindel, Charles H., and C. Edwin Vaughan
   1978   "A multidimensional approach to religiosity and disengagement." *Journal of Gerontology* 33:103–108.

Mojzes, Paul
   1987   "The impact of the Eastern European church upon their own societies." Pp. 258–279 in R. L. Rubenstein (ed.), *Spirit Matters: The Worldwide Impact of Religion on Contemporary Politics*. New York: Paragon.

Mol, Hans J. (ed.)
   1976   *Identity and the Sacred*. New York: Free Press.

Moodie, T. Dunbar
   1978   "The Afrikaner civil religion." Pp. 203–228 in H. Mol (ed.), *Identity and Religion*. London: Sage.

Moody, Edward
   1971   "Urban witches." Pp. 280–290 in J. Spradley and D. McCurdy (eds.), *Conformity and Conflict: Readings in Cultural Anthropology*. Boston: Little, Brown.

1974   "Magical therapy: An anthropological investigation of contemporary Satanism."
        Pp. 355–382 in I. Zaretsky and M. Leone (eds.), *Religious Movements in Contemporary America*. Princeton, N.J.: Princeton University Press.

Mooney, James
1965   *The Ghost-Dance Religion and the Sioux Outbreak of 1890*. Chicago: University of
        Chicago Press.

Moore, R. Laurence
1994   *Selling God: American Religion in the Marketplace of Culture*. New York: Oxford
        University Press.

Moorhead, James H.
1994   "The American Israel: Protestant tribalism and the universal mission."
        Pp. 145–166 in W. Hutchinson and H. Lehmann (eds.), *Many Are Chosen: Divine Election and Western Nationalism*. Minneapolis: Fortress/Harvard Theological
        Studies.

Morrow, Duncan
1995   "Church and religion in the Ulster crisis." Pp. 151–167 in S. Dunn (ed.), *Facets
        of the Conflict in Northern Ireland*. New York: St. Martin's Press.

Mörth, Ingo
1987   "Elements of religious meaning in science-fiction literature." *Social Compass* 34,
        1:87–108.

Moses, Wilson Jeremiah
1993   *Black Messiahs and Uncle Toms: Social and Literary Manipulations of a Religious
        Myth*. University Park: Pennsylvania State University Press.

Mosqueda, Lawrence
1986   *Chicanos, Catholicism and Political Ideology*. New York: University Press of
        America.

Moxon-Browne, Edward
1983   *Nation, Class and Creed in Northern Ireland*. Aldershot, England: Gower.

Muller, Jessica H., and Barbara A. Koenig
1988   "On the boundary of life and death: The definitions of dying by medical residents." Pp. 351–374 in M. Lock and D. R. Gordon (eds.), *Biomedicine Examined*. Dordrecht, Netherlands: Kluwer.

Murphy, Joseph M.
1988   *Santería: An African Religion in America*. Boston: Beacon.

Murray, Dominic
1995   "Culture, religion and violence in Northern Ireland." Pp. 213–229 in S. Dunn
        (ed.), *Facets of the Conflict in Northern Ireland*. New York: St. Martin's Press.

Murvar, Vatro
1971   "Messianism in Russia: Religious and revolutionary." *Journal for the Scientific
        Study of Religion* 10, 4:277–338.

Myerhoff, Barbara
1974   *The Peyote Hunt: The Sacred Journey of the Huichol Indians*. Ithaca, N.Y.: Cornell
        University Press.

1978   "Bobbes and Zeydes: Old and new roles for elderly Jews." Pp. 207–244 in
        J. Hoch-Smith and A. Spring (eds.), *Women in Ritual and Symbolic Roles*. New
        York: Plenum.

Nash, Manning
1991   "Islamic resurgence in Malaysia and Indonesia." Pp. 691–739 in M. E. Marty
        and R. S. Appleby (eds.), *Fundamentalisms Observed*. Chicago: University of
        Chicago Press.

Nason-Clark, Nancy
  1987a   "Are women changing the image of ministry? A comparison of British and
          American realities." *Review of Religious Research* 28, 4:330–340.
  1987b   "Ordaining women as priests: Religious vs. sexist explanations for clerical atti-
          tudes." *Sociological Analysis* 48, 3:259–273.
  1995   "Conservative Protestants and violence against women." *Religion and the Social
          Order* 5:109–130.

Neitz, Mary Jo
  1986   *Charisma and Community: A Study of Religious Commitment Within the Charismatic
          Renewal.* New Brunswick, N.J.: Transaction.
  1990   "In goddess we trust." Pp. 353–372 in T. Robbins and D. Anthony (eds.), *In
          Gods We Trust: New Patterns of Religious Pluralism in America.* New Brunswick,
          N.J.: Transaction.
  1994   "Quasi-religions and cultural movements: Contemporary witchcraft as a
          churchless religion." *Religion and the Social Order* 4:127–149.
  1995   "Constructing women's rituals: Roman Catholic women and 'Limina.'"
          Pp. 283–304 in N. T. Ammerman and W. C. Roof (eds.), *Work, Family, and
          Religion in Contemporary Society.* New York: Routledge.

Neitz, Mary Jo, and James V. Spickard
  1990   "Steps toward a sociology of religious experience: The theories of Mihaly Csik-
          szentmihalyi and Alfred Schutz." *Sociological Analysis* 51, 1:15–33.

Nelsen, Hart M., Neil H. Cheek, Jr., and Paul Au
  1985   "Gender differences in images of God." *Journal for the Scientific Study of Religion*
          24, 4:396–402.

Nelsen, Hart M., and Conrad L. Kanagy
  1993   "Churched and unchurched black Americans." Pp. 311–323 in D. Roozen and
          K. Hadaway (eds.), *Church and Denominational Growth.* Nashville: Abingdon.

Nelsen, Hart M., and Anne Kusner Nelsen
  1975   *The Black Church in the Sixties.* Lexington: University of Kentucky Press.

Nelson, Benjamin
  1949   *The Idea of Usury: From Tribal Brotherhood to Universal Otherhood.* Princeton, N.J.:
          Princeton University Press.

Nelson, Geoffrey
  1969   *Spiritualism and Society.* New York: Schocken.

New York Times, The
  1986   "South Africa's main white Dutch Reformed Church, in break from its support
          of racial separation, denounces apartheid," October 23.
  1989   "For mothers, elderly present second burden." May 13.
  1995   "In a faded Chicago area, a full-time church shoulders full-time burdens." Feb-
          ruary 3.

Niebuhr, H. Richard
  1929   *The Social Sources of Denominationalism.* New York: Meridian.

Niebuhr, R. Gustav
  1992   "The Lord's name." *Wall Street Journal,* April 27.
  1995   "Where shopping-mall culture gets a big dose of religion." *New York Times,*
          April 16.

Noonan, John T., Jr.
  1965   *Contraception: A History of Its Treatment by Catholic Theologians and Canonists.*
          Cambridge, Mass.: Harvard University Press.

Obelkevich, James
  1979   "Introduction." Pp. 3–7 in J. Obelkevich (ed.), *Religion and the People,
          800–1700.* Chapel Hill: University of North Carolina Press.

Oberoi, Harjot
  1993   "Sikh Fundamentalism: Translating history into theory." Pp. 256–285 in
          M. Marty and R. S. Appleby (eds.), *Fundamentalisms and the State: Remaking Poli-
          ties, Economies, and Militance.* Chicago: University of Chicago Press.

O'Brien, Conor Cruise
  1974   *States of Ireland.* London: Granada.

  1987   *The Siege: The Saga of Israel and Zionism.* New York: Simon & Schuster.

  1995   *Ancestral Voices: Religion and Nationalism in Ireland.* Chicago: University of
          Chicago Press.

O'Brien, Jay
  1993   "Ethnicity, national identity, and social conflict." *Nordic Journal of African Studies*
          2, 2:60–82.

O'Connor, Mary
  1989   "The Virgin of Guadalupe and the economics of symbolic behavior." *Journal for
          the Scientific Study of Religion* 28, 2:105–119.

O'Dea, Thomas
  1961   "Five dilemmas in the institutionalization of religion." *Journal for the Scientific
          Study of Religion* 1:30–39.

O'Leary, Stephen
  1994   *Arguing the Apocalypse: A Theory of Millennial Rhetoric.* New York: Oxford Uni-
          versity Press.

Oomen, T. K.
  1989   "State and religion in multi-religious nation-states: The case of South Asia."
          Unpublished paper presented to the Conference Internationale de Sociologie
          des Religions.

"Orthodox Eastern Church"
  1943   *Encyclopaedia Brittanica* 16:938–942.

O'Toole, Roger
  1976   "Underground traditions in the study of sectarianism: Non-religious uses of the
          concept 'sect.'" *Journal for the Scientific Study of Religion* 15, 2:145–156.

  1984   *Religion: Classic Sociological Approaches.* Toronto: McGraw-Hill-Ryerson.

Ozorak, Elizabeth Weiss
  1989   "Social and cognitive influences on the development of religious beliefs and
          commitment in adolescence." *Journal for the Scientific Study of Religion* 28,
          4:448–463.

Pace, Enzo
  1989   "Pilgrimage as spiritual journey: An analysis of pilgrimage using the theory of
          V. Turner and the resource mobilization approach." *Social Compass* 36, 2:229–244.

Page, Ann L., and Donald A. Clelland
  1978   "The Kanawha County textbook controversy: A study of the politics of lifestyle
          concern." *Social Forces* 57:265–281.

Palmer, Susan J.
  1993   "Women's 'cocoon work' in new religious movements: Sexual experimentation
          and feminine rites of passage." *Journal for the Scientific Study of Religion* 32,
          4:343–355.

  1994   *Moon Sisters, Krishna Mothers, Rajneesh Lovers: Women's Roles in New Religions.*
          Syracuse, N.Y.: Syracuse University Press.

Palmer, Susan J., and Frederick Bird
  1992   "Therapy, charisma, and social control in the Rajneesh movement." *Sociological Analysis* 53(S):S71–S85.

Parker, Cristián
  1993   *Otra Lógica en América Latina: Religión Popular y Modernización Capitalista.* Santiago, Chile: Funda de Cultura Económica.

  1994   "La sociología de la religión y la modernidad: por una revisión crítica de las categorías durkheimianas desde América Latina." *Revista Mexicana de Sociología* 55:229–254.

Parsons, Talcott
  1944   "The theoretical development of the sociology of religion." *Journal of the History of Ideas* 5:176–190.

  1969   "Family and church as 'boundary' structures." Pp. 423–429 in N. Birnbaum and G. Lenzer (eds.), *Sociology and Religion.* Englewood Cliffs, N.J.: Prentice-Hall.

  1971   "Belief, unbelief and disbelief." Pp. 207–245 in R. Caporale and A. Grumelli (eds.), *The Culture of Unbelief.* Berkeley: University of California Press.

  1972   "Definitions of health and illness in the light of American values and social structure." Pp. 107–127 in E. G. Jaco (ed.), *Patients, Physicians and Illness.* New York: Macmillan.

Parsons, Talcott, Edward Shils, Kaspar D. Naegele, and Jesse R. Pitts
  1961   *Theories of Society: Foundations of Modern Sociological Theory.* New York: Free Press.

Pepper, George B.
  1988   *The Boston Heresy Case in View of the Secularization of Religion: A Case Study in the Sociology of Religion.* Lewiston, N.Y.: Edwin Mellen.

Peshkin, Alan
  1986   *God's Choice: The Total World of a Fundamentalist Christian School.* Chicago: University of Chicago Press.

Peter, Karl A.
  1987   *The Dynamics of Hutterite Society.* Edmonton: University of Alberta Press.

Peters, Victor
  1971   *All Things Common: The Hutterite Way of Life.* New York: Harper & Row.

Pfeffer, Leo
  1974   "The legitimation of marginal religions in the United States." Pp. 9–26 in I. Zaretsky and M. Leone (eds.), *Religious Movements in Contemporary America.* Princeton, N.J.: Princeton University Press.

  1987   "Religious exemption." Pp. 103–114 in T. Robbins and R. Robertson (eds.), *Church–State Relations: Tensions and Transitions.* New Brunswick, N.J.: Transaction.

  1992   "Issues that divide: The triumph of secular humanism." Pp. 428–443 in M. Marty (ed.), *Civil Religion, Church and State.* Munich: K. G. Saur.

Pickering, W. S. F.
  1984   *Durkheim's Sociology of Religion: Themes and Theories.* London: Routledge & Kegan Paul.

Pilarzyk, Thomas
  1978   "The origin, development, and decline of a youth culture religion: An application of sectarianization theory." *Review of Religious Research* 20, 1:23–43.

Pitts, Walter F.
  1993   *Old Ship of Zion: The Afro-Baptist Ritual in the African Diaspora.* New York: Oxford University Press.

Poll, Solomon
  1969   *The Hasidic Community in Williamsburg: A Study in Sociology of Religion.* New
          York: Schocken.

Poloma, Margaret
  1989   *The Assemblies of God at the Crossroads: Charisma and Institutional Dilemmas.*
          Knoxville: University of Tennessee Press.

Poloma, Margaret, and Brian F. Pendleton
  1989   "Religious experiences, evangelism, and institutional growth within the Assem-
          blies of God." *Journal for the Scientific Study of Religion* 28, 4:415–431.

Pope, Barbara Corrado
  1985   "Immaculate and powerful: The Marian revival in the nineteenth century."
          Pp. 173–200 in C. W. Atkinson, C. H. Buchanan, and M. R. Miles (eds.), *Im-
          maculate and Powerful: The Female in Sacred Image and Social Reality.* Boston: Bea-
          con Press.

Pope, Liston
  1942   *Millhands and Preachers.* New Haven, Conn.: Yale University Press.

Posterski, Donald C., and Irwin Barker
  1993   *Where's a Good Church?* Winfield, B.C.: Wood Lake Books.

Potvin, Raymond H., and C. F. Lee
  1982   "Adolescent religion: A developmental approach." *Sociological Analysis* 43,
          2:131–144.

Prell, Riv-Ellen
  1989   *Prayer and Community: The Havurah in American Judaism.* Detroit: Wayne State
          University Press.

Priesand, Sally
  1975   *Judaism and the New Woman.* New York: Behrman.

Prokesch, Steven
  1990   "With the world made over, can even Belfast change?" *New York Times,*
          April 2.

Raab, Earl (ed.)
  1964   *Religious Conflict in America.* New York: Doubleday.

Raboteau, Albert J.
  1978   *Slave Religion: The "Invisible Institution" in the Antebellum South.* New York: Ox-
          ford University Press.

  1995   *A Fire in the Bones: Reflections on African-American Religious History.* Boston:
          Beacon.

Ramet, Pedro
  1989   "The interplay of religious policy and nationalistic policy in the Soviet Union
          and Eastern Europe." Pp. 3–41 in P. Ramet (ed.), *Religion and Nationalism in
          Soviet and East European Politics.* Durham, N.C.: Duke University Press.

Ray, Melissa L.
  1994   "Partial alienation as organizational parent-member accommodation: An urban,
          midwestern Catholic parish." *Sociology of Religion* 55, 1:53–64.

Regan, Daniel
  1989   "Islam as a new religious movement in Malaysia." Pp. 124–146 in J. A. Beck-
          ford and T. Luckmann (eds.), *The Changing Face of Religion.* London: ISA/Sage.

Report of the National Advisory Commission on Civil Disorders
  1968   New York: Bantam.

Ribeiro de Oliveira, Pedro A.
  1979   "The 'Romanization' of Catholicism and agrarian capitalism in Brazil." *Social
          Compass* 26, 2–3:309–329.

Ribuffo, Leo P.
  1983   *The Old Christian Right: The Protestant Far Right from the Great Depression to the Cold War.* Philadelphia: Temple University Press.
  1988   "Religious prejudice and nativism." Pp. 1525–1546 in C. H. Lippy and P. W. Williams (eds.), *Encyclopedia of the American Religious Experience,* Vol. III. New York: Scribner.

Richardson, James T.
  1979   "From cult to sect: Creative eclecticism in new religious movements." *Pacific Sociological Review* 22, 2:139–166.
  1981   (ed.), *The Deprogramming Controversy: Sociological, Psychological, Legal and Historical Perspectives.* New Brunswick, N.J.: Transaction.
  1985   "The active vs. passive convert: Paradigm conflict in conversion/recruitment research." *Journal for the Scientific Study of Religion* 24, 2:119–236.
  1991   "Cult/brainwashing cases and freedom of religion." *Journal of Church and State* 33, 1:55–74.
  1993a   "Definitions of cult: From sociological-technical to popular-negative." *Review of Religious Research* 34, 4:348–356.
  1993b   "Mergers, 'marriages,' coalitions and denominationalization: The growth of Calvary Chapel." *Syzygy: Journal of Alternative Religion and Culture* 2, 3–4:205–223.
  1993c   "A social psychological critique of 'brainwashing' claims about recruitment to new religions." *Religion and the Social Order* 3B:75–97.
  1995a   "Legal status of minority religions in the United States." *Social Compass* 42, 2:249–264.
  1995b   "Minority religions, religious freedom, and the new pan-European political and judicial institutions." *Journal of Church and State* 37, 1:39–58.

Richardson, James T., and John Dewitt
  1992   "Christian Science spiritual healing, the law, and public opinion." *Journal of Church and State* 34, 3:549–561.

Richardson, James T., Mary Harder, and Robert B. Simmonds
  1979   *Organized Miracles: A Sociological Study of a Jesus Movement Organization.* New Brunswick, N.J.: Transaction.

Richardson, James T., Jan van der Lans, and Frans Derks
  1986   "Leaving and labeling: Voluntary and coerced disaffiliation from religious social movements." *Research in Social Movements, Conflicts and Change* 9:97–126.

Riis, Ole
  1989   "The role of religion in legitimating the modern structuration of society." *Acta Sociologica* 32, 2:137–153.
  1993   "Recent developments in the study of religion in modern society." *Acta Sociologica* 36, 3:371–383.

Robbins, Thomas
  1981   "Church, state, and cult." *Sociological Analysis* 42, 3:209–226.
  1984   "Constructing cultist 'mind control.'" *Sociological Analysis* 45, 3:241–256.
  1985   "Government regulatory powers and church autonomy: Deviant groups as test cases." *Journal for the Scientific Study of Religion* 24, 3:237–252.
  1987   "Church–state tensions and marginal movements." Pp. 135–149 in T. Robbins and R. Robertson (eds.), *Church–State Relations: Tensions and Transitions.* New Brunswick, N.J.: Transaction.
  1993   "The intensification of church-state conflict in the United States." *Social Compass* 40, 4:505–527.

Robbins, Thomas, and Dick Anthony
  1972   "Getting straight with Meher Baba." *Journal for the Scientific Study of Religion* 11, 2:122–140.

  1979   "Cults, brainwashing, and counter-subversion." *Annals of the American Academy of Political and Social Science* 446 (Nov.):78–90.

  1990   (eds.), *In Gods We Trust: New Patterns of Religious Pluralism in America*. New Brunswick, N.J.: Transaction.

  1995   "Sects and violence: Factors enhancing the volatility of marginal religious movements." Pp. 236–259 in S. A. Wright (ed.), *Armageddon in Waco: Critical Perspectives on the Branch Davidian Conflict*. Chicago: University of Chicago Press.

Robbins, Thomas, Dick Anthony, and Thomas Curtis
  1975   "Youth culture religious movements: Evaluating the integrative hypothesis." *Sociological Quarterly* 16, 1:48–64.

Robbins, Thomas, Dick Anthony, and James Richardson
  1978   "Theory and research on today's 'new religions.'" *Sociological Analysis* 39, 2:95–122.

Robbins, Thomas, and James Beckford
  1993   "Religious movements and church-state issues." *Religion and the Social Order* 3B:199–218.

Robbins, Thomas, and David Bromley
  1991   "New religious movements and the sociology of religion." *Religion and the Social Order* 1:183–205.

  1993a "New religious movements in the United States." *Archives de Sciences Sociales des Religions* 83 (July–Sept.):91–106.

  1993b "Research on new religious movements: A bibliography." *Archives de Sciences Sociales des Religions* 83 (July–Sept.):107–125.

Robbins, Thomas, William Shepherd, and J. McBride (eds.)
  1986   *Cults, Culture and the Law*. Chico, Calif.: Scholars Press.

Robertson, Roland
  1967   "The Salvation Army: The persistence of sectarianism." Pp. 49–105 in B. Wilson (ed.), *Patterns of Sectarianism*. London: Heinemann.

  1970   *The Sociological Interpretation of Religion*. New York: Schocken.

  1977   "Individualism, societalism, worldliness, universalism: Thematizing theoretical sociology of religion." *Sociological Analysis* 38, 4:281–308.

  1978   *Meaning and Change: Explorations in the Cultural Sociology of Modern Societies*. New York: New York University Press.

  1979   "Religious movements and modern societies: Toward a progressive problem-shift." *Sociological Analysis* 40, 4:297–314.

  1981   "Consideration from within the American context on the significance of church-state tension." *Sociological Analysis* 42, 3:193–208.

  1985   "The sacred and the world system." Pp. 347–358 in P. Hammond (ed.), *The Sacred in a Secular Age*. Berkeley: University of California Press.

  1987   "Church-state relations and the world system." Pp. 39–51 in T. Robbins and R. Robertson (eds.), *Church-State Relations: Tensions and Transitions*. New Brunswick, N.J.: Transaction.

  1988   "Christian Zionism and Jewish Zionism: Points of contact." Pp. 239–258 in A. Shupe and J. Hadden (eds.), *The Politics of Religion and Social Change*. New York: Paragon.

  1989   "Globalization, politics and religion." Pp. 10–23 in J. Beckford and T. Luckmann (eds.), *The Changing Face of Religion*. London: ISA/Sage.

1991a "Globalization, modernization, and postmodernization: The ambiguous position of religion." Pp. 281–191 in R. Robertson and W. R. Garrett (eds.), *Religion and the Global Order*. New York: Paragon House.

1991b "The globalization paradigm: Thinking globally." *Religion and the Social Order* 1 (1991):207–224.

1992 *Globalization: Social Theory and Global Culture*. Newbury Park, Calif.: Sage.

Robertson, Roland, and William R. Garrett, eds.
1991 *Religion and the Global Order*. New York: Paragon House.

Robinson, John
1974 "A song, a shout, and a prayer." Pp. 213–235 in C. E. Lincoln (ed.), *The Black Experience in Religion*. Garden City, N.Y.: Doubleday.

Robinson, John A.
1963 *Honest to God*. Philadelphia: Westminster.

Rochford, E. Burke, Jr.
1985 *Hare Krishna in America*. New Brunswick, N.J.: Rutgers University Press.

1989 "Factionalism, group defection, and schism in the Hare Krishna movement." *Journal for the Scientific Study of Religion* 28, 2:162–179.

Rokeach, Milton
1960 *The Open and Closed Mind*. New York: Basic Books.

Roof, Wade Clark
1978 *Commitment and Community: Religious Plausibility in a Liberal Protestant Church*. New York: Elsevier.

1989 "Multiple religious switching: A research note." *Journal for the Scientific Study of Religion* 28, 4:530–535.

1993 *A Generation of Seekers: The Spiritual Journeys of the Baby Boom Generation*. San Francisco: Harper.

Roof, Wade Clark, and Lyn Gesch
1995 "Boomers and the culture of choice: Changing patterns of work, family and religion." Pp. 61–79 in N. T. Ammerman and W. C. Roof (eds.), *Work, Family, and Religion in Contemporary Society*. New York: Routledge.

Roof, Wade Clark, and William McKinney
1987 *American Mainline Religion: Its Changing Shape and Future*. New Brunswick, N.J.: Rutgers University Press.

Roof, Wade Clark, and Jennifer L. Roof
1984 "Review of the polls: Images of God among Americans." *Journal for the Scientific Study of Religion* 23, 2:201–205.

Roozen, David A.
1993 "Empty nest; empty pew: The Boomers continue through the life-cycle." Unpublished paper presented to the Religious Research Association.

Roozen, David A., William McKinney, and Wayne Thompson
1990 "The 'Big Chill' generation warms to worship." *Review of Religious Research* 31, 1:314–322.

Rosaldo, Renato
1989 *Culture and Truth: The Remaking of Social Analysis*. Boston: Beacon Press.

Rose, Richard
1971 *Governing Without Consensus*. London: Faber and Faber.

Rose, Susan D.
1988 *Keeping Them Out of the Hands of Satan: Evangelical Schooling in America*. New York: Routledge, Chapman and Hall.

Rosenberg, Ellen M.
   1989   *The Southern Baptists: A Subculture in Transition.* Knoxville: University of Tennessee Press.

Rosenthal, Andrew
   1992   "President strikes a religious note." *New York Times,* January 28.

Rowthorn, Bob, and Naomi Wayne
   1988   *Northern Ireland: The Political Economy of Conflict.* Boulder, Colo.: Westview.

Rudy, David R., and Arthur L. Greil
   1987   "Taking the pledge: The commitment process in Alcoholics Anonymous." *Sociological Focus* 20, 1:45–60.

   1989   "Is Alcoholics Anonymous a religious organization? Meditations on marginality." *Sociological Analysis* 50, 1:41–51.

Ruether, Rosemary R.
   1974   (ed.), *Religion and Sexism: Images of Women in Jewish and Christian Traditions.* New York: Simon & Schuster.

   1975   *New Woman/New Earth.* New York: Seabury Press.

   1989   "Sexism and God-language." Pp. 151–162 in J. Plaskow and C. P. Christ (eds.), *Weaving the Visions: New Patterns in Feminist Spirituality.* San Francisco: Harper & Row.

Ruether, Rosemary R., and Rosemary Skinner Keller (eds.)
   1981   *Women and Religion in America: The Nineteenth Century.* New York: Harper & Row.

Ruether, Rosemary R., and Eleanor McLaughlin
   1979   *Women of Spirit: Female Leadership in the Jewish and Christian Traditions.* New York: Simon & Schuster.

Ruether, Rosemary R., and Herman Ruether
   1989   *The Wrath of Jonah: The Crisis of Religious Nationalism in the Israeli-Palestinian Conflict.* San Francisco: Harper & Row.

Ryan, Mary P.
   1983   *Womanhood in America: From Colonial Times to the Present.* New York: Franklin Watts.

Sacks, Karen
   1974   "Engels revisited: Women, the organization of production and private property." Pp. 207–222 in M. S. Rosaldo and L. Lamphere (eds.), *Woman, Culture and Society.* Stanford, Calif.: Stanford University Press.

Sacred Congregation for the Doctrine of the Faith
   1977   "Declaration on the question of the admission of women to the ministerial priesthood." *Origins* 6, Feb. 3:524.

Samuelsson, Kurt
   1964   *Religion and Economic Action.* New York: Harper & Row.

Sargant, William
   1957   *Battle for the Mind.* Garden City, N.Y.: Doubleday.

Sarna, Jonathan
   1986   *The American Jewish Experience.* New York: Holmes and Meier.

Scanzoni, John and Cynthia Arnett
   1987   "Enlarging the understanding of marital commitment via religious devoutness, gender role preferences, and locus of control." *Journal of Family Issues* 8, 1:135–156.

Schmitt, David
   1973   *The Irony of Irish Democracy.* Lexington, Mass.: Heath.

Schneider, Louis and Sanford M. Dornbusch
  1957   "Inspirational religious literature: From latent to manifest functions of religion."
          American Journal of Sociology 62:476–481.

Schor, Juliet B.
  1992   *The Overworked American*. New York: Basic Books.

Schumann, Howard
  1971   "The religious factor in Detroit: Review, replication and reanalysis." *American
          Sociological Review* 36, 1:30–48.

Schutz, Alfred
  [1951] "Making music together: A study in social relationship." Pp. 135–158 in
  1964   A. Broderson (ed.), *Collected Papers II: Studies in Social Theory*. The Hague: Mar-
          tinus Nijhoff.

Schweitzer, Albert
  1936   *Indian Thought and Its Development*. Boston: Beacon Press.

Scribner, Bob
  1984   "Cosmic order and daily life: Sacred and Secular in pre-industrial German soci-
          ety." Pp. 17–32 in K. von Greyerz (ed.), *Religion and Society in Early Modern
          Europe, 1500–1800*. London: The German Historical Institute/Allen & Unwin.

Seidler, John, and Katherine Meyer
  1989   *Conflict and Change in the Catholic Church*. New Brunswick, N.J.: Rutgers Uni-
          versity Press.

Shaffir, William
  1974   *Life in a Religious Community: Lubavitcher Chassidim in Montreal*. Toronto: Holt,
          Rinehart & Winston.

  1978   "Witnessing as identity consolidation." Pp. 39–57 in H. Mol (ed.), *Identity and
          Religion*. London: Sage.

  1995   "Leaving ultra-Orthodoxy: The experiences of haredi Jews." Unpublished paper
          presented to the Association for the Sociology of Religion.

Shannon, William V.
  1963   *The American Irish*. New York: Macmillan.

Shaw, Rosalind, and Charles Stewart, eds.
  1994   *Syncretism/Anti-Syncretism: The Politics of Religious Synthesis*. London: Routledge.

Shupe, Anson D., Jr., and David G. Bromley
  1980   *The New Vigilantes: Deprogrammers, Anti-Cultists and the New Religions*. Beverly
          Hills, Calif.: Sage.

Shupe, Anson D., Jr., and William Stacey
  1982   *Born Again Politics and the Moral Majority: What Social Surveys Really Show*. New
          York: Edwin Mellen.

Silber, Laura, and Allan Little
  1994   *The Death of Yugoslavia*. London: Penguin.

Simmel, Georg
  1906   "The sociology of secrecy and of secret societies." *American Journal of Sociology*
          11:441–498.

  1955   *Conflict: The Web of Group Affiliations*. Glencoe, Ill.: Free Press.

  [1906] *Sociology of Religion*. New York: Philosophical Library.
  1959

  [1908] *On Individuality and Social Forms*. Chicago: University of Chicago Press.
  1971

Simms, J. G.
1967　"The restoration and the Jacobite war." Pp. 204–216 in T. W. Moody and F. X. Martin (eds.), *The Course of Irish History*. Cork, Ireland: Mercier.

Simpson, George E.
1965　*The Shango Cult of Trinidad*. Puerto Rico: University of Puerto Rico Press.

1978　*Black Religions in the New World*. New York: Columbia University Press.

Simpson, John H.
1983　"Moral issues and status politics." Pp. 187–205 in R. C. Liebman and R. Wuthnow (eds.), *The New Christian Right*. New York: Aldine.

Sizer, Sandra
1979　*Gospel Hymns and Social Religion: The Rhetoric of Nineteenth Century Revivalism*. Philadelphia: Temple University Press.

Slater, Philip
1966　*Microcosm: Structural, Psychological and Religious Evolution in Groups*. New York: Wiley.

Smith, Alan
1995　"Education and the conflict in Northern Ireland." Pp. 168–186 in S. Dunn (ed.), *Facets of the Conflict in Northern Ireland*. New York: St. Martin's Press.

Smith, Brian H.
1979　"Churches and human rights in Latin America: Recent trends on the subcontinent." Pp. 155–193 in D. Levine (ed.), *Churches and Politics in Latin America*. Beverly Hills, Calif.: Sage.

1982　*The Church and Politics in Chile: Challenges to Modern Catholicism*. Princeton, N.J.: Princeton University Press.

Smith, Christian
1994　"The spirit and democracy: Base communities, Protestantism, and democratization in Latin America." *Sociology of Religion* 55, 2:119–143.

Smith, David J., and Gerald Chambers
1991　*Inequality in Northern Ireland*. Oxford: Clarendon.

Smith, Jonathan Z.
1987　*To Take Place: Toward a Theory in Ritual*. Chicago: University of Chicago Press.

Smith, Timothy L.
1983　"My rejection of a cyclical view of 'Great Awakenings.'" *Sociological Analysis* 44:97–102.

Snow, David A.
1987　"Organization, ideology and mobilization: The case of Nichiren Shoshu of America." Pp. 153–172 in D. Bromley and P. Hammond (eds.), *The Future of New Religious Movements*. Macon, Ga.: Mercer University Press.

Snow, David A., and Richard Machalek
1983　"The convert as a social type." Pp. 259–289 in R. Collins (ed.), *Sociological Theory: 1983*. San Francisco: Jossey-Bass.

1984　"The sociology of conversion." *Annual Review of Sociology* 10:167–190.

Snow, David A., Louis Zurcher, Jr., and Sheldon Ekland-Olson
1980　"Social networks and social movements: A micro-structural approach to differential recruitment." *American Sociological Review* 45:787–801.

Sollors, Werner (ed.)
1989　*The Invention of Ethnicity*. New York: Oxford University Press.

Solomon, Trudy
1981　"Integrating the 'Moonie' experience: A survey of ex-members of the Unification Church." Pp. 275–294 in T. Robbins and D. Anthony (eds.), *In Gods We Trust* (first edition). New Brunswick, N.J.: Transaction.

Spence, Jonathan D.
1996	*God's Chinese Son: The Taiping Heavenly Kingdom of Hong Xiuquan.* New York: W. W. Norton.

Spickard, James V.
1991	"Experiencing religious rituals: A Schutzian analysis of Navajo ceremonies." *Sociological Analysis* 52, 2:191–204.

1993	"For a sociology of religious experience." Pp. 109–128 in W. Swatos (ed.), *A Future for Religion?* Newbury Park, Calif.: Sage.

1994	"Are there universal human rights?: Reflections on the ethical implications of a multi-cultural world." Unpublished paper presented to the Jameson Associates Lecture, University of Redlands.

Spickard, James V., and Meredith B. McGuire
1994	"Religion and social activism: Narratives of commitment." Unpublished paper presented to the International Sociological Association.

Spilka, Bernard, Richard L. Gorsuch, and Ralph W. Hood, Jr.
1985	*The Psychology of Religion: An Empirical Approach.* Englewood Cliffs, N.J.: Prentice-Hall.

Spillers, Hortense J.
1971	"Martin Luther King and the style of the black sermon." *Black Scholar* 3, 1:14–27.

Spiro, Melford
1966	"Religion: Problems of definition and explanation." Pp. 85–126 in M. Banton (ed.), *Anthropological Approaches to the Study of Religion.* London: Tavistock.

1970	*Kibbutz.* New York: Schocken.

Sprenger, Reverend J., and H. Kramer
[1486]	*Malleus Mallificarum.* Tr. M. Summers. London: Pushkin Press.
1948

Sprinzak, Ehud
1993	"Three models of religious violence: The case of Jewish fundamentalism in Israel." Pp. 462–490 in M. E. Marty and R. S. Appleby (eds.), *Fundamentalisms and the State.* Chicago: University of Chicago Press.

Stacey, William, and Anson Shupe
1982	"Correlates of support for the electronic church." *Journal for the Scientific Study of Religion* 21, 4:291–303.

Staples, Clifford L., and Armand L. Mauss
1987	"Conversion or commitment? A reassessment of the Snow and Machalek approach to the study of conversion." *Journal for the Scientific Study of Religion* 26, 2:133–147.

Stark, Rodney, and William Sims Bainbridge
1979	"Of churches, sects, and cults: Preliminary concepts for a theory of religious movements." *Journal for the Scientific Study of Religion* 18, 2:117–133.

1980	"Toward a theory of religion: Religious commitment." *Journal for the Scientific Study of Religion* 19, 2:114–128.

Stark, Rodney, and Charles Glock
1968	*American Piety: The Nature of Religious Commitment.* Berkeley: University of California Press.

Stauffer, Robert E.
1974	"Radical symbols and conservative functions." Unpublished paper presented to Society for the Scientific Study of Religion.

Steeman, Theodore M.
1975   "Church, sect, mysticism, denomination: Periodological aspects of Troeltsch's types." *Sociological Analysis* 36, 3:181–204.

Steinberg, Stephen
1965   "Reform Judaism: The origin and evolution of a 'church movement.'" *Journal for the Scientific Study of Religion* 5:117–129.

Steinfels, Peter
1990   "Mormons drop rites opposed by women." *New York Times,* May 3.

1992   "Pastoral letter on women's role fails in vote of Catholic bishops." *New York Times,* November 19.

1995   "Wariness greets Vatican doctrinal claim." *New York Times,* November 26.

Stevens-Arroyo, Anthony M.
1994   "The emergence of a social identity among Latino Catholics: An appraisal." Pp. 77–130 in J. P. Dolan and A. F. Deck (eds.), *Hispanic Catholic Culture in the U.S.: Issues and Concerns.* Notre Dame, Ind.: University of Notre Dame Press.

Stevens-Arroyo, Anthony M., and Andres I. Pérez y Mena (eds.)
1995   *Enigmatic Powers: Syncretism with African and Indigenous Peoples' Religions Among Latinos.* New York: PARAL/City University of New York.

Stevenson, Richard W.
1994a   "Belfast, mired in poverty, pins its hopes on peace." *New York Times,* April 26.

1994b   "Peace on Irish horizon doesn't spell prosperity." *New York Times,* September 4.

Stewart, Charles
1994   "Syncretism as a dimension of nationalist discourse in modern Greece." Pp. 127–144 in C. Stewart and R. Shaw (eds.), *Syncretism/Anti-Syncretism: The Politics of Religious Synthesis.* London: Routledge.

Stoll, David
1990   *Is Latin America Turning Protestant? The Politics of Evangelical Growth.* Berkeley: University of California Press.

Straus, Roger
1979   "Religious conversion as a personal and collective accomplishment." *Sociological Analysis* 40, 2:158–165.

Stuard, Susan Mosher
1989   "Women's witnessing: A new departure." Pp. 3–25 in E. P. Brown and S. M. Stuard (eds.), *Witnesses for Change: Quaker Women Over Three Centuries.* New Brunswick, N.J.: Rutgers University Press.

Swatos, William, Jr.
1975   "Monopolism, pluralism, acceptance, and rejection: An integrated model for church–sect theory." *Review of Religious Research* 16, 3:174–185.

1979   *Into Denominationalism: The Anglican Metamorphosis.* Monograph Series 2. Storrs, Conn.: Society for the Scientific Study of Religion.

Szasz, Thomas S.
1970   *The Manufacture of Madness.* New York: Dell.

Talmon, Yonina
1966   "Millenarian movements." *Archives Europeenes de Sociologie* 7, 2:159–200.

Tambiah, Stanley J.
1992   *Buddhism Betrayed? Religion, Politics and Violence in Sri Lanka.* Chicago: University of Chicago Press.

1993   "Buddhism, politics, and violence in Sri Lanka." Pp. 589–619 in M. E. Marty and R. S. Appleby (eds.), *Fundamentalisms and the State.* Chicago: University of Chicago Press.

Tamney, Joseph, and Stephen Johnson
   1983   "The Moral Majority in Middletown." *Journal for the Scientific Study of Religion*
          22, 2:145–157.

Taussig, Michael T.
   1980   *The Devil and Commodity Fetishism in South America*. Chapel Hill: University of
          North Carolina Press.

Taves, Ann
   1986   *The Household of Faith: Roman Catholic Devotions in Mid-Nineteenth Century Amer-
          ica*. Notre Dame, Ind.: University of Notre Dame Press.

Tawney, R. H.
   1926   *Religion and the Rise of Capitalism*. New York: Harcourt Brace.

Tepper, Sheri S.
   1990   *Raising the Stones*. New York: Doubleday.

Terkel, Studs
   1975   *Working*. New York: Random House.

Thompson, E. P.
   1968   *The Making of the English Working Class*. Harmondsworth, England: Penguin
          Books.

Thompson, Kenneth
   1986   *Beliefs and Ideology*. Chichester, England: Ellis Horwood.

Thompson, Leonard
   1990   *A History of South Africa*. New Haven, Conn.: Yale University Press.

Thompson, William Irwin
   1967   *The Imagination of an Insurrection: Dublin, Easter, 1916*. New York: Harper &
          Row.

Tippett, Alan R.
   1973   "The phenomenology of worship, conversion and brotherhood." Pp. 92–109 in
          W. Clark (ed.), *Religious Experience: Its Nature and Function in the Human Psyche*.
          Springfield, Ill.: Thomas.

Tipton, Steven M.
   1982a *Getting Saved from the Sixties: Moral Meaning in Conversion and Cultural Change*.
          Berkeley: University of California Press.

   1982b "The moral logic of alternative religions." Pp. 79–107 in M. Douglas and
          S. Tipton (eds.), *Religion and America: Spirituality in a Secular Age*. Boston: Bea-
          con Press.

Tiryakian, Edward A.
   1974   "Toward the sociology of esoteric culture." Pp. 257–280 in E. Tiryakian (ed.),
          On the Margins of the Visible. New York: Wiley.

Tobey, Alan
   1976   "Summer solstice of the Happy-Healthy-Holy Organization." Pp. 5–30 in
          C. Glock and R. Bellah (eds.), *The New Religious Consciousness*. Berkeley: Uni-
          versity of California Press.

Towler, Robert and Audrey Chamberlain
   1973   "Common religion." Pp. 1–27 in M. Hill (ed.), *Sociological Yearbook of Religion in
          Britain,* Vol. 6. London: SCM.

Trexler, Richard C.
   1984   "Reverence and profanity in the study of early modern religion." Pp. 245–269
          in K. von Greyerz (ed.), *Religion and Society in Early Modern Europe, 1500–1800*.
          London: Allen & Unwin/The German Historical Institute.

Troeltsch, Ernst
[1931] *The Social Teachings of the Christian Churches*. Vols. 1 and 2. Tr. O. Wyon. New
1960   York: Harper & Row.

Truzzi, Marcello
1972   "The occult revival as popular culture: Some random observations in the old
        and nouveau witch." *Sociological Quarterly* 13:16–36.

1974a   "Definition and dimension of the occult: Towards a sociological perspective."
        Pp. 243–255 in E. Tiryakian (ed.), *On the Margins of the Visible*. New York:
        Wiley.

1974b   "Witchcraft and Satanism." Pp. 215–222 in E. Tiryakian (ed.), *On the Margins of
        the Visible*. New York: Wiley.

Turner, Bryan
1977   "Confession and social structure." *Annual Review of the Social Sciences of Religion*
        1:29–58.

1980   "The body and religion: Towards an alliance of medical sociology and sociology
        of religion." *Annual Review of the Social Sciences of Religion* 4:247–286.

1991   *Religion and Social Theory: A Materialistic Perspective*. London: Sage.

Turner, Victor
1969   *The Ritual Process*. Chicago: Aldine.

1974a   *Dramas, Fields, and Metaphors*. Ithaca, N.Y.: Cornell University Press.

1974b   "Metaphors of anti-structure in religious culture." Pp. 63–84 in A. Eister (ed.),
        Changing Perspectives in the Scientific Study of Religion. New York: Wiley.

1979   "Betwixt and between: The liminal period in 'rites de passage.'" Pp. 234–243 in
        E. Vogt (ed.), *Reader in Comparative Religion: An Anthropological Approach*. New
        York: Harper & Row.

Turner, Victor, and Edith Turner
1978   *Image and Pilgrimage in Christian Culture: Anthropological Perspectives*. New York:
        Columbia University Press.

Tyson, Ruel W.
1988   "The testimony of Sister Annie Mae." Pp. 105–125 in R. W. Tyson, J. L. Pea-
        cock, and D. W. Patterson (eds.), *Diversities of Gifts: Field Studies in Southern Re-
        ligion*. Urbana: University of Illinois Press.

Umansky, Ellen M.
1985   "Feminism and the reevaluation of women's roles within American Jewish life."
        Pp. 477–494 in Y. Y. Haddad and E. B. Findly (eds.), *Women, Religion and So-
        cial Change*. Albany: State University of New York Press.

Underhill, Evelyn
1961   *Mysticism*. New York: Dutton.

Van Gennep, A.
1960   *The Rites of Passage*. Tr. M. Vikedom and G. Coffee. Chicago: University of
        Chicago Press.

Vauchez, André
1993   *Laity in the Middle Ages: Religious Beliefs and Devotional Practices*. Notre Dame,
        Ind.: University of Notre Dame.

Vaughn, William Preston
1983   *The Antimason Party in the United States, 1826–1843*. Lexington: University of
        Kentucky Press.

Verdesi, Elizabeth Howell
1976   *In But Still Out: Women in the Church*. Philadelphia: Westminster.

Voll, John Obert
1982　*Islam: Continuity and Change in the Modern World.* Boulder, Colo.: Westview.

Vonnegut, Kurt, Jr.
1963　*Cat's Cradle.* Baltimore: Penguin Books.

Wach, Joachim
1944　*Sociology of Religion.* Chicago: University of Chicago Press.

Wagner, Melinda Bollar
1983　*Metaphysics in Midwestern America.* Columbus: Ohio State University Press.

1990　*God's Schools: Choice and Compromise in American Society.* New Brunswick, N.J.: Rutgers University Press.

1994　"Metaphysics: A practical religious philosophy." *Religion and the Social Order* 4:111–125.

Wald, Kenneth D.
1987　*Religion and Politics in the United States.* New York: St. Martins.

Wald, Kenneth D., Dennis E. Owen, and Samuel S. Hill, Jr.
1989　"Evangelical politics and status issues." *Journal for the Scientific Study of Religion* 28, 1:1–16.

Walker, Andrew, and James S. Atherton
1971　"An Easter Pentecostal convention: The successful management of a 'time of blessing.'" *Sociological Review* 19, 3:367–387.

Wallace, Anthony
1956　"Revitalization movements." *American Anthropologist* 58:264–281.

1957　"Mazeway disintegration: The individual's perception of socio-cultural disorganization." *Human Organization* 16:23–27.

Wallace, Ruth A.
1975　"Bringing women in: Marginality in the churches." *Sociological Analysis* 36, 4:291–303.

1988　"Catholic women and the creation of a new social reality." *Gender and Society* 2, 1:24–38.

1991　*They Call Her Pastor.* Albany: State University of New York Press.

1993　"The social construction of a new leadership role: Catholic women pastors." *Sociology of Religion* 54, 1:31–42.

Wallis, Roy
1973　"A comparative analysis of problems and processes of change in two manipulationist movements: Christian Science and Scientology." Pp. 407–422 in *Contemporary Metamorphosis of Religion?: Acts of the 12th Conference Internationale de Sociologie Religieuse.* Lille, France: CISR.

1974　"Ideology, authority and the development of cultic movements." *Social Research* 41:299–327.

1977　*The Road to Total Freedom: A Sociological Analysis of Scientology.* New York: Columbia University Press.

Wallis, Roy, and Steve Bruce
1984　"The Stark-Bainbridge theory of religion: A critical analysis and counter proposal." *Sociological Analysis* 45, 1:11–28.

1986　*Sociological Theory, Religion and Collective Action.* Belfast: Queens University.

Wardwell, Walter
1972　"Orthodoxy and heterodoxy in medical practice." *Social Science and Medicine* 6:759–763.

Warner, R. Stephen
1988   *New Wine in Old Wineskins: Evangelicals and Liberals in a Small-Town Church.*
Berkeley: University of California Press.

1994   "The place of the congregation in the contemporary American religious configuration." Pp. 54–99 in J. P. Wind and J. W. Lewis (eds.), *American Congregations,* Vol. 2. Chicago: University of Chicago Press.

Warner, W. Lloyd
1953   *American Life: Dream and Reality.* Chicago: University of Chicago Press.

Washington, Joseph
1972   *Black Sects and Cults.* Garden City, N.J.: Doubleday.

Waters, Mary C.
1996   "Optional ethnicities: For Whites only?" Pp. 444–454 in S. Pederanza and
R. G. Rumbart (eds.), *Origins and Destinies.* Belmont, Calif.: Wadsworth.

Weber, Max
[1920–21] *From Max Weber: Essays in Sociology.* Tr. and ed. H. H. Gerth and C. W.
1946   Mills. New York: Oxford University Press.

[1925] *The Theory of Social and Economic Organization.* Tr. A. M. Henderson and T.
1947   Parsons. New York: Oxford University Press.

[1920–21] *The Religion of China.* Tr. H. H. Gerth. New York: Free Press.
1951

[1920 21] *Ancient Judaism.* Tr. H. H. Gerth and D. Martindale. New York: Free Press.
1952

[1904] *The Protestant Ethic and the Spirit of Capitalism.* Tr. T. Parsons. New York:
1958a   Scribner.

[1920–21] *The Religion of India.* Tr. H. H. Gerth and D. Martindale. New York: Free
1958b   Press

[1922] *The Sociology of Religion.* Tr. E. Fischoff. Boston: Beacon Press.
1963

[1925] *Economy and Society,* Vol. 3. Ed. G. Roth and C. Wittich. New York:
1968   Bedminster.

Weil, Andrew
1973   *The Natural Mind.* Boston: Houghton Mifflin.

Welter, Barbara
1966   "The cult of true womanhood: 1820–1860." *American Quarterly* (Summer), Part
I:151–174.

1976   *Dimity Convictions: The American Woman in the Nineteenth Century.* Athens: Ohio
University Press.

Wessinger, Catherine (ed.)
1993   *Women's Leadership in Marginal Religions: Explorations Outside the Mainstream.* Urbana: University of Illinois Press.

Westhues, Kenneth
1973   "The established church as an agent of change." *Sociological Analysis* 34,
2:106–123.

1976   "The church in opposition." *Sociological Analysis* 37, 4:299–314.

1978   "Stars and stripes, the maple leaf, and the papal coat of arms." *Canadian Journal of
Sociology* 3, 2:245–261.

Westin, Alan
1964   "The John Birch Society." Pp. 239–268 in D. Bell (ed.), *The Radical Right.* Garden City, N.Y.: Doubleday.

Westley, Frances
 1978 "The cult of man: Durkheim's predictions and new religious movements." *Sociological Analysis* 39, 2:135–145.

 1983 *The Complex Forms of the Religious Life: A Durkheimian View of New Religious Movements.* Chico, Calif.: Scholars Press.

Whyte, J. H.
 1980 *Church and State in Modern Ireland, 1923–1979.* Dublin: Gill.

Whyte, John
 1990 *Interpreting Northern Ireland.* Oxford: Clarendon.

Wilcox, Clyde
 1991 "Religion and electoral politics among black Americans in 1988." Pp. 159–172 in J. L. Guth and J. C. Green (eds.), *The Bible and the Ballot Box: Religion and Politics in the 1988 Election.* Boulder: Westview Press.

Wilcox, Clyde, Sharon Linzey, and Ted. G. Jelen
 1991 "Reluctant warriors: Premillennialism and politics in the Moral Majority." *Journal for the Scientific Study of Religion* 30, 3:245–258.

Willen, Richard S.
 1983 "Religion and law: The secularization of testimonial procedures." *Sociological Analysis* 44, 1:53–64.

Williams, Melvin D.
 1974 *Community in a Black Pentecostal Church: An Anthropological Study.* Pittsburgh: University of Pittsburgh Press.

Williams, Rhys H.
 1988 "American political culture and religion: Conflict and convergence." Paper presented to the Society for the Scientific Study of Religion.

 1995 "Breaching the 'Wall of Separation': The balance between religious freedom and social order." Pp. 299–322 in S. A. Wright (ed.), *Armageddon in Waco: Critical Perspectives on the Branch Davidian Conflict.* Chicago: University of Chicago Press.

Williams, Rhys H., and Susan M. Alexander
 1994 "Religious rhetoric in American Populism: Civil religion as movement ideology." *Journal for the Scientific Study of Religion* 33, 1:1–15.

Williams, Rhys H., and N. J. Demerath III
 1991 "Religion and political process in an American city." *American Sociological Review* 56 (August):417–431.

Williams, Richard N., and James E. Faulconer
 1994 "Religion and mental health: A hermeneutic reconsideration." *Review of Religious Research* 35, 4:335–349.

Willits, Fern K., and Donald M. Crider
 1989 "Church attendance and traditional religious beliefs in adolescence and young adulthood: A panel study." *Review of Religious Research* 31, 1:68–81.

Wilmore, Gayraud S., Jr.
 1972 *Black Religion and Black Radicalism.* Garden City, N.Y.: Doubleday.

Wilson, Bryan R.
 1967 (ed.), *Patterns of Sectarianism: Organization and Ideology in Social and Religious Movements.* London: Heinemann.

 1970 *Religious Sects.* New York: McGraw-Hill.

 1976 *Contemporary Transformations of Religion.* London: Oxford University Press.

 1989 "Sects and society in tension." Pp. 159–184 in P. Badham (ed.), *Religion, State and Society in Modern Britain.* Lewiston, N.Y.: Edwin Mellen Press.

1990   *The Social Dimensions of Sectarianism: Sects and New Religious Movements in Contemporary Society.* Oxford: Clarendon.

1993   "Historical lessons in the study of sects and cults." *Religion and the Social Order* 3A:53–73.

Wilson, Bryan R., and Karel Dobbelaere
1994   *A Time to Chant: The Soka Gakkai Buddhists in Britain.* Oxford: Clarendon.

Wilson, John, and Sharon Sandomirsky
1991   "Religious affiliation and the family." *Sociological Forum* 6, 2:289–309.

Wilson, William Julius
1980   *The Declining Significance of Race: Blacks and Changing American Institutions.* Chicago: University of Chicago Press.

1987   *The Truly Disadvantaged: The Inner City, the Underclass and Public Policy.* Chicago: University of Chicago Press.

Wimberley, Ronald, Thomas Hood, C. M. Lipsey, Donald Clelland, and M. Hay
1975   "Conversion in a Billy Graham crusade: Spontaneous event or ritual performance." *Sociological Quarterly* 16:162–170.

Winter, J. Alan
1996   "Symbolic ethnicity or religion among Jews in the United States: A test of Gansian hypotheses." *Review of Religious Research* 37, 3:137–151.

Winter, Miriam Therese, Adair Lummis, and Allison Stokes
1994   *Defecting in Place: Women Claiming Responsibility for Their Own Spiritual Lives.* New York: Crossroad.

Wogaman, J. Philip
1988   "The use of government infiltration in religious groups: A new/old issue." Pp. 123–137 in J. E. Wood (ed.), *Ecumenical Perspectives on Church and State.* Waco, Tex.: Baylor University, J. M. Dawson Institute on Church–State Studies.

Wood, James E., Jr.
1988   "Religious pluralism and American society." Pp. 1–18 in J. E. Wood (ed.), *Ecumenical Perspectives on Church and State.* Waco, Tex.: Baylor University, J. M. Dawson Institute on Church–State Studies.

Wood, James R.
1970   "Authority and controversial policy: The churches and civil rights." *American Sociological Review* 35, 6:1057–1069.

1981   *Leadership in Voluntary Organizations: The Controversy over Social Action in Protestant Churches.* New Brunswick, N.J.: Rutgers University Press.

Woodcock, George and Ivan Avakumonic
1977   *The Doukhobors.* Toronto: McClelland and Stewart.

Woodward, C. Vann
1957   *The Strange Career of Jim Crow.* New York: Oxford University Press.

Worsley, Peter
1968   *The Trumpet Shall Sound: A Study of "Cargo" Cults in Melanesia.* New York: Schocken.

Wright, Stuart A.
1987   *Leaving Cults: The Dynamics of Defection.* Washington, D.C.: Society for the Scientific Study of Religion, Monograph Series.

1991   "Reconceptualizing cult coercion and withdrawal: A comparative analysis of divorce and apostasy." *Social Forces* 70, 1:125–145.

Wright, Stuart A., and Helen Rose Ebaugh
1993   "Leaving new religions." *Religion and the Social Order* 3B:117–138.

Wuthnow, Robert
  1976a "Astrology and marginality." *Journal for the Scientific Study of Religion* 15,
          2:157–168.

  1976b *The Consciousness Reformation.* Berkeley: University of California Press.

  1978   "Religious movements and the transition in world culture." Pp. 63–79 in
          J. Needleman and G. Baker (eds.), *Understanding the New Religions.* New York:
          Seabury Press.

  1980   "World order and religious movements." Pp. 57–75 in A. Bergesen (ed.), *Studies of the Modern World System.* New York: Academic Press.

  1988   *The Restructuring of American Religion: Society and Faith Since World War II.*
          Princeton, N.J.: Princeton University Press.

  1994   *Producing the Sacred: An Essay on Public Religion.* Urbana, Ill.: University of Illinois Press.

Yinger, J. Milton
  1946   *Religion in the Struggle for Power.* Durham, N.C.: Duke University Press.

  1963   "The 1962 H. Paul Douglass lectures, I. Religion and social change: Functions
          and dysfunctions of sects and cults among the disprivileged." *Review of Religious
          Research* 4, 2:65–84.

  1970   *The Scientific Study of Religion.* New York: Macmillan.

Young, Allan
  1976   "Some implications of medical beliefs and practices for social anthropology."
          American Anthropologist 78:5–24.

Zablocki, Benjamin
  1971   *The Joyful Community.* Baltimore: Penguin Books.

Zald, Mayer
  1970   *Organization Change: The Political Economy of the Y.M.C.A.* Chicago: University
          of Chicago Press.

Author Index

Subject Index